THE
CASIMIR EFFECT IN
CRITICAL SYSTEMS

THE

CASIMIR EFFECT IN
CRITICAL SYSTEMS

Michael Krech

World Scientific
Singapore • New Jersey • London • Hong Kong

Published by

World Scientific Publishing Co. Pte. Ltd.

P O Box 128, Farrer Road, Singapore 9128

USA office: Suite 1B, 1060 Main Street, River Edge, NJ 07661

UK office: 73 Lynton Mead, Totteridge, London N20 8DH

THE CASIMIR EFFECT IN CRITICAL SYSTEMS

ISBN 981-02-1845-1

Printed in Singapore by Utopia Press.

Foreword

Almost half a century ago Casimir* showed that two uncharged perfectly conducting plates placed parallel to each other in vacuum experience an attractive force proportional to the inverse fourth power of their separation. The effect is due to the change, brought about by the presence of the plates, of the zero-point modes of the electromagnetic field. For conductors of arbitrary shape (rather than plates), these modes depend on the boundary conditions, on the shape of the conductors, and on their location relative to each other. Thus a contribution to the zero-point energy depending on the distance between them exists. The resulting tiny force could be verified in a number of experiments.*

From the underlying mechanism it is clear that a Casimir effect should also occur in other physical situations; the generic case to be considered is a field theory with a fluctuating field restricted by surfaces on which it is constrained to satisfy certain boundary conditions. The fluctuations may be quantum mechanical or thermal in origin. Whenever this field is massless, the resulting Casimir force between the surfaces should be of a long-range nature, decaying algebraically rather than exponentially (as does in the case of massive fields).

Posed in this general form, the problem of the Casimir effect is of considerable interest in both condensed matter physics and quantum field theory. In quantum field theory it arises quite naturally in studies of field theories on manifold with boundaries. Another source of motivation for investigations has been the idea that it should afford insight into the bag model for hadrons.

In condensed matter physics most of the early activities focused on variants of the original Casimir problem; the vacuum energy of the electromagnetic field in the presence of dielectric and conducting surfaces, and a calculation of the fluctuation-induced effective interactions. One important result has been the general theory of van der Waals forces developed by Dzyaloshinskii, Lifshitz, and Pitaevskii. The interested reader may find references and a more detailed account of this work in chapter 3.1 and the review articles cited there.

A more recent development has been the interest in the Casimir effect in *critical* systems — the central subject of this book. This trend may be viewed as a logical consequence of a number of impressive developments that took place in the area of critical phenomena during the last 25 years. Building on Wilson's pioneering

---

*References are given in subsequent chapters of this book.

work on the renormalization group, the modern field-theoretic approach to bulk critical behavior was developed in the 1970s. Aside from providing a conceptually satisfactory theoretical framework, it has led to quantitatively accurate predictions, many of which were carefully checked by experiments. Its spectacular successes have led to a much broader awareness of the many intimate relations between Euclidean field theories and problems of condensed matter physics and of statistical physics.

In the 1980s the field theoretic renormalization group approach was extended to systems with surfaces. This paved the way to its application in the study of surface critical phenomena. While it became clear already at that time how the Casimir effect could be handled within the framework of renormalized field theory, detailed investigations along these lines came only much later. These advances were paralleled by enormous activity and progress in three other, but related, research areas: the physics of wetting phenomena, the theory of finite size effects, and conformal field theory. As the research in any of these fields progressed, more and more interrelations among these and the Casimir effect were discovered, providing further impetus to the study of the latter.

Thus an overview of the current knowledge on the Casimir effect in critical systems and problems related to it involves a fairly broad spectrum of topics. In this book, M. Krech gives a detailed and authoritative review of this field, describing both the necessary background as well as the latest results, in many of which he has been involved himself. It should be equally useful for the novice seeking an introduction into the field, and for research workers as a standard reference and coherent account of the presently available information. Containing detailed discussions of many experimental results as well as of possible future experiments, it should also be very valuable for the interested experimentalist. It is my expectation and hope that the book will make the field accessible to an even broader community and stimulate vigorous further experimental and theoretical research in this fascinating field.

H. W. Diehl
Essen, April 1994

Contents

———THE———
CASIMIR EFFECT IN
CRITICAL SYSTEMS

1. Introduction

1.1 Bulk Critical Phenomena

Macroscopic bodies consist of a very large number of particles, which belong to only very few different species and interact basically through electromagnetic forces. The appearance of such a body on a macroscopic scale heavily depends on the collective behavior of its microscopic constituents, which in turn depends on the thermodynamic state of the system. The thermodynamic state is characterized by the values of a few thermodynamic variables like temperature, pressure, concentrations of other species, or electromagnetic fields applied from outside. At high temperature and low pressure the separation between the particles is very large compared to their diameter and the particles themselves move rather freely through the available space. This is a simple picture of the *vapor phase* of the system, which is homogeneous and isotropic. At some intermediate temperature and pressure the particles form small aggregates according to regular patterns over distances of a few particle diameters. On macroscopic scales these small aggregates move still rather freely exchanging particles from their outermost shells among each other. This is a rough description of the *liquid phase* of the system, which on *macroscopic* scales is still homogeneous and isotropic. At low temperature and high pressure the particles can be "frozen" into regular arrangements over *macroscopic* distances. The system then displays discrete lattice symmetries instead of continuous homogeneity and isotropy. The mean distance between two neighboring particles has decreased to the order of the magnitude of a particle diameter. This lattice picture is the most common visualization of the *solid phase* of the system. Besides vapor, liquid, and solid there are many more other phases possible in real systems. Especially the liquid and solid phases may be subdivided into additional phases with different macroscopic behavior, e.g., superfluids, ferromagnets, or liquid crystals [1, 2, 3] (see below).

The phase transition from one phase to another takes place abruptly, i.e., the one–phase regions in the *phase diagram* are bounded by sharp lines. These lines are known as *coexistence curves* between the two adjacent phases and they indicate the locations of *first–order* phase transitions in the phase diagram. The typical appearance of a phase diagram of a simple nonpolar fluid in the temperature–pressure (T, p) plane is shown in Fig.1.1(a) [1, 2]. The three coexistence curves intersect in a point of

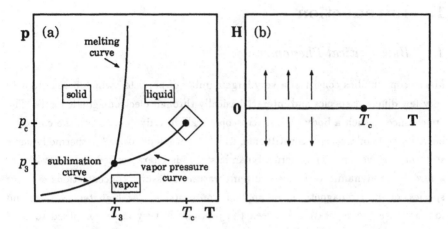

Fig. 1.1:   (a) Schematic bulk phase diagram of a simple system in the temperature–pressure (T, p) plane. (T_3, p_3) denotes the triple point and (T_c, p_c) denotes the
critical point. Beyond the critical point any difference between the liquid and the vapor phase has vanished. (b) Schematic bulk phase diagram of a uniaxial ferromagnet
in the temperature–magnetic field (T, H) plane. The critical point $(T = T_c, H = 0)$
terminates the line $H = 0$ of first–order transitions between the two possible ferromagnetic states. Note the similarity of (b) to the region close to (T_c, p_c) indicated by the
box in (a).

three–phase coexistence, which is known as the *triple point* (T_3, p_3). The vapor pressure curve *terminates* in a *critical point* (T_c, p_c), whereas to present knowledge there
is no critical point terminating the melting curve. The density jump across the vapor
pressure curve diminishes as the critical point is approached and it vanishes right
at the critical point. For $T > T_c$ or $p > p_c$ liquid and vapor have become indistin-
guishable. The critical point indicates the location of a *second–order* phase transition
between the liquid and the vapor phase, where the density changes *continuously* from
its value in the liquid to its value in the vapor.

The critical point is a remarkable exception point in the phase diagram shown in
Fig.1.1 (a). Beyond that point vapor and liquid merge to form a single phase. This
phenomenon is imtimately linked to the fact that the liquid and the vapor phase are
both spatially homogeneous and isotropic on macroscopic scales. The solid phase,
which often forms a regular lattice structure, only exhibits *discrete* spatial symmetries.

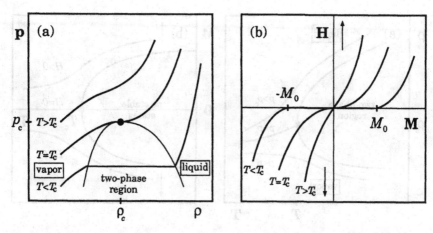

Fig. 1.2: (a) Isotherms in the schematic phase diagram of a simple fluid in the (ρ, p) plane for $T < T_c$, $T = T_c$, and $T > T_c$. The two–phase coexistence region is indicated by the flat portion of the isotherms for $T < T_c$. The critical isotherm $(T = T_c)$ displays a turning point with zero slope at (ρ_c, p_c). (b) Isotherms of a uniaxial ferromagnet in the (M, H) plane for $T < T_c$, $T = T_c$, and $T > T_c$ (see (a)). $M_0(T)$ and $-M_0(T)$ indicate the possible values of the spontaneous magnetization for $T < T_c$ and $H = 0$ (see Refs.[1], [2], and Sec.1.4).

Provided, that there is *no* external field which mimics the spatial symmetry of the solid phase, an argument given by L. D. Landau holds according to which the phase transition between the liquid and the solid state is always first order due to the absence of a *continuous* crossover between *different* spatial symmetries. For a bulk phase diagram like Fig.1.1(a) this statement seems to hold, but, however, it does not hold for layers adsorbed on solid substrates (see Refs.[4] and [5]).

As pointed out above the solid phase of a system may itself be subdivided into solid phases with different properties. One important example is *ferromagnetism*. A simplified phase diagram of a *uniaxial* ferromagnet in the plane spanned by the temperature T and the external magnetic field H is shown in Fig.1.1(b). The other thermodynamic variables characterizing the solid phase have been disregarded here. In a uniaxial ferromagnet the magnetization is either zero or parallel (antiparallel) to the magnetic axis. The phase diagram contains a critical point at $(T, H) = (T_c, 0)$, below which the transition between the two possible ferromagnetic phases indicated

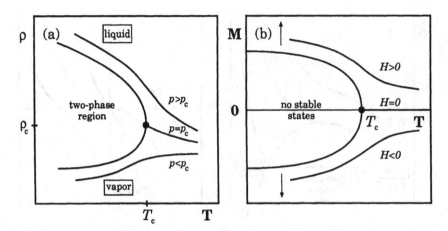

Fig. 1.3:   (a) Schematic phase diagram of a simple fluid in the (T, ρ) plane in the vicinity of a critical point. The equation of state $\rho = \rho(T, p)$ bifurcates at the critical temperature for $p = p_c$. (b) Schematic phase diagram of a uniaxial ferromagnet in the (T, M) plane (see (a)). Note, that the density gap in the two–phase coexistence region for $T < T_c$ in (a) is not symmetric around ρ_c, whereas the magnetization gap for $T < T_c$ in (b) is symmetric around $M = 0$ [1, 2].

by the arrows in Fig.1.1(b) is first order. Above the critical temperature T_c these two phases have merged to a single *paramagnetic* phase. *At* the critical point the ferromagnetic transition has become second order. Note, that the phase diagram shown in Fig.1.1(b) looks like an enlargement of the boxlike region around the critical point (T_c, p_c) indicated in Fig.1.1(a) in rotated coordinates.

It is very illustrative to look at the region near (T_c, p_c) in Fig.1.1(a) and $(T_c, 0)$ in Fig.1.1(b) in different thermodynamic variables [1, 2]. The isotherms in the density–pressure (ρ, p) plane for a simple fluid are shown in Fig.1.2(a) and the isotherms in the magnetization–magnetic field (M, H) plane for a uniaxial ferromagnet are shown in Fig.1.2(b). In either case the isotherms show a flat portion below the critical temperature T_c, which indicates the jump in the density ρ and the magnetization M, respectively, at the first–order phase transition. At $T = T_c$ the jump has disappeared and the plateau in the isotherms has shrunk to a single point at $\rho = \rho_c$ $(M = 0)$, through which the isotherms pass with zero slope. For $T > T_c$ the isotherms have a positive slope for all ρ (or M). Note, that below T_c and without an external field

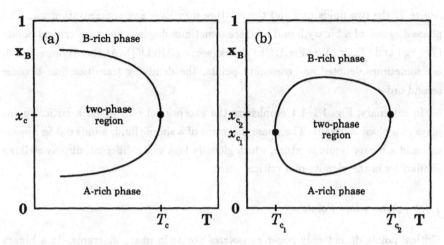

Fig. 1.4: (a) Schematic phase diagram of a binary liquid mixture in the vicinity of the
critical (consolute) point (T_c, x_c) in the (T, x_B) plane. Below the critical temperature T_c
the liquid demixes into an A–rich and a B–rich liquid. The concentration gap vanishes
at the critical point. (b) Schematic phase diagram of a binary liquid mixture with two
consolute points. The fluids demix in the temperature range $T_{c_1} < T < T_{c_2}$.

$M_0(T)$ and $-M_0(T)$ indicate the possible values of the *spontaneous magnetisation*
(see Sec.1.4).

In order to complete our picture of the vicinity of the critical point we show the
phase diagram in the (T, ρ) plane for a simple fluid in Fig.1.3(a) and in the (T, M)
plane for a uniaxial ferromagnet in Fig.1.3(b). The curves show $\rho(T, p)$ and $M(T, H)$,
respectively, for $p > p_c$ $(H > 0)$, $p = p_c$ $(H = 0)$, and $p < p_c$ $(H < 0)$ in the vicinity
of the critical point. For $p = p_c$ $(H = 0)$ the critical temperature T_c indicates a
bifurcation in the critical equation of state $\rho = \rho(T, p_c)$ $(M = M(T, 0))$.

A very common example for systems with more than one liquid phase is provided
by binary liquid mixtures, which consist of two species A and B. Depending on the
temperature and the mole fraction or concentration $x_B = 1 - x_A$ of B in the mixture
the A–rich and B–rich fluid may or may not be miscible. A typical phase diagram
of a binary liquid mixture in the vicinity of the critical point (T_c, x_c) is displayed
in Fig.1.4(a), which resembles Figs.1.3(a) and (b). Below a critical temperature T_c
the fluids *demix* into an A–rich phase $(x_B < x_c)$ and a B–rich phase $(x_B > x_c)$.

Above T_c the two fluids mix and the mixture may have any concentration x_B. The phase diagram of a binary liquid mixture sometimes displays two such critical points (T_{c_1}, x_{c_1}) and (T_{c_2}, x_{c_2}), as Fig.1.4(b) shows (see, e.g., Ref.[5]). At these points, which are sometimes denoted as consolute points, the demixing transition has become second order.

In summary, Figs.1.1–1.4 emphasize the exceptional role which a critical point plays in a phase diagram. The phase diagrams of a simple fluid, a uniaxial ferromagnet, and a binary liquid mixture, which globally look quite different, display striking similarities in the vicinity of a critical point.

1.1.1 Critical End Points

Critical points do not only occur as *isolated* points in phase diagrams. In a binary liquid mixture there are three intensive thermodynamic variables at our disposal, the two chemical potentials μ_A and μ_B of the species A and B and the temperature T. In terms of these variables the analogue of the phase diagram shown in Fig.1.1(a) restricted to the vicinity of the critical point becomes three–dimensional. A possible structure of such a phase diagram is shown in Fig.1.5 [1] [6]. The one–phase regions are now separated by two–phase coexistence *surfaces* S_1 and S_2, which indicate the locations of first–order phase transitions. The liquid–vapor coexistence surface and the coexistence surface between the A–rich and the B–rich liquid shown in Fig.1.5 terminate in *lines* L_1 and L_2 of critical points, where the liquid–vapor and the demixing transitions, respectively, become second order (see Fig.1.4). Starting from low temperatures first the A–rich and the B–rich liquid phases merge to form a single liquid phase and then this liquid phase and the vapor phase merge to a single fluid (see Fig.1.1(a)). The two coexistence surfaces intersect in a *line* of three–phase coexistence, which is denoted as the *triple line*. The triple line is terminated by the *critcal end point* of the line L_2 of the second–order demixing transitions in the liquid–vapor coexistence surface S_1. (The presence of the critical end point in the liquid–vapor coexistence surface can be observed in a wetting experiment (see Sec.6.3)). Two widely used binary liquid mixtures are methanol–cyclohexane and cyclohexane–aniline (see Ref.[1]).

A second very prominent system which exhibits critical end points is liquid ^{4}He.

---

[1] We disregard the solid phases here.

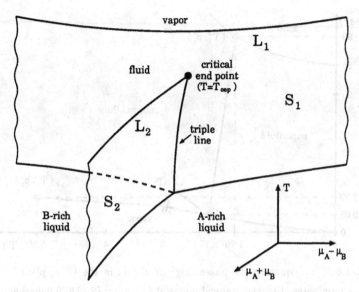

Fig. 1.5: Schematic phase diagram of a binary liquid mixture in the space spanned by the temperature T and the sum and the difference of the two chemical potentials μ_A and μ_B of the species A and B. $\mu_A + \mu_B$ basically represents the pressure. The two surfaces S_1 and S_2 of first–order phase transitions are terminated by the critical lines L_1 and L_2, respectively. The surfaces S_1 and S_2 intersect in the triple line, which ends in the critical end point of the line L_2 of the second order demixing transitions (see Ref.[6]).

A semiquantitative phase diagram is displayed in Fig.1.6 [7]. The line of the second–order (critical) superfluid phase transitions in liquid ^{4}He, which is usually denoted as the λ–line, terminates in *two* critical end points, one on the vapor pressure curve at $(T_\lambda, p_\lambda) = (2.17K, 0.05\text{bar})$ and the other on the melting curve at $(T_\lambda^+, p_\lambda^+) = (1.76K, 30\text{bar})$. The former critical end point is sometimes referred to as the *lower* λ–*point* and the latter is denoted as the *upper* λ–*point*. The vapor pressure curve terminates in a usual critical point (see Fig.1.1) at $(T_c, p_c) = (5.20K, 2.29\text{bar})$. Besides superfluidity a second outstanding property of ^{4}He is visible in the phase diagram shown in Fig.1.6: ^{4}He has no triple point. Below a pressure of 25bar it is impossible to freeze ^{4}He solely by cooling. (Again, an effect of the lower λ–point on a wetting

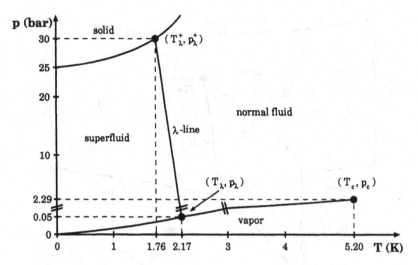

Fig. 1.6:   Semiquantitative phase diagram of ^{4}He in the (T,p) plane. The
λ–line terminates at two critical end points at $(T_\lambda, p_\lambda) = (2.17K, 0.05\text{bar})$ and at
$(T_\lambda^+, p_\lambda^+) = (1.76K, 30\text{bar})$. The liquid–vapor coexistence line ends in the critical
point $(T_c, p_c) = (5.20K, 2.29\text{bar})$. Note the break in scale on the pressure axis
above $p_\lambda = 0.05\text{bar}$.

layer of ^{4}He near $T = T_\lambda$ can be expected (see Sec.6.3).)

Some metallic elements like nickel show *ferromagnetic* phases. Without external
field ($H = 0$) the transition to the ferromagnetic state is second order (see Fig.1.1(b)).
In the (T,p) plane of the phase diagram of, e.g., nickel the second–order ferromagnetic
transitions are located on a line of critical points as shown schematically in Fig.1.7
[8]. This line terminates in the critical end point (T_{cep}, p_{cep}) of the ferromagnetic
transitions on the sublimation curve.

The above examples may illustrate that the occurrence of critical points as critical
end points is a rather common feature in systems with more than one component (see
Fig.1.5) or in one–component systems, in which additional internal degrees of freedom
give rise to a subdivision of the liquid or the solid one–phase region.

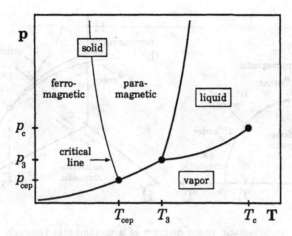

Fig. 1.7: Schematic phase diagram of a ferromagnetic
element like nickel in the (T, p) plane $(H = 0)$. The critical line indicates the locations of the second–order phase
transitions to the ferromagnetic state. It terminates in
a critical end point (T_{cep}, p_{cep}) on the sublimation curve.

1.1.2 Tricritical Points

In phase diagrams of systems with several internal degrees of freedom it may happen that several critical lines meet in a point, where critical behavior of different
origin prevails simulaneously. Such points are called *multicritical points*. The critical
behavior observed near a multicritical point depends on the *path* in the phase diagram along which this point is approached [2]. Depending on the number of critical
lines which meet at a multicritical point one distinguishes bicritical, tricritical, and
tetracritical points [2]. Here we will restrict ourselves to tricritical points.

One of the simplest systems which shows a tricritical point is a strongly anisotropic
antiferromagnet, often denoted as a *metamagnet* (see Eq.(1.19)) [2]. A schematic
phase diagram of a metamagnet in the (T, H) plane is shown in Fig.1.8(a) (see also
Fig.1.1(b) and Ref.[2]). The antiferromagnetic phase is characterized by an arrangement of spins with alternating orientation from one lattice point to another. A strong
magnetic field H which is homogeneous on the scale of the lattice spacing forces the
spins to align along the direction of the field thus forming a ferromagnetic type of spin
alignment. Upon releasing the field H at $T < T_c$ a phase transition to the antiferro-

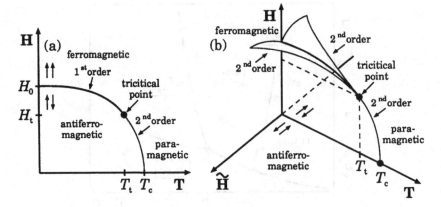

Fig. 1.8: (a) Schematic phase diagram of a metamagnet (strongly anisotropic antiferromagnet) in the (T, H) plane. The transition from the low field paramagnetic phase to the antiferromagnetic phase is second order (critical), the transition from the high field ferromagnetic phase to the antiferromagnetic phase is first order. The critical line meets the first–order line in the tricritical point (T_t, H_t). (b) Theoretical phase diagram of a metamagnet in the space spanned by T, H, and the *staggered field* $\widetilde{H}$. In this representation the tricritical point becomes visible as the intersection of three critical lines (see main text).

magnetic phase occurs. Below the *tricritical temperature* T_t this phase transition is first order, between T_t and the critical temperature T_c the transition is second order. A corresponding behavior can be observed as a function of temperature for magnetic fields H below H_0 (see Fig.1.8(a)). The *tricritical point* (T_t, H_t) separates the line of first order transitions from the line of second order (critical) transitions. According to the introductory remarks one should expect that *three* critical lines meet at the tricritical point, but only one is visible in Fig.1.8(a). In order to resolve this apparent contradiction we recall that a uniaxial antiferromagnet on a *cubic lattice* (see Sec.1.4) has the same phase diagram as a uniaxial ferromagnet (see Fig.1.1(b)), if the magnetic field H which imposes a ferromagnetic type of spin alignment on the system is replaced with a field $\widetilde{H}$ which imposes an antiferromagnetic (alternating) type of spin alignment on the system. Such a field must reverse its sign from one lattice point to another and it is therefore called the *staggered* field. There is no experimental realization of a staggered field, but it provides a helpful theoretical tool to reveal the missing two critical lines in the phase diagram of a metamagnet as Fig.1.8(b) shows. First, we observe that the $(T, \widetilde{H})$ plane in fact resembles the phase diagram of a uniaxial ferromagnet in Fig.1.1(b). Second, the additional critical lines terminate the two winglike surfaces (tricritical wings) of first–order transitions between the antiferromagnetic and the ferromagnetic state. Finally, the first–order line in Fig.1.8(a) is in fact the *triple line* in the completed phase diagram in Fig.1.8(b). The tricritical point is then located at $(\widetilde{H}, T, H) = (0, T_t, H_t)$.

The phase diagram of a binary liquid mixture shown in Fig.1.5 implicitly contains the potential to form a tricritical point, which can be visualized without the introduction of an artificial field. Depending on the microscopic interaction potentials w_{AA}, w_{BB}, and w_{AB} between the particles of species A or B, the critical end point moves to different positions on the liquid–vapor coexistence (vapor pressure) surface. For a special arrangement of w_{AA}, w_{BB}, and w_{AB} the critical end point falls onto the critical line L_1 of liquid–vapor transitions. In such a case the critical end point has become a tricritical point, which joins the three critical lines of the vapor–liquid A, liquid A–liquid B, and liquid B–vapor transitions. Note, that the phase diagram in Fig.1.5 then resembles Fig.1.8(b) in the vicinity of the tricritical point.

Like critical points tricritical points do not always occur as isolated exception points in a phase diagram. In a mixture of ^{3}He and ^{4}He the additional concentration degree of freedom extends the λ–line of superfluid phase transitions in ^{4}He to a *critical surface* of superfluid transitions. A schematic phase diagram of a ^{3}He–^{4}He mixture in the space spanned by the fugacity $z = \exp(\mu_{3He}/k_B T)$ of ^{3}He, the temperature T, and the pressure p is shown in Fig.1.9(a). For increasing fugacity or concentration of ^{3}He the superfluid transition temperature decreases until the transition switches to first order. The first– and second–order superfluid transition surfaces join in a *tricritical line*, which terminates in a lower and an upper *tricritical end point* on the vapor pressure surface and the melting surface, respectively. Near the lower tricritical end point the quantitative phase diagram *in* the vapor pressure surface of the ^{3}He–^{4}He mixture is shown in Fig.1.9(b) versus the mole fraction x of ^{3}He and the temperature T [9]. The curve $T_\lambda(x)$ is the intersection of the critical surface of the superfluid transitions with the vapor pressure surface. At the tricritical end point $(x_t, T_t) = (0.675, 0.867K)$ a gap in the mole fraction opens indicating that the superfluid transition has become first order. In a wetting experiment with ^{3}He–^{4}He mixtures close to the vapor pressure coexistence surface a tricritical effect on the

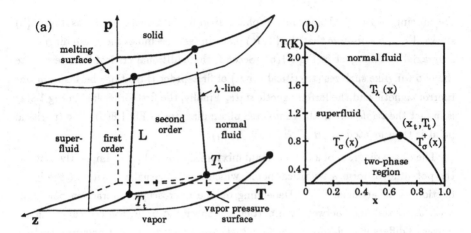

Fig. 1.9: (a) Schematic bulk phase diagram of a ^{3}He–^{4}He mixture versus the fugacity z of ^{3}He, the temperature T, and the pressure p. At $z = 0$ the phase diagram of pure ^{4}He is recovered (see Fig.1.6). The tricritical line L joins the surfaces of the first–order (front) and the second–order superfluid transitions. (b) Phase diagram of a liquid ^{3}He–^{4}He mixture at coexistence with the vapor phase versus the mole fraction x of ^{3}He and the temperature T. The line of the second order superfluid transitions $T_\lambda(x)$ ends in the tricritical point $(x_t, T_t) = (0.675, 0.867K)$, below which the transition becomes first order. The resulting two–phase region is bounded by $T_\sigma^-(x)$ and $T_\sigma^+(x)$ [9].

wetting layer is visible near the lower tricritical end point (see Sec.6.3).

1.1.3 Universality and Critical Exponents

The behavior of a simple fluid, a uniaxial ferromagnet, or a binary liquid mixture in the vicinity of the critical point (see Figs.1.1–1.4) is governed by the competition between two macroscopic states of different internal symmetry. For $T > T_c$ the system is in a high symmetry state which means that there is no difference between liquid and vapor in a simple fluid, that there is no spontaneous magnetization in a uniaxial ferromagnet, and that the two fluids A and B in a binary liquid mixture are completely miscible. Below T_c this symmetry is sponaneously *broken*, and phase separation takes place into liquid and vapor (simple fluid), into magnetized phases of opposite orientation (uniaxial ferromagnet), or into an A–rich and a B–rich liquid (binary liquid mixture). The competition between these opposing tendencies in the critical region leads to *critical fluctuations* of the density, the spontaneous magnetization, or the concentration, respectively, both in space and time. An experimental manifestation of these fluctuations is the so–called *critical opalescence* in simple fluids in which light is scattered at density fluctuations which vary typically on a scale of several thousand Angstroms. (see, e.g., Refs.[1] and [2]). Upon approaching the critical point the fluctuations become more violent, and their spectrum extends to larger and larger length and time scales. In the idealized picture of the thermodynamic limit the critical fluctuations loose any characteristic length and time scale as the critical point is reached so that to a certain extent the system becomes *scale invariant*. Molecular diameters or lattice spacings of course impose a lower bound on the length scales on which critical fluctuations can occur. However, the concept of scale invariance at the critical point matches the real physical situation very well and leads to valuable conclusions. The distance from the critical point is conveniently measured in terms of the reduced temperature $t \equiv (T - T_c)/T_c$ in combination with the reduced pressure $\delta p \equiv (p - p_c)/p_c$ for fluids, the magnetic field H for ferromagnets, or the reduced chemical potential $\delta \mu \equiv (\mu - \mu_c)/\mu_c$ with $\mu \equiv \mu_A - \mu_B$ for binary liquid mixtures. In terms of these variables thermodynamic quantities are *predominantly* governed by *power laws* as a consequence of approximate scale invariance.

Apart from thermodynamical quantities *correlation functions* contain important structural informations about a system near criticality. For very large distances correlation functions do not decay faster than the microscopic interaction potential $w(r)$ between two particles, which becomes important if $w(r)$ decays like a *power law*. At intermadiate but still macroscopic distances the scale of the (usually exponential) decay is set by the *correlation length* ξ. At the critical point ξ diverges due to scale invariance (in the thermodynamic limit) and the decay of the correlation function is governed by a *power law* as function of the distance. The exponents of these various power laws governing the critical behavior of the system are called *critical indices* or *critical exponents*. A summary of the definitions of the (bulk) critical exponents is given in Table 1.1 [2] (see Ref.[1]). The power laws shown in Table 1.1 should be understood as the *leading* terms in an expansion of the indicated quantity in powers of t, H, or δp, respectively. Therefore Table 1.1 shows the leading *singular* behavior of thermodynamic quantities, the correlation length, and the pair correlation

---

[2]Here the exponents above and below T_c are assumed to be equal.

Table 1.1: Definitions of the critical exponents of a bulk system in d dimensions (see Ref.[1]). As paradigms the definitions are given for a ferromagnet and a simple fluid. Amplitudes denoted by $A^\pm$ take *different* values A^+ for $t > 0$ and A^- for $t < 0$.

Quantitiy	Expo-nent	Definition	Variables t	H^a	δp^b
specific heat	α	$C = C_0^\pm |t|^{-\alpha}$	$\neq 0$	0	0
spontaneous magnetization[a]	β	$M = M_0(-t)^\beta$	< 0	0	
liquid–vapor density difference[b]	β	$\rho_l - \rho_g = \rho_0(-t)^\beta$	< 0		0
susceptibility[a]	γ	$\chi = \chi_0^\pm |t|^{-\gamma}$	$\neq 0$	0	
compressibility[b]	γ	$\kappa = \kappa_0^\pm |t|^{-\gamma}$	$\neq 0$		0
critical isotherm[a]	δ	$|M| = A_H |H|^{1/\delta}$	0	$\neq 0$	
critical isotherm[b]	δ	$|\rho_l - \rho_g| = A_p |\delta p|^{1/\delta}$	0		$\neq 0$
correlation length	ν	$\xi_\pm = \xi_0^\pm |t|^{-\nu}$	$\neq 0$	0	0
pair correlation function[c]	η	$G(r) = g_0 r^{-(d-2+\eta)}$	0	0	0

[a] ferromagnet
[b] fluid
[c] see Ref.[10]

function close to or at the critical point. Depending on the sign of the exponent the singularities are either cusps or algebraic divergences. The corrections to the leading singular behavior are often denoted as *corrections to scaling* [11]. The singular part of the specific heat for example reads

$$C(t) = C_0^\pm |t|^{-\alpha} \left(1 + c_1^\pm |t|^{x_1} + c_2^\pm |t|^{x_2} + \ldots \right) \quad \text{with} \quad 0 < x_1 < x_2 < \ldots, \qquad (1.1)$$

if the correction terms characterized by the correction exponents x_1, x_2, ... are included. The regular (background) part of the specific heat has the form of a Taylor series around $t = 0$.

The critical exponents introduced in Table 1.1, certain amplitude combinations, and the correction exponents x_1, x_2, ... do usually not depend on the specific microscopic structure of the system under consideration [12]. This property is known as *universality*. Both real and model systems can be assigned to *universality classes* within which the critical exponents take certain values. The universality classes have

Table 1.2:   Critical exponents for the Ising universality class in $d = 2$ (exact, see Ref.[1]) and in $d = 3$ [14] for short–ranged interactions. The classical exponents according to van–der–Waals[a] and Ornstein–Zernicke[b] theory are correct for long–ranged interactions (see main text).

	α	β	γ	δ	ν	η
$d = 2$	$0(\ln|t|)$	$\frac{1}{8}$	$\frac{7}{4}$	15	1	$\frac{1}{4}$
$d = 3$	0.11	0.33	1.24	4.79	0.63	0.037
classical	0^a	$\frac{1}{2}^a$	1^a	3^a	$\frac{1}{2}^b$	0^b

been named after their most prominent representative. The uniaxial ferromagnet (see Figs.1.1(b)–1.3(b)), which can be described by the *Ising model* (see Eq.(1.17)), a simple liquid (see Figs.1.1(a)–1.3(a)), and a binary liquid mixture (see Figs.1.4 and 1.5) belong to the *Ising universality class*. Liquid ^{4}He (see Fig.1.6) and the *XY–model* are members of the *XY universality class*. An isotropic ferromagnet like nickel (see Fig.1.7) is described by the Heisenberg model, which represents the *Heisenberg universality class*. Note, that the aforementioned three models describe *spin* systems (see Sec.1.4) in which the magnetization M has one component (Ising), two components (XY), or three components (Heisenberg). The high symmetry (paramagnetic) state of these spin systems in zero external field is $O(N)$–symmetric with respect to the magnetization, where $N = 1$ (Ising), $N = 2$ (XY), or $N = 3$ (Heisenberg) (see, e.g., Ref.[2]). Within a given universality class the critical exponents depend on the spatial dimension d and on the range of the interaction as demonstrated in Table 1.2 for the Ising universality class. An interaction potential is *short–ranged* concerning critical behavior, if $w(r)$ decays *faster* than $r^{-(d+2)}$ for $r \to \infty$ [13]. In this case the exponents given in the first and second row of Table 1.2 apply to $d = 2$ [1] and $d = 3$ [14], respectively. If the potential $w(r)$ decays *slower* than $r^{-(d+\sigma)}$ with $\sigma < d/2$ for $r \to \infty$, the potential is *long–ranged*. In this case the exponents denoted as *classical* in Table 1.2 apply [13]. In the interval $d/2 < \sigma < 2$ the exponents depend on σ [13]. Note, that $\alpha = 0$ in $d = 2$ indicates a logarithmic law for the specific heat C, whereas the classical $\alpha = 0$ indicates a discontinuity of C at $t = 0$. Note further, that a constraint acting on an extensive variable x can lead to a "renormalization" of the critical exponents according to [15]

$$\begin{aligned} \alpha &\rightarrow \alpha_x = -\alpha/(1-\alpha), \\ \beta &\rightarrow \beta_x = \beta/(1-\alpha), \\ \gamma &\rightarrow \gamma_x = \gamma/(1-\alpha), \end{aligned} \tag{1.2}$$

where $\alpha > 0$ has been assumed.

1.1.4   The Free Energy

The critical exponents introduced in Table 1.1 are not independent of each other. From thermodynamical stability requirements *inequalities* among the critical exponents can be derived [16]. In fact, both theoretical calculations and experiments have shown that these inequalities all hold as *equalities*. The origin of these equations is the fact that the thermodynamic quantities in Table 1.1 can be represented as *derivatives* of the (Gibbs) *free energy* in the critical region. In the spirit of a *leading singular* behavior as displayed in Table 1.1 the leading singular contribution $\mathcal{F}^{sing}$ to the free energy $\mathcal{F}$ in units of $k_B T_c$ and per volume V for $V \rightarrow \infty$ (thermodynamic limit) can be written in the *scaling form*

$$\lim_{V \to \infty} \frac{\mathcal{F}^{sing}(t,H)}{k_B T_c V} \equiv F^{sing}(t,H) = |t|^{2-\alpha} f_\pm \left(H|t|^{-\Delta} \right), \tag{1.3}$$

where the magnetic "language" has been used for convenience. The critical exponent Δ is the *gap exponent*, which can be related to other exponents defined in Table 1.1 (see below). The scaling relation given by Eq.(1.3) had first been formulated as a *scaling hypothesis* [17] and was later confirmed by the renormalization group treatment of critical behavior (see, e.g., Ref.[18]).

A first remarkable consequence of Eq.(1.3) is *data collapse*. If the *scaled* singular part of the free energy density $y \equiv F^{sing}(t,H)|t|^{\alpha-2}$ is plotted versus the *scaling argument* $x \equiv H|t|^{-\Delta}$ the data points collapse onto the two curves $y = f_+(x)$ for $t > 0$ and $y = f_-(x)$ for $t < 0$. The functions $f_+(x)$ and $f_-(x)$ are therefore called *scaling functions*. A corresponding scaling relation for the pair correlation function $G(r,t)$ for $H = 0$ and short–ranged interactions reads [1, 2]

$$G(r,t) = r^{-(d-2+\eta)} g_\pm(r/\xi_\pm), \tag{1.4}$$

where the scaling functions $g_\pm(x)$ have the property $g_\pm(0) = g_0$ and $\xi_\pm$ is the correlation length (see Table 1.1 and Eq.(5.10) in Sec.5.1). As a second consequence of Eqs.(1.3) and (1.4) relations among the critical exponents, the so–called *scaling laws*

$$\Delta = \beta\delta \quad ,$$
$$\alpha + 2\beta + \gamma = 2 \quad ,$$
$$\alpha + (\delta + 1)\beta = 2 \quad ,$$
$$\gamma = \nu(2 - \eta) \quad , \tag{1.5}$$

and the *hyperscaling law*

$$\alpha + d\nu = 2 \tag{1.6}$$

can be obtained for $2 \leq d \leq 4$ [1, 2]. The hyperscaling law plays an outstanding role among the scaling laws, because it explicitly involves the spatial dimension d. The exponents in Table 1.2 for $d = 2$ and $d = 3$ are in accordance with both Eq.(1.5) and Eq.(1.6) and the classical values for the critical exponents fulfill Eq.(1.5). Hyperscaling, however, for the classical exponents only holds in $d = 4$ which defines the so–called *upper critical dimension* (see Sec.1.4).

1.2   Surface Critical Phenomena

The thermodynamic limit is a valuable theoretical picture for the description of thermodynamic properties of macroscopic bodies, which are usually dominated by bulk (volume) effects. Especially in the vicinity of a critical point the simple power laws in Table 1.1 and the scaling properties of thermodynamic potentials (see Eq.(1.3)) and pair correlation functions (see Eq.(1.4)) are only valid in strictly homogeneous and isotropic systems. However, experimental samples differ from this idealized picture in the sense that they are of finite size, have a certain shape, and inevitably have *surfaces*.

On the level of a microscopic lattice model a surface in the simplest case is a plane of unsaturated (missing) bonds. In the magnetic language, which has been so successful in classifying bulk critical behavior, we can think of a uniaxial ferromagnet (Ising model) on a lattice which is bounded by a single plane surface on one side and otherwise extends to infinity. We will refer to such a system as a *halfspace* of

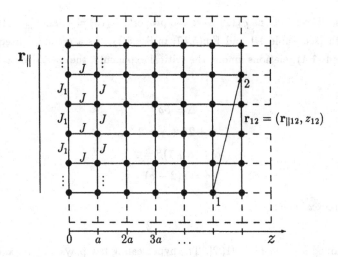

Fig. 1.10: Schematic view of a semi–infinite uniaxial ferromagnet
in d dimensions, where the surface is located at $z = 0$. The interaction is restricted to nearest neighbours with the strengths J inside
the lattice and J_1 in the surface. Position vectors $\mathbf{r}_{12} = \mathbf{r}_2 - \mathbf{r}_1$
are decomposed according to $\mathbf{r}_{12} = (\mathbf{r}_{\|12}, z_{12})$.

semi–infinite system in the following. This is again an idealized picture, which permits the identification of surface contributions to critical behavior. A schematic view
of a semi–infinite uniaxial ferromagnet on a simple cubic lattice in d dimensions is
shown in Fig.1.10 [19] (for the corresponding Ising model see Eq.(1.22) in Sec.1.4).
The direction perpendicular to the surface is denoted by z, the other $d - 1$ direc-
tions parallel to the surface are combined to the $d - 1$-dimensional vector $\mathbf{r}_\|$. The
interactions sketched as bonds connecting the dots in Fig.1.10 only couple nearest
neighbours, i.e., the interactions are *short–ranged*. In order to keep the picture as
simple as possible the interaction strengths only take the values J inside the lattice
and J_1 in the surface. Surface and bulk interactions can be of different nature. For
example, the surface interactions may be antiferromagnetic for an otherwise ferromagnetic system [19]. For a liquid system a surface in the spirit of Fig.1.10 can be
imposed by confining walls.

At a surface the spatial symmetry of the bulk system is broken. The quantitative effect of the presence of a surface on, say, the overall magnetization of a ferromagnetic sample depends on the penetration depth of the symmetry breaking effect of the surface into the volume. For short–range interactions such an effect will be damped out after a few lattice spacings so that the influence of a surface on overall properties of the system will be minute. However, there are two phenomena which increase the surface effect: *long–range interactions* and *long–range correlations* [20]. For short-ranged interactions the penetration depth of surface effects due to correlations is set by the correlation length $\xi_\pm$, which becomes large in the vicinity of a *critical point* (see Table 1.1). We can therefore expect that surface effects on the *critical* behavior of a system are particularly enhanced. In experiments *finite–size effects* on the critical behavior may accompany the surface effects (see Sec.1.3). Note, that long–range correlations prevail for *all* $T < T_c$ in $d = 3$ due to spin waves, if the system belongs to the XY– or Heisenberg universality class.

If the dimensionality of the semi–infinite system in Fig.1.10 is sufficiently large (depending on the range of the interactions) the surface can exhibit phase transitions and critical behavior of its own [19, 20]. A schematic summary of all possible combinations of bulk and surface phase transitions of a semi–infinite uniaxial ferromagnet without external fields is shown in Fig.1.11 versus the reduced coupling constants $K = J/k_B T$ and $K_1 = J_1/k_B T$ [20]. For a fixed ratio J_1/J the system follows a straight line of type 1,2, or 3 upon cooling starting from the origin with a slope J_1/J. The other lines in Fig.1.11 indicate the locations of second–order phase transitions. Three scenarios of successive surface and bulk transitions can be distinguished. For $J_1/J < K_1^c/K^c$ (path 1) a surface transition does not occur. After the bulk transition has taken place the spontaneous bulk magnetization imposes a ferromagnetic type of spin alignment on the surface. This is the so–called *ordinary* (O) transition. For $J_1/J > K_1^c/K^c$ (path 2) the surface undergoes a ferromagnetic surface phase transition indicated by the line (S) *before* the bulk transition occurs. The transition in the presence of a magnetized surface is called *extraordinary* (E) transition. For $J_1/J = K_1^c/K^c$ (path 3), finally, the surface and the bulk transition to the ferromagnetic state occur at the same temperature. This is called the *special* or *surface–bulk* (SB) transition. The three critical lines of the O, S, and E transitions intersect in the point (K^c, K_1^c) which identifies the location of the SB transition as a *multicritical* point [19, 20]. In two dimensions the surface in Fig.1.10 is a line which does not show any phase transition at finite temperatures for short–ranged interactions. In this case

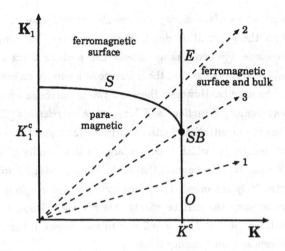

Fig. 1.11:   Schematic phase diagram of a semi–infinite
uniaxial ferromagnet in $d > 2$ versus $K = J/k_BT$ and
$K_1 = J_1/k_BT$. The lines O, S, E indicate the locations
of the second–order ordinary, surface, and extraordinary
phase transitions, respectively. The three critical lines intersect in a multicritical point, which ind'cates the special
or surface–bulk (SB) transition. The paths 1,2,3 correspond to trajectories of the system upon cooling for fixed
$J_1/J < K_1^c/K^c$, $J_1/J > K_1^c/K^c$, and $J_1/J = K_1^c/K^c$, respectively. In $d = 2$ the point SB does not exist [20].

only the O critical line is realized in the phase diagram, i.e., K_1^c is shifted to infinity.

In order to account for the surface critical behavior quantitatively various surface
quantities must be studied. The surface magnetizaticn for example can be probed
experimentally by spin polarized low–energy electron d'ffraction (SPLEED) (see, e.g.,
Ref.[20] for more examples). One finds that the leading behavior of surface quantities near *bulk* criticality (O, SB, or E) is again described by power laws. The
critical exponents, however, are *different* from the corresponding exponents in the
bulk system.

1.2.1 The Surface Tension

The presence of a surface increases the number of thermodynamically relevant variables for, e.g., a uniaxial ferromagnet. Besides the reduced bulk coupling constant $K = J/k_B T$ and the external bulk field H the *surface specific* reduced coupling $K_1 = J_1/k_B T$ and an external *surface* field H_1, which acts on the surface spins only, must be taken into account (see Refs.[19, 20]). The free energy $\mathcal{F}$ therefore depends on four variables which we choose according to $\mathcal{F} = \mathcal{F}(T, H; J_1/J, H_1)$. In order to study critical behavior of a semi–infinite system the SB multicritical point must be treated on the same footing as a bulk critical point. This is achieved by defining a *surface enhancement* $c = K_1^c/K^c - K_1/K$ [19, 20], which can be interpreted as the surface analogue of the reduced temperature t. Accordingly, the surface field H_1 is kept as the surface analogue of H. The O, SB, and E transitions in Fig.1.11 are then characterized by $H = H_1 = 0$, $t = 0$, and $c > 0$, $c = 0$, or $c < 0$, respectively [19, 20]. The *singular* part $\mathcal{F}^{sing}$ of the free energy $\mathcal{F}$ is conveniently written as a function of t, H, c, and H_1 according to $\mathcal{F}^{sing} = \mathcal{F}^{sing}(t, H; c, H_1)$. In order to obtain the surface contribution to $\mathcal{F}^{sing}$ the thermodynamic limit must be performed in two successive steps. The first step yields the singular bulk free energy density $F_b^{sing} \equiv F^{sing}$ according to Eq.(1.3), which contains no surface information. The singular part F_s^{sing} of the *surface tension*, i.e., the surface free energy per area A, can be extracted from $\mathcal{F}^{sing}$ in the limit $A \to \infty$. For a simple regular geometry the limit $A \to \infty$ leads to $V \to \infty$ and one finds

$$\lim_{A \to \infty} \left(\frac{\mathcal{F}^{sing}(t, H; c, H_1)}{A k_B T_c} - \frac{V}{A} F_b^{sing}(t, H) \right) = F_s^{sing}(t, H; c, H_1). \qquad (1.7)$$

Using Eqs.(1.3) and (1.7) one can set up a geometrical expansion of $\mathcal{F}^{sing}$, in which F_b^{sing} governs the leading (bulk) contribution and the singular surface tension F_s^{sing} governs the next to leading (surface) contribution. Disregarding higher order terms one has for a finite volume V [19, 20]

$$\mathcal{F}^{sing}(t, H; c, H_1) = k_B T_c \left(V F_b^{sing}(t, H) + A F_s^{sing}(t, H; c, H_1) + \dots \right). \qquad (1.8)$$

Note, that Eq.(1.8) holds for systems of arbitrary shape, where A is the total surface area. Higher order terms involve edge contributions (line tensions), curvature terms (shape tensions), and finite–size contributions for example.

As indicated above the limit $V \to \infty$ yields only the bulk free energy density which does not depend on the surface variables. However, the surface free energy

Table 1.3:   Definitions for the singular parts of surface and surface-induced themodynamical quantities (taken from Ref.[20]).

Type	Definition	Quantity
excess quantities	$M_s = -\partial F_s^{sing}/\partial H$	excess magnetization
	$U_s = -\partial F_s^{sing}/\partial t$	excess internal energy
	$\chi_s = \partial M_s/\partial H$	excess susceptibility
	$C_s = \partial U_s/\partial t$	excess specific heat
local quantities	$M_1 = -\partial F_s^{sing}/\partial H_1$	surface magnetization
	$U_1 = -\partial F_s^{sing}/\partial c$	surface internal energy
	$\chi_{11} = \partial M_1/\partial H_1$	local susceptibility
	$C_{11} = \partial U_1/\partial c$	local specific heat
mixed quantities	$\chi_1 = \partial M_1/\partial H$	layer susceptibility
	$C_1 = \partial U_1/\partial t$	layer specific heat

depends on the whole set of variables, which gives rise to a large variety of surface or surface–induced quantities. A summary of the definitions of their *singular* parts is given in Table 1.3 (see especially Ref.[20]). The difference between excess and local (surface) quantities can easily be demonstrated for the magnetization $M_{\infty/2}(z)$ in a semiinfinite geometry, which is a function of z due to the presence of a surface at $z = 0$ (see Fig.1.10). The bulk magnetization M (see Table 1.1) is then given by $M = M_{\infty/2}(z = \infty)$ and the surface and the excess magnetization have the representation [20]

$$
\begin{aligned}
M_1 &= M_{\infty/2}(z = 0), \\
M_s &= \int_0^\infty (M_{\infty/2}(z) - M)dz,
\end{aligned}
\tag{1.9}
$$

respectively.

In order to obtain a systematic overview over the leading singularities of the quantities defined in Table 1.3 a generalization of the scaling relation given by Eq.(1.3) for $F_s^{sing}(t, H; c, H_1)$ in a semiinfinite geometry is desirable. For the ordinary (O) and the extraordinary (E) transition (see Fig.1.11) $c \neq 0$ and only t, H, and H_1 remain as relevant scaling arguments near these transitions. The scaling relation then reads

[19, 20, 21]

$$F_s^{sing}(t, H; c \neq 0, H_1) = |t|^{2-\alpha_s} f_{s\pm} \left(H|t|^{-\Delta}, H_1|t|^{-\Delta_1} \right), \qquad (1.10)$$

where Δ is the gap exponent from Eq.(1.3) and Δ_1 is its surface counterpart. The exponent α_s is the critical exponent of the excess specific heat C_s (see below). Near the multicritical point (SB) in Fig.1.11 c manages the crossover from the ordinary transition to the surface (S) and the extraordinary transition. The scaling relation for F_s^{sing} near the multicritical point SB then involves a third scaling argument according to [19, 20]

$$F_{s,SB}^{sing}(t, H; c, H_1) = |t|^{2-\alpha_s} f_{s\pm}^{SB} \left(H|t|^{-\Delta}, H_1|t|^{-\Delta_1^{SB}}, c|t|^{-\phi} \right), \qquad (1.11)$$

where Δ_1^{SB} corresponds to Δ_1 in Eq.(1.10) and ϕ is the crossover exponent. For $H = H_1 = 0$ and fixed $t > 0$ the right hand side of Eq.(1.11) has a singularity at a finite negative value of c which marks the position of the surface transition S in Fig.1.11. Therefore the *scaling function* $f_{s+}^{SB}(0, 0, x)$ is singular at a certain value $x_s = c|t|^{-\phi} < 0$ of the argument x. In the vicinity of the multicritical point SB the line S can then be parameterized according to

$$t_s(c) = \frac{T_s(c) - T_{c,b}}{T_{c,b}} = \left(\frac{c}{x_s} \right)^{1/\phi}, \qquad (1.12)$$

where $T_s(c)$ denotes the critical temperature of the surface transition and $T_{c,b} \equiv T_c$ is the bulk critical temperature.

1.2.2 Surface Universality Classes

The scaling relations given by Eqs.(1.10) and (1.11) make a clear distinction between the ordinary and the extraordinary transition on one hand and the surface–bulk multicritical point on the other hand. Surface critical behavior near $t = 0$ can therefore not be expected to depend only on the bulk universality class. In fact, the type O, SB, or E of the surface, which can be characterized by the sign of the surface enhancement c, generates a subdivision of the bulk universality classes into *surface universality classes*. For $O(N)$–symmetric spin systems Fig.1.11 already indicates the three possible surface universality classes in $d = 3$. They can be characterized by $c = +\infty$ for the O surface universality class, $c = 0$ for the SB surface universality class, and $c = -\infty$ for the E surface universality class [20]. Each of these values of

c yields the *leading* singular behavior of the quantities defined in Table 1.3 at the associated O, SB, or E transition. *Finite* values of c generate *corrections to scaling* analogous to Eq.(1.1). The fact that $c = 0$ represents a surface universality class of its own emphasizes the meaning of the SB transition as a multicritical point.

The extraordinary transition is characterized by the presence of a magnetized surface, which can be magnetized either *spontaneously* due to a surface transition or externally by a surface field H_1. It has been shown that the origin of the surface magnetization is irrelevant for the leading singular behavior of the quantities in Table 1.3 at the extraordinary transition [22]. A semi–infinite system with a surface field can therefore be regarded as a representative of the E surface universality class. Surface fields then give access to the extraordinary transition in *two dimensions*, where a surface transition usually does not exist. In this respect the O and SB surface universality classes are *symmetry conserving* and the E surface universality class is *symmetry breaking*.

The singular behavior of the surface–induced quantities given by Table 1.3 is governed by power laws in the reduced temperature t and the fields H and H_1, where the three surface universality classes must be considered separately.

1.2.3 Surface Critical Exponents

According to Table 1.3 a large variety of surface critical exponents can be expected. However, the scaling relations for the singular surface free energy given by Eqs.(1.10) and (1.11) guarantee the existence of scaling laws like those in Eq.(1.5) among the surface critical exponents. The definitions of the exponents for the excess, local, and mixed quantities (see Table 1.3) and the exponents for the pair correlation function are collected in Table 1.4 (taken from Ref.[20]). The pair correlation function G is no longer a function of only the distance between the points $\mathbf{r} = (\mathbf{r}_{\parallel}, z)$ and $\mathbf{r}' = (\mathbf{r}'_{\parallel}, z')$ as in the fully isotropic bulk case. For a semi–infinite geometry spatial isotropy only holds in planes parallel to the surface so that $G = G(|\mathbf{r}_{\parallel} - \mathbf{r}'_{\parallel}|, z, z')$. Note further, that only for $|\mathbf{r} - \mathbf{r}'| \to \infty$ in a direction *strictly parallel* to the surface one has $G \sim |\mathbf{r}_{\parallel} - \mathbf{r}'_{\parallel}|^{-(d-2+\eta_{\parallel})}$. In *any* other case the decay of G is governed by the exponent $\eta_{\perp}$ [20, 23].

The singularity of the surface free energy F_s^{sing} for $H = H_1 = 0$ and $2 \leq d \leq 4$ is already fixed by dimensional analysis according to $F_s^{sing} \sim \xi_{\pm}^{-(d-1)} \sim |t|^{2-\alpha_s}$. From the hyperscaling relation Eq.(1.6) and the scaling relations given by Eqs.(1.10) and (1.11) the scaling laws

$$\alpha_s = \alpha + \nu,$$
$$\beta_s = \beta - \nu,$$
$$\gamma_s = \gamma + \nu,$$
$$\delta_s = \Delta/\beta_s, \tag{1.13}$$

follow [19, 20] so that the exponents of the excess quantities are solely given by bulk exponents. Therefore only Δ_1 or ϕ appear as new exponents in the scaling relations for the singular surface free energy. Furthermore, the following scaling laws hold for any surface universality class [19, 20]

$$\Delta_1 = (\nu/2)(d - \eta_{\parallel}),$$
$$\eta_{\perp} = (\eta + \eta_{\parallel})/2,$$
$$\beta_1 = (\nu/2)(d - 2 + \eta_{\parallel}),$$
$$\gamma_1 = \nu(2 - \eta_{\perp}),$$
$$\delta_1 = \Delta/\beta_1,$$
$$\gamma_{11} = \nu(1 - \eta_{\parallel}),$$
$$\delta_{11} = \Delta_1/\beta_1. \tag{1.14}$$

The scaling laws for α_1 and α_{11} explicitly depend on the surface universality class. For the SB surface universality class one finds (see Eq.(1.11) and Table 1.3)

$$\alpha_1^{SB} = \alpha_s - 1 + \phi,$$
$$\alpha_{11}^{SB} = \alpha_s - 2 + 2\phi, \tag{1.15}$$

whereas the O surface universality class yields [20, 24]

$$\alpha_1^{O} = \alpha - 1,$$
$$\alpha_{11}^{O} = \alpha - 2 - \nu. \tag{1.16}$$

The critical exponents for local and mixed quantities and $\eta_{\parallel}$ and $\eta_{\perp}$ can therefore be expressed in terms of Δ_1, ϕ, and bulk exponents. For the E surface universality class there is evidence that *all* surface exponents can be expressed in terms of bulk exponents [19, 20].

Table 1.4: Definition of the critical exponents for excess, local, and mixed quantities according to Table 1.3 and for the pair correlation function parallel and perpendicular to the surface (taken from Ref.[20]). Amplitudes have not been defined for brevity.

Type of quantitiy	Expo– nent	Definition	t	H	H_1
excess	α_s	$C_s \sim |t|^{-\alpha_s}$	$\neq 0$	0	0
	β_s	$M_s \sim (-t)^{\beta_s}$	< 0	0	0
	γ_s	$\chi_s \sim |t|^{-\gamma_s}$	$\neq 0$	0	0
	δ_s	$M_s \sim |H|^{1/\delta_s}$	0	$\neq 0$	0
local	α_{11}	$C_{11} \sim |t|^{-\alpha_{11}}$	$\neq 0$	0	0
	β_1	$M_1 \sim (-t)^{\beta_1}$	< 0	0	0
	γ_{11}	$\chi_{11} \sim |t|^{-\gamma_{11}}$	$\neq 0$	0	0
	δ_1	$M_1 \sim |H|^{1/\delta_1}$	0	$\neq 0$	0
	δ_{11}	$M_1 \sim |H_1|^{1/\delta_{11}}$	0	0	$\neq 0$
mixed	α_1	$C_1 \sim |t|^{-\alpha_1}$	$\neq 0$	0	0
	γ_1	$\chi_1 \sim |t|^{-\gamma_1}$	$\neq 0$	0	0
correlation	$\eta_{\|}$	$G(|\mathbf{r}_{\|} - \mathbf{r}'_{\|}|, z, z) \sim |\mathbf{r}_{\|} - \mathbf{r}'_{\|}|^{-(d-2+\eta_{\|})}$	0	0	0
	$\eta_\perp$	$G(0, z, z') \sim |z - z'|^{-(d-2+\eta_\perp)}$	0	0	0

Finally, we mention that the *surface* transition (line S in Fig.1.11) is governed by the $d - 1$–dimensional counterparts of the bulk critical exponents in d dimensions (see Table 1.1). The reduced temperature for the surface transition is given by $t_s = (T - T_s(c))/T_s(c)$ (see Eq.(1.12)) and the relevant external field is a linear combination of H and H_1 [20].

1.3   Finite Size Systems

Experimental samples of critical systems are not only characterized by the presence of surfaces, they are also always of finite size and have a certain shape. The critical behavior discussed in Secs.1.1 and 1.2 was based on the thermodynamic limit in which the volume and, in the case of a semi–infinite geometry, the surface area has become infinite at constant particle density in the bulk and in the surface. The bulk free energy F_b^{sing} and the surface free energy F_s^{sing} both display critical singularities in the form of power laws as shown in Tables 1.1 and 1.4. A finite size system naturally has *curved* surfaces which give rise to additional curvature contributions (shape tensions) to the free energy. Edges and corners and the corresponding line tensions and corner contributions to the free energy in this respect are only extreme cases of curvature. Moreover, the *finite distance* of two opposing surfaces generates specific *finite-size* contributions to the free energy, which decrease if the geometry is enlarged without changing its shape. The free energy of finite size systems therefore displays a whole geometrical hierarchy of contributions of which Eq.(1.8) only gives a first impression.

The finite size of experimental samples has even more drastic repercussions on the structure of the free energy. The partition function of a strictly finite system, which in reality always consists of a finite number of constituents, is a finite sum of exponentials and therefore becomes neither zero nor infinite. The free energy will then never show singularities so that the critical behavior in the sense of Secs.1.1 and 1.2 will not exist. However, in practice the laboratory samples are usually so large and consist of so many constituents that the power laws given in Tables 1.1 and 1.4 are excellent approximations of the actual "critical" behavior except for the singularity right *at* the critical point. To which extent the experimental samples can be regarded to be inifinitely large depends on the size which the *correlation length* $\xi_{\pm}$ actually reaches in comparison with the linear extensions L of the system. If even very close to a critical point $\xi_{\pm}$ remains far smaller than L, any finite-size effect on the critical behavior will be invisible. If, however, $\xi_{\pm}$ becomes comparable with L, strong deviations from the bulk critical behavior will be observed. In this respect geometries with a large extension L' in one direction and a small extension L in another direction are of particular interest. If an experiment can be designed such that close to a second-order phase transition $L < \xi_{\pm} \ll L'$, critical behavior in *reduced spatial dimension* will be observed. Whether the "remaining" dimensionality is still large enough to sustain a phase transition depends on the range of the interaction potential and on the internal symmetry of the high temperature state (see Secs.1.4 and 5.5). The simplest example for such an arrangement is given by a system confined to a *film geometry*.

1.3.1   Critical Films

Liquid films form in a natural way due to *wetting* transitions. In the simplest case a
homogeneous flat substrate with a smooth surface is exposed to the vapor of a fluid.
Close to liquid–vapor coexistence a liquid layer of macroscopic thickness forms on
the substrate if the substrate potential is stronger than the corresponding interaction
between the fluid particles [25]. For binary liquid mixtures (see Fig.1.5) or liquid
^{4}He (see Fig.1.6) critical phenomena in the film occur near $T = T_{cep}$ or $T = T_\lambda$,
respectively (see Sec.6.3). In the case of a critical wetting film one of the boundaries
is in fact a liquid–vapor *interface* which exhibits fluctuations (capillary waves) of its
own. If the interface undergoes a *roughness transition* the film thickness will fluctuate
on macroscopic length scales [19, p.64]. In particular, the wetting behavior of liquid
^{4}He near the λ–transition on various substrates has attracted some attention in the
past years (see Sec.6.3 for details).

In an experiment a film geometry can also be prepared artificially by confining
the system between two external parallel plane walls at a prescribed distance L. Such
a sharply defined geometry allows the experimental observation of critical behavior
crossing over from three to two dimensions as the correlation length grows (see also
Sec.5.5). One of the best studied systems with respect to bulk critical phenomena
and critical surface and finite–size effects is liquid ^{4}He near the lower λ–point [26],
because ^{4}He allows the preparation of very pure samples and, moreover, temperature
gradients in ^{4}He are strongly suppressed in the superfluid state. The measurements in
particular probe the *specific heat* (see Sec.6.1) and the superfluid density (see Sec.6.2).
The onset of superfluidity in a film of thickness L is expected to be located at an L–
dependent temperature $T_c(L)$ below T_λ governed by the power law $T_\lambda - T_c(L) \sim L^{-\Lambda}$
(see Eq.(2.5) in Sec.2.1). The critical behavior of the helium film itself is governed by
the XY universality class ($N = 2$) in two dimensions (see Secs.1.4 and 5.5 and Fig.2.1).
For a detailed comparison between theory and experiment we refer to Secs.6.1 and
6.2.

The derivative of the free energy of the film with respect to L can be interpreted as
a *force* acting between the boundaries. Especially near the bulk critical temperature
this force bears a very close resemblance with the *Casimir force* which has first been
discovered in electrodynamics, and we will therefore refer to this force as the (critical)
Casimir force (see Chap.3). Near bulk criticality the critical Casimir force enters the

force balance in a fluid film which is governed by van–der–Waals forces outside the critical region. Experimentally the critical modification of the force balance can either be observed directly using a suitably adapted atomic force microscope (see Sec.6.4) or indirectly by measuring the change of a wetting layer thickness near a critical point (see Sec.6.3).

1.3.2 Critical Fluids in Capillaries

Capillaries or tubes are typical realizations of *cylindrical* geometries. Experimental investigations have primarily been performed on liquid ^{4}He confined in Nuclepore filters which provide a cylindrical confinement in capillaries between a few hundred and a few thousand Angstroms in diameter [26]. As for the film geometry much effort has been spent on the specific heat and the superfluid density concerning their scaling behavior with the diameter D of the cylindrical channels in Nuclepore filters. The length of these capillaries is usually much larger than their diameter so that a crossover from three–dimensional to one–dimensional "critical" behavior is observed. For ^{4}He $d = 1$ is not large enough to sustain a phase transition and therefore the specific heat exhibits only a rounded maximum at a certain temperature T_m instead of a sharp peak (see Fig.2.1). The difference $T_\lambda - T_m$ is expected to scale with $D^{-\Lambda}$, which has also been checked against the theoretical prediction $\Lambda = 1/\nu$. The experimental data, however, first indicated a violation of the expected scaling behavior (see Ref.[26] and Secs.6.1 and 6.2).

For investigations concerning the wetting behavior of liquid ^{4}He on solid substrates the *vibrating–wire microbalance technique* provides a very useful tool [27] (see Sec.6.3). The mass load on the wire due to the adsorbed film changes the frequency of the wire oscillation which in turn gives a measure of the thickness of the adsorbed layer. This arrangement represents an "inverted" capillary in the sense that the surface of the wire provides an inner wall for the liquid ^{4}He, whereas the outer boundary is a liquid–vapor interface. Provided, that the layer thickness is much smaller than the circumference of the wire, intermediate two–dimensional critical behavior will occur during the crossover from $d = 3$ to $d = 1$. In this situation the geometry resembles a film, because the wire radius at this intermediate stage is much larger than the correlation length so that curvature effects only form corrections to the finite–size contribution to the free energy. Of course a corresponding filmlike geometry can be prepared inside a capillary, but as the film thickness increases beyond a certain threshold value the whole capillary will be filled by the liquid at once due to the onset of *capillary condensation* (see, e.g., Ref.[28]).

1.3.3 Critical Fluids in Pores

Fluids in porous media are confined to a large number of small cavities each about 100Å in diameter in, e.g., carbon black powder. Altogether the interior of a porous medium provides a rather irregularly shaped geometry which can hardly be reproduced from one experiment to another [26]. Unlike the geometries discussed before a pore in the above sense provides a complete confinement, i.e., the fluid is bounded by solid walls in any direction. Depending on the shape of the pore a cascade of crossovers in the critical behavior from $d = 3$ via $d = 2$ (films) to $d = 1$ (capillaries) and finally $d = 0$ (closed cavities) can be studied. However, for quantitative tests of scaling laws uniform confinement is required in order to avoid the disturbing influence of a broad pore size distribution on the experimental data [26].

Liquid ^{4}He near the lower λ–point again seems to be by far the best–studied system. As for confinement in capillaries no real phase transition occurs for complete confinement so that the specific heat again exhibits a maximum at a temperature T_m which is shifted with respect to T_λ (see Fig.2.1). Scaling of the maximum and its position can be obtained, but it is hard to define a unique confining size (see Secs.6.1, 6.2, and Ref.[26] for details).

1.4 *Theoretical Concepts*

The quantitative theoretical understanding of critical phenomena has been a rather challenging task due to the lack of characteristic length and time scales in critical fluctuations. The most fundamental approach is based on microscopic models of the system under consideration. In two dimensions the problem of finding a theoretical understanding of critical phenomena has been very successfully attacked this way. The highly developed methods for the solution of two–dimensional models (see Ref.[29]) have led to many exact results for the critical exponents one of which is displayed in Table 1.2. For the construction of model systems which exhibit critical behavior the concept of universality can be exploited in the sense that a detailed representation of the microscopic physics is not essential. Unfortunately the theoretical tools developed for two–dimensional models do not apply in three dimensions so that exact results for critical exponents in $d = 3$ for the most relevant universality classes do not exist.

A very useful approximate treatment of model systems in statistical mechanics is provided by *mean–field theory*. The microscopic interaction of one particle in the system with its neighbours is replaced with an averaged (mean) field which self–consistently accounts for the interaction potential between the particles. Mean–field theory succeeds in predicting phase transitions and critical points, but, however, collective phenomena responsible for critical behavior are overestimated so that phase transitions are predicted even in one dimension for short–ranged interactions. The critical exponents predicted by mean–field theory for the Ising-, XY-, and Heisenberg universality class turn out to be the classical ones (see Table 1.2) in contrast to experimental findings. Anyway, mean–field theory as a theoretical tool must not be rejected on the basis of the above shortcomings, because the global structure of phase diagrams is often reproduced correctly by a mean–field analysis. Moreover, mean–field theory already bears the *order parameter* concept and the more abstract *Landau theory* in it which is the starting point for the modern view of critical behavior, the *field–theoretical renormalization group*.

Computer simulations of microscopic model systems based on the Monte–Carlo technique have become a very powerful tool for the investigation of critical phenomena [30]. The lattices on which these microscopic models are defined and simulated are inevitably *finite*. In order to obtain reliable estimates on, say, bulk critical exponents the lattices must of course be sufficiently large. However, instead of simulating only the largest possible system a *series* of simulations on smaller lattices of different sizes is performed so that the bulk critical exponents can be extracted from a *finite–size scaling* analysis. Surface effects on the critical behavior (see Sec.1.2) are usually eliminated by imposing periodic boundary conditions on the lattice.

1.4.1   Lattice Models and Universality

A large variety of critical phenomena can be conveniently described in magnetic terms. It is therefore not surprising that those lattice models which are usually studied concerning their critical behavior are *spin systems*. The most common of these spin systems is given by the spin-$\frac{1}{2}$ *Ising* model defined by the Hamiltonian [1, 2]

$$\mathcal{H}(\{s\}) = -\tfrac{1}{2}\sum_{i,j} J_{ij}s_i s_j - g\mu_B H \sum_i s_i, \qquad (1.17)$$

where $J_{ij} \equiv J(|\mathbf{r}_i - \mathbf{r}_j|)$ is the pair exchange interaction strength and H is the external magnetic field. The spin variable $s_i \equiv s(\mathbf{r}_i)$ at the lattice site $\mathbf{r}_i$ can take the values $\pm\frac{1}{2}$ so that $g\mu_B s_i$ is the magnetic moment associated with the spin variable s_i in units of the Bohr magneton μ_B. For $J_{ij} > 0$ the model Hamiltonian defined by Eq.(1.17) describes a *uniaxial ferromagnet* which belongs to the *Ising universality class* like simple fluids or binary liquid mixtures. The corresponding critical exponents are displayed in Table 1.2. For $J_{ij} < 0$ the *antiparallel* orientation of neighbouring spins is energetically favored and Eq.(1.17) represents a *uniaxial antiferromagnet* in an external magnetic field. On a cubic lattice the antiferromagnetic and the ferromagnetic models are equivalent, i.e., the uniaxial antiferromagnet belongs to the Ising universality class. If, however, the underlying lattice is triangular ($d = 2$), the antiferromagnetic interaction leads to *frustration* which totally changes the critical behavior. Although the frustrated uniaxial antiferromagnet is still represented by an *Ising* model, it does no longer belong to the Ising universality class [5] (see below).

The XY– and Heisenberg universality classes are represented by classical spin models as well. In these cases the spin variable $\mathbf{s}_i \equiv \mathbf{s}(\mathbf{r}_i)$ at a lattice site $\mathbf{r}_i$ is chosen as a unit vector with two (XY) or three (Heisenberg) components. Reducing the degree of freedom to one component leads back to the Ising model. With a pair interaction $J_{ij} = J(|\mathbf{r}_i - \mathbf{r}_j|)$ the classical XY– and Heisenberg models in an external field $\mathbf{H}$ have the form

$$\mathcal{H}(\{\mathbf{s}\}) = -\tfrac{1}{2}\sum_{i,j} J_{ij}\mathbf{s}_i \cdot \mathbf{s}_j - g\mu_B \mathbf{H} \cdot \sum_i \mathbf{s}_i. \qquad (1.18)$$

The classical nature of the spin variable $\mathbf{s}_i$ in Eq.(1.18) is irrelevant for the critical behavior of XY– or Heisenberg models. In order to illustrate the differences between the critical behavior of the ferromagnetic ($J_{ij} > 0$) Ising–, XY–, and Heisenberg model estimates for their bulk critical exponents in $d = 3$ are listed in Table 1.5 [14]. Note, that the singularity in the specific heat changes its character from the Ising model (simple fluid near T_c, see Figs.1.1–1.3) to the XY model (liquid ^{4}He near T_λ, see Fig.1.6).

Table 1.5: Bulk critical exponents for a ferromagnetic Ising–, XY–, and Heisenberg model in three dimensions (see Ref.[14]).

	α	β	γ	δ	ν	η
Ising	0.11	0.33	1.24	4.79	0.63	0.037
XY	-0.01	0.35	1.31	4.77	0.67	0.040
Heisenberg	-0.13	0.37	1.39	4.78	0.71	0.040

A lattice model for an *anisotropic* antiferromagnet ($J_{ij} < 0$) in terms of the classical spin variables s_i of the Heisenberg model is given by [2]

$$\mathcal{H}(\{s\}) = -\tfrac{1}{2} \sum_{i,j} J_{ij} s_i \cdot s_j - K \sum_i (s_i^z)^2 - g\mu_B \mathbf{H} \cdot \sum_i s_i, \qquad (1.19)$$

where s_i^z denotes the z–component of s_i. For sufficiently strong anisotropy $K < 0$ the Hamiltonian in Eq.(1.19) provides a model for a *metamagnet* which exhibits a *tricritical* point in the phase diagram (see Fig.1.8 and Ref.[2]).

For the theoretical description of critical behavior only the identification of the universality class is essential. In this perspective a model representative of a certain universality class does not need to incorporate any correspondence to microscopic physics. The most famous example in this respect is given by the q–state Potts model which is defined by the Hamiltonian [31]

$$\mathcal{H}(\{p\}) = -J \sum_{<ij>} \delta_{p_i p_j}, \qquad (1.20)$$

where the summation over i and j is only carried out over nearest neighbour pairs $< ij >$. The state of a Potts "spin" at the lattice site i is characterized by an integer $p_i \in \{1, 2, .., q\}$, and according to Eq.(1.20) two spins at neighbouring sites i and j only interact with a strength $-J$, if they are in the same state. For $J > 0$ the ground state is given by the q–fold degenerate uniform configuration $p_i = p$ of the Potts spins on the lattice. In the case $q = 2$ Eq.(1.20) is equivalent to the spin-$\tfrac{1}{2}$ Ising model without an external field (see Eq.(1.17)). The Potts model always shows a phase transition in $d \geq 2$, but whether for a given q and d the transition is of first or second order turns out to be a rather delicate question. The present knowledge about the answer is summarized in Fig.1.12 [31] which shows the conjectured *critical*

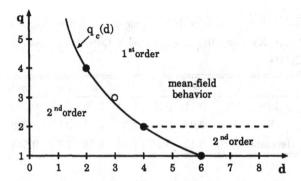

Fig. 1.12:   Critical curve $q = q_c(d)$ for the q–state Potts model (taken from Ref.[31]). Below the critical curve (solid line) the Potts model shows a second–order phase transition. Above the critical curve the behavior is as expected from mean–field theory, i.e., the phase transition is first–order for $q > 2$ and otherwise second–order. Only the points $(d, q) = (2, 4), (4, 2)$, and $(6, 1)$ (full circles) on the critical curve are rigorosly known. To present knowledge the point $(d, q) = (3, 3)$ (open circle) belongs to the first–order regime.

curve $q = q_c(d)$ as a solid line interpolating the exactly known values $q_c(2) = 4$, $q_c(4) = 2$, and $q_c(6) = 1$. Above the critical curve the order of the phase transition is as expected from mean–field theory, it is first order for $q > 2$ (dashed line) and second order for $q \leq 2$. Below the critical curve and at the known three points the transition is always second order. To present knowledge the three–state Potts model in $d = 3$ (open circle) shows a weak first–order transition and therefore has been assigned to the first–order regime in Fig.1.12 [31]. For $q \leq \max(q_c(d), 2)$ the Potts model defines its own, the q–state Potts, universality class which has widespread realizations for phase transitions in adsorbed layers on solid substrates [4] and beyond Eq.(1.20) for $q = 1$ (percolation) or $q = 0$ (random linear resistor networks) [31]. In two dimensions the Potts model has been exactly solved for $q = 3$ and $q = 4$ [29, 31] and the critical exponents are summarized in Table 1.6. The antiferromagnetic spin–$\frac{1}{2}$ Ising model on a triangular lattice belongs to the *three–state Potts* universality class in $d = 2$ [5]. The transition of an atomic layer adsorbed on the (111)–surface of a solid substrate to the

Table 1.6:   Critical exponents for the q–state Potts
model in $d = 2$ ($J > 0$) (taken from Ref.[31]). The
q–state Potts universality class incorporates various
critical phenomena in terms of the parameter q.

q	α	β	γ	δ	ν	η
0^a	$-\infty$	1/6	∞	∞	∞	0
1^b	-2/3	5/36	43/18	91/5	4/3	5/24
2^c	0	1/8	7/4	15	1	1/4
3^d	1/3	1/9	13/9	14	5/6	4/15
4^e	2/3	1/12	7/6	15	2/3	1/4

^{a} random resistor network
^{b} percolation
^{c} ferromagnetic Ising model (see Table 1.2)
^{d} transition to $\sqrt{3} \times \sqrt{3}$–phase in an adsorbate [4]
^{e} transition to 2×2–phase in an adsorbate [4]

$\sqrt{3} \times \sqrt{3}$–phase is another representative of the three–state Potts universality class
in $d = 2$. The corresponding transition to the 2×2–phase represents the four–state
Potts universality class in $d = 2$ (see Table 1.6) [4].

Keeping at least some microscopic aspects in a statistical lattice model does not
neccessarily lead to a representative of an experimentally relevant universality class.
An example for this is the so–called *spherical model*. Although the universality class
of the spherical model does not seem to be relevant for real systems, the model itself
has the advantage to be rigorously solvable in any spatial dimension of interest. This
ability to find exact solutions assigns a prominent role to the spherical model in the
theoretical analysis of critical finite–size properties (see Sec.2.2). The definition of
the model is based on the Ising Hamiltonian (see Eq.(1.17)), where the spin variables
$s_i = s(\mathbf{r}_i)$ are allowed to vary in the continuous range of the real numbers rather
than on a finite set of discrete values. This Ising–like Hamiltonian together with the
spherical constraint $\sum_i s_i^2 = N$, where N is the number of lattice sites, defines the
spherical model. A widely used variant of the spherical model relaxes the spherical
constraint to the *mean spherical constraint* $\sum_i \langle s_i^2 \rangle = N$ which in a *bulk* model is
equivalent to $\langle s_i^2 \rangle = 1$. This second variant of the spherical model is called *mean*

spherical model and it displays the same critical behavior as the spherical model. The pair interaction potential $J_{ij} = J(|\mathbf{r}_i - \mathbf{r}_j|)$ is usually chosen to be positive. For the asymptotic behavior $J(r \to \infty) \sim r^{-(d+\sigma)}$ of the pair interaction potential the following values for the bulk critical exponents can be obtained [29]

$$\alpha = \frac{d - 2\sigma^*}{d - \sigma^*} \, , \, \beta = \frac{1}{2} \, , \, \gamma = \frac{\sigma^*}{d - \sigma^*} \, , \, \delta = \frac{d + \sigma^*}{d - \sigma^*} \, , \, \nu = \frac{1}{d - \sigma^*} \, , \, \eta = 2 - \sigma^* \, , \quad (1.21)$$

where $\sigma^* < d < 2\sigma^*$ and $\sigma^* = \min(\sigma, 2)$. For short–ranged interaction ($\sigma > 2$, see Ref.[13]) the exponents given by Eq.(1.21) reduce to the classical (mean–field) exponents (see Table 1.2) in $d = 4$. A phase transition occurs for the spherical model, if $d \geq \sigma^*$.

On the level of microscopic lattice models surfaces can be introduced by adding a surface contribution to the bulk Hamiltonian. The semi–infinte uniaxial ferromagnet sketched in Fig.1.10 is described by an Ising model according to the Hamiltonian [19, 20]

$$\mathcal{H}(\{s\}) = -\frac{1}{2} \sum_{i,j \in V} J_{ij} s_i s_j - g\mu_B H \sum_{i \in V} s_i - \frac{1}{2} \sum_{i,j \in S} J_{1,ij} s_i s_j - g\mu_B H_1 \sum_{i \in S} s_i, \quad (1.22)$$

where the summation indices i and j symbolically represent lattice points $\mathbf{r}_i$ and $\mathbf{r}_j$ in the bulk ($i, j \in V$) or in the surface ($i, j \in S$). The surface pair interaction $J_{1,ij}$ and the surface field H_1 are the counterparts of the corresponding bulk quantities J_{ij} and H. For *ferromagnetic* bulk and surface interactions ($J_{ij} > 0, J_{1,ij} > 0$) and $H = H_1 = 0$ the model Hamiltonian defined by Eq.(1.22) displays the phase diagram shown in Fig.(1.11). Semi–infinite versions of the XY– or Heisenberg model are constructed by direct analogy with Eq.(1.22), where the surface field $\mathbf{H}_1$ has become a two– or three–component vector. For a film geometry a second surface contribution of the above type completes the Hamiltonian. For periodic boundary conditions such surface contributions do not occur.

1.4.2   The Order Parameter Concept

The theoretical picture of a second–order phase transition is closely related to the process of *spontaneous symmetry breaking*. This picture has already been used in Sec.1.1 in order to visualize the origin of critical fluctuations as a competition between two macroscopic states of low ($T < T_c$) and high ($T > T_c$) internal symmetry. From a slightly different point of view the low symmetry and the high symmetry state can be characterized as *ordered* and *disordered*, respectively. In a simple liquid the degree of order (phase separation) is indicated by a finite gap $\rho_l - \rho_g$ between the liquid and the vapor density which opens below T_c. In a uniaxial ferromagnet the amount of order is indicated by a nonzero spontaneous magnetization, and in a binary liquid mixture a concentration gap opens between the A–rich and the B–rich liquid phase (see Figs.1.1–1.4). The density difference, the spontaneous magnetization, and the concentration difference therefore serve as *order parameters* which indicate disorder by vanishing identically and order by taking nonzero values. The broken symmetry here is the symmetry of the disordered state under *sign reversal* of the order parameter which characterizes the *Ising universality class*. The order parameter in an XY– or Heisenberg system is again given by a spontaneous magnetization which in this case is a two– or three–component vector. In the ordered state the spontaneous magnetization defines a (spontaneously chosen) preferred direction in the system which breaks the continuous $O(N)$–symmetry characterizing the disordered state of the XY– ($N = 2$) or Heisenberg ($N = 3$) universality class. In this sense the disordered state of the Ising universality class has the discrete $O(N = 1)$–symmetry. A general definition of the order parameter **M**, the *spontaneous magnetization*, for an $O(N)$–symmetric spin system is given by the thermodynamical average

$$\mathbf{M} = \left\langle \sum_i \mathbf{s}_i \right\rangle \tag{1.23}$$

for $\mathbf{H} = 0$. For a semiinfinte uniaxial ferromagnet (see Eq.(1.22) and Fig.1.10) one can define a spontaneous layer magnetization by

$$M_{\infty/2,n} = \left\langle \sum_{i \in S_n} s_i \right\rangle, \tag{1.24}$$

where S_n is the n–th layer of spins below the surface layer S_0. In the continuum limit $M_{\infty/2,n}$ becomes a continuous function $M_{\infty/2}(z)$ of the coordinate $z = an$ (see Fig.1.10). The spontaneous *surface magnetization* $M_1 = M_{\infty/2}(z = 0)$ then provides the order parameter for the surface transition (see Fig.1.11, line S) and the function $M_{\infty/2}(z)$ defines the *order parameter profile* (see Eq.(1.9)).

In a uniaxial antiferromagnet the ordered state is characterized by an alternating spin alignment. The lattice can be subdivided into two sublattices A and B such that all spins on sublattice A point "up" and all spins on sublattice B point "down".

The appropiate order parameter is the *staggered magnetization* $\widetilde{M}$ which is defind by

$$\widetilde{M} = \left\langle \sum_{i \in A} s_i \right\rangle - \left\langle \sum_{i \in B} s_i \right\rangle. \tag{1.25}$$

In the examples described so far the definition of the order parameter is more or less obvious. In general, however, there is no recipe according to which an order parameter can be constructed. At least some hints concerning the type of order preferred in the ordered state can be obtained from an analysis of the ground state of the system, *provided* there are no other phase transitions between the one under consideration and $T = 0$. A second source of information is mean–field theory which is expected to have the order parameter already incorporated, provided, it reasonably reproduces the structure of the phase diagram in the vicinity of the critical point of interest. The order parameter itself is not restricted to be a scalar or a vector. In a liquid crystal rodlike molecules sitting at *random* positions can exhibit *orientational order*. The order parameter in this case is the *tensor*

$$S_{ij}^{\alpha\beta} = \tfrac{1}{2} \left\langle 3 e_i^\alpha e_j^\beta - \delta_{ij} \delta^{\alpha\beta} \right\rangle, \tag{1.26}$$

where $\mathbf{e}_i = (e_i^1, e_i^2, e_i^3)$ for $i = 1, 2, 3$ forms a local coordinate system in which the orientation of that molecule is fixed [3, 5].

The order parameter for the Potts model is an analogue of the spontaneous magnetization M in an Ising system. However, the symmetry group is not $O(N)$ but Z_q, the group of all permutations of q elements, which characterizes the disordered macroscopic state of the Potts model. In order to determine the structure of the order parameter in this case we introduce external fields which enable us to impose that type of order on the disordered state from *outside* which *spontaneously* occurs below the critical temperature. The nearest neighbour Potts model in an external field reads ($J > 0$, see Eq.(1.20))

$$\mathcal{H}(\{p\}) = -J \sum_{<ij>} \delta_{p_i p_j} - \sum_{p=1}^{q-1} h_p \sum_i \delta_{p_i p}. \tag{1.27}$$

In order to favor the homogeneous (ordered) configuration $p_i = p \in \{1, 2, \ldots, q-1\}$ the field component h_p can be turned positive with all other components kept zero. The last possible ordered configuration $p_i = q$ is then energetically favored if *all* field components $h_1, h_2, \ldots, h_{q-1}$ are turned *negative*. The order parameter which couples linearly to the external field therefore has $q - 1$ independent components $M_1, M_2, \ldots, M_{q-1}$ which can be arranged in $q - 1$–dimensional space (see below).

1.4.3   Continuum Models

The microscopic lattice Hamiltonian of a model system can be interpreted as a functional on the space of the configurations of microscopic states on the lattice. In continuum theory for critical systems this microscopic Hamiltonian is replaced with an *effective Hamiltonian* which is a functional of a coarse grained but still fluctuating field $\Phi(\mathbf{r})$ describing, e.g., the local magnetization in a spin system. In absence of external fields the thermodynamical average $\langle\Phi(\mathbf{r})\rangle$ with respect to the effective Hamiltonian is the *order parameter*. In the vicinity of a critical point the order parameter varies continuously between zero $(T > T_c)$ and small finite values $(T < T_c)$ so that in the spirit of *Landau theory* the effective Hamiltonian is expanded in powers of $\Phi(\mathbf{r})$. The expansion is truncated if a finite minimum of the effective Hamiltonian is guaranteed in the whole critical region. The resulting expression is known as the *Ginzburg–Landau Hamiltonian* $\mathcal{H}\{\Phi(\mathbf{r})\}$, and it represents the standard continuum model for the theoretical description of critical behavior.

Below a second–order phase transition the system is in an ordered state which has been spontaneously chosen from a set of equivalent (degenerate) ordered states. This sudden preference for one of the ordered states breaks the internal symmetry which characterizes the disordered state. The equivalence of the ordered states below T_c without external fields is incorporated into the functional form of the Ginzburg–Landau Hamiltonian by the requirement that $\mathcal{H}\{\Phi(\mathbf{r})\}$ is *invariant* under the symmetry group of the disordered state. For short–ranged interactions a gradient expansion is combined with the expansion of $\mathcal{H}\{\Phi(\mathbf{r})\}$ in powers of $\Phi(\mathbf{r})$, where only the lowest order term compatible with the symmetry is kept. Long–ranged pair interaction potentials generate *nonlocal* contributions to the Ginzburg–Landau functional. The construction of Ginzburg–Landau functionals especially under this symmetry aspect is discussed in Ref.[3] in great detail so that we quote only a few examples here.

We restrict the following discussion to the case of short–ranged interactions which can be incorporated in a *local* Ginzburg–Landau Hamiltonian. In the case of $O(N)$–symmetry, which captures the Ising–, XY–, and the Heisenberg universality classes, $\mathcal{H}\{\Phi(\mathbf{r})\}$ is only allowed to depend on $|\Phi(\mathbf{r})|$ and $|\nabla\Phi(\mathbf{r})|$ in an analytical way apart from contributions due to external fields. The Hamiltonian reads

$$\mathcal{H}\{\Phi(\mathbf{r})\} = \int d^d r \left[\tfrac{1}{2}\left(\nabla\Phi(\mathbf{r})\right)^2 + \tfrac{\tau}{2}\Phi^2(\mathbf{r}) + \tfrac{g}{4!}\left(\Phi^2(\mathbf{r})\right)^2 - \mathbf{h}(\mathbf{r})\cdot\Phi(\mathbf{r})\right], \qquad (1.28)$$

where

$$\Phi = (\phi_1, \ldots, \phi_N) \quad , \quad (\nabla \Phi)^2 \equiv \sum_{i=1}^{N} (\nabla \phi_i)^2 ,$$

and $\mathbf{h(r)}$ is a smoothly varying external (magnetic) field which couples linearly to (the magnetization) $\Phi(\mathbf{r})$ and thus breaks the $O(N)$–symmetry. The parameter τ is the *bare* reduced temperature and the parameter g is assumed to be a *positive* constant in the whole critical region. Neglecting fluctuations the Ginzburg–Landau Hamiltonian can be interpreted as the *Landau* free energy which upon *minimalization* with respect to $\Phi(\mathbf{r})$ yields the order parameter and the *leading singular* behavior of the free energy near the critical point (see Eq.(2.54) in Sec.2.3). The critical exponents predicted by this minimum principle (Landau theory) for the $O(N)$–symmetric Hamiltonian in Eq.(1.28) are again the classical ones given in Table 1.2 for any N. In this respect Landau theory and mean–field theory are equivalent. For *semi–infinite* $O(N)$–symmetric systems an effective surface Hamiltonian must be added to Eq.(1.28) which is typically of the form [20]

$$\mathcal{H}_1\{\Phi(\mathbf{r}_{\|}, 0)\} = \int d^{d-1} r_{\|} \left[\frac{c}{2} \Phi^2(\mathbf{r}_{\|}, 0) - \mathbf{h}_1(\mathbf{r}_{\|}) \cdot \Phi(\mathbf{r}_{\|}, 0) \right], \qquad (1.29)$$

where $\mathbf{r} = (\mathbf{r}_{\|}, z)$, c is the surface enhancement [20] and $\mathbf{h}_1(\mathbf{r}_{\|})$ is the surface field. Landau theory here leads to *order parameter profiles* $M_{\infty/2}(z)$ which are of different shapes in the various regions of the phase diagram shown in Fig.1.11 [20]. Note, that Eq.(1.29) does not represent the most general form of $\mathcal{H}_1$. A cubic surface contribution like $\int d^{d-1} r_{\|} \frac{w_1}{3!} \Phi^3(\mathbf{r}_{\|}, 0)$ may become relevant in the context of *critical adsorption* [32].

Tricritical behavior in an $O(N)$–symmetric system is captured by the effective Hamiltonian

$$\mathcal{H}\{\Phi(\mathbf{r})\} = \int d^d r \left[\frac{1}{2} (\nabla \Phi(\mathbf{r}))^2 + \frac{\tau}{2} \Phi^2(\mathbf{r}) + \frac{g}{4!} \left(\Phi^2(\mathbf{r}) \right)^2 + \frac{w}{6!} \left(\Phi^2(\mathbf{r}) \right)^3 - \mathbf{h}(\mathbf{r}) \cdot \Phi(\mathbf{r}) \right],$$
$$(1.30)$$

where τ and g are smooth functions of the temperature T and a second parameter x (for examples see Figs.1.8 and 1.9). *At the tricritical point* $(T, x) = (T_t, x_t)$ one has $\tau = g = 0$ (see Ref.[2]). The parameter w is assumed to be a positive constant in the tricritical region in order to provide stability. A surface contribution to Eq.(1.30) for a semi–infinite tricritical $O(N)$–symmetric system allows more terms than Eq.(1.29). It reads [33]

$$\mathcal{H}_1\{\Phi(\mathbf{r}_{\|}, 0)\} = \int d^{d-1} r_{\|} \left[\frac{c}{2} \Phi^2(\mathbf{r}_{\|}, 0) + \frac{v}{4!} \left(\Phi^2(\mathbf{r}_{\|}, 0) \right)^2 \right] \qquad (1.31)$$

in absence of symmetry–breaking external fields.

The order–parameter space of the 3–state Potts model is two–dimensional ($\Phi = (\phi_1, \phi_2)$), and due to the Z_3–symmetry the three equivalent ordered states can be visualized as the corner points of an equilateral triangle centered at the origin $\phi_1 = \phi_2 = 0$. Without loss of generality $\langle \phi_2 \rangle = 0$ can be assumed for one of the ordered states so that the Z_3–invariance becomes equivalent to invariance under reflection at the ϕ_1–axis and rotations by 120 and 240 degrees. The effective Hamiltonian for the 3–state Potts model then reads (see Ref.[5])

$$\mathcal{H}\{\Phi(\mathbf{r})\} = \int d^d r \left[\tfrac{1}{2} (\nabla \Phi(\mathbf{r}))^2 + \tfrac{\tau}{2} \Phi^2(\mathbf{r}) - \tfrac{f}{3!} \left(\phi_1^3(\mathbf{r}) - 3\phi_1(\mathbf{r}) \phi_2^2(\mathbf{r}) \right) + \tfrac{g}{4!} \left(\Phi^2(\mathbf{r}) \right)^2 \right],$$
(1.32)

where $f, g > 0$ is assumed and symmetry–breaking fields are omitted. The order–parameter space for the 4–state Potts model is three–dimensional ($\Phi = (\phi_1, \phi_2, \phi_3)$) so that the four equivalent ordered states can be visualized as the corner points of a tetrahedron centered at the origin. By similar arguments the Z_4–operations can be mapped onto reflections and rotations in three–dimensional space which leads to the effective Hamiltonian [5]

$$
\begin{aligned}
\mathcal{H}\{\Phi(\mathbf{r})\} &= \int d^d r \left[\tfrac{1}{2} (\nabla \Phi(\mathbf{r}))^2 + \tfrac{\tau}{2} \Phi^2(\mathbf{r}) - \tfrac{f}{3!} \phi_1(\mathbf{r}) \phi_2(\mathbf{r}) \phi_3(\mathbf{r}) \right. \\
&\quad \left. + \tfrac{g}{4!} \left(\Phi^2(\mathbf{r}) \right)^2 + \tfrac{u}{4!} \left(\phi_1^4(\mathbf{r}) + \phi_2^4(\mathbf{r}) + \phi_3^4(\mathbf{r}) \right) \right],
\end{aligned}
$$
(1.33)

where symmetry–breaking fields are again omitted and $f, g, u > 0$. In contrast to $O(N)$–symmetry the Z_3– and Z_4–symmetries allow the construction of *cubic* invariants, and therefore Landau (mean–field) theory predicts a *first–order* phase transition for the 3–state and 4–state Potts model in absence of external fields. According to Fig.1.12 this prediction is correct in $d \geq 3$, but it fails in $d = 2$.

Due to the Ginzburg–Landau criterion critical behavior is captured correctly by Landau theory if the spatial dimension d exceeds a certain threshold value which is known as the *upper critical dimension* d_c. For the $O(N)$–symmetric systems near critical points (see Eqs.(1.28) and (1.29)) one has $d_c = 4$, near tricritical points (see Eqs.(1.30) and (1.31)) one has $d_c = 3$. In $d \leq d_c$ a systematic treatment of the critical fluctuations is needed which is usually performed within the framework of the *field–theoretical renormalization group* [34, 35, 36, 37]. The deviation $\varepsilon = d_c - d$ of the spatial dimension d from the upper critical dimension d_c is a useful parameter which controls the modification of critical behavior by critical fluctuations.   For $\varepsilon = 0$

($d = d_c$) fluctuations at most lead to *logarithmic corrections* of the asymptotic power laws which are still governed by Landau theory. For $\varepsilon > 0$ ($d < d_c$) the fluctuations become so violent that the critical exponents depart from their classical values given by Landau theory. *Field theory* provides the key to the mechanism that brings about universality for critical exponents (see Table 1.1 and Refs.[34, 35, 36, 37]), certain amplitude combinations (see Ref.[38], Chap.3, and Chap.5), and scaling functions (see Chap.4 and Chap.5).

Conformal field theory in two dimensions provides a general framework in which certain two–dimensional models are characterized by the value of a single parameter, the *conformal anomaly number* or *cental charge c* (see Eq.(2.61) in Sec.2.3). The spectrum of possible values of the critical exponents for these models is completely determined by general properties of the corresponding scaling operators in conformal field theory (see Ref.[39]). The principle of *conformal invariance* relates correlation functions and scaling density profiles in various infinite and finite geometries (see Sec.2.3, Chap.3, and Chap.4).

2.   Finite Size Scaling

2.1   Phenomenology

In the thermodynamic limit the concept of scale invariance leads to power–law behavior of thermodynamical quantities in the vicinity of critical or multicritical points. These power laws can be derived from scaling relations for the bulk free energy density (see Eq.(1.3)) and the surface tensions (see Eq.(1.7)) which characterize the leading singular behavior of the total free energy of the system in the vicinity of a critical point (see Eq.(1.8)). As already pointed out in Sec.1.3 the theoretical description of the critical behavior in real finite systems by singular thermodynamical quantities can only be approximate. The accuracy of this approximation is governed by the size of the correlation length $\xi_{\pm}$ in comparison with the smallest spatial extension of the system. Using modern experimental techniques correlation lengths can be achieved up to several thousand Angstroms so that system sizes of this order of magnitude become clearly visible in the experimental data of, e.g., the specific heat (see Sec.1.3 and Ref.[26]). On the theoretical side a decomposition of the free energy of a finite system *only* into bulk and surface contibutions according to Eq.(1.8) is clearly not sufficient although Eq.(1.8) may be quite accurate away from the critical point. This is simply due to the singularities in F_{bulk}^{sing} and F_{s}^{sing} which remain uncompensated in Eq.(1.8). So besides the geometrical hierarchy of bulk, surface, edge, or curvature contributions a distinguished finite–size term must be included in Eq.(1.8) in order to obtain analytic behavior at bulk criticality. This will be illustrated below for one of the simplest finite geometries one can think of, the *film geometry*.

A system confined in a film geometry is bounded by two plane walls a and b which are a finite distance L apart. A film does not exhibit edges or curved surfaces so that the decomposition of the singular part of the free energy in units of $k_B T_c$ and per cross–section area A reads in the limit $A \to \infty$

$$\lim_{A \to \infty} \frac{\mathcal{F}^{sing}(t, H; c_a, H_a; c_b, H_b; L)}{k_B T_c A} = L F_{bulk}^{sing}(t, H)$$

$$+ F_s^{sing}(t, H; c_a, H_a) + F_s^{sing}(t, H; c_b, H_b) + \delta F^{sing}(t, H; c_a, H_a; c_b, H_b; L), \quad (2.1)$$

where δF^{sing} is the finite–size contribution. As in Eq.(1.8) we use the magnetic language for convenience. Due to the presence of two surfaces two surface fields H_a

and H_b and two surface enhancements c_a and c_b must be considered which appear in the list of arguments of δF^{sing}. According to Eq.(1.3), (1.10), and (1.11) the bulk and surface contributions in Eq.(2.1) can be written in a *scaling form*. The basic idea of *finite–size scaling theory* is that the size dependence incorporated in the finite–size correction δF^{sing} and therefore the size dependence of $\mathcal{F}^{sing}$ exhibits scaling as well. The number of scaling arguments depends on the surface universality classes which the two walls a and b represent. As shown in Eq.(1.11) the case of SB surfaces is the most complicated one, because the six scaling fields t, H, c_a, H_a, c_b, and H_b in this case generate macroscopic length scales according to

$$\xi_t \sim |t|^{-\nu}, \xi_H \sim |H|^{-\frac{\nu}{\Delta}}, \xi_{c_a} \sim |c_a|^{-\frac{\nu}{\phi}}, \xi_{H_a} \sim |H_a|^{-\frac{\nu}{\Delta_1^{SB}}}, \xi_{c_b} \sim |c_b|^{-\frac{\nu}{\phi}}, \xi_{H_b} \sim |H_b|^{-\frac{\nu}{\Delta_1^{SB}}}.$$

$$(2.2)$$

For short–ranged interactions ξ_t can be chosen to be the bulk correlation length $\xi_\pm$, whereas for long–ranged interactions this is no longer possible (see below). The film thickness L forms an additional length scale so that for $d < d_c$ $\mathcal{F}^{sing}$ can be written in the scaling form [40, 41, 42, 43]

$$\lim_{A\to\infty} \frac{\mathcal{F}^{sing}(t, H; c_a, H_a; c_b, H_b; L)}{k_B T_c A} = L^{-(d-1)} \left\{ (|t|^\nu L)^d f_{bulk\pm} \left(H L^{\Delta/\nu} \right) \right.$$

$$+ (|t|^\nu L)^{d-1} \left[f_{s\pm}^{SB} \left(H L^{\Delta/\nu}; c_a L^{\phi/\nu}, H_a L^{\Delta_1^{SB}/\nu} \right) + f_{s\pm}^{SB} \left(H L^{\Delta/\nu}; c_b L^{\phi/\nu}, H_b L^{\Delta_1^{SB}/\nu} \right) \right]$$

$$\left. + \delta f_\pm \left(|t| L^{1/\nu}, H L^{\Delta/\nu}; c_a L^{\phi/\nu}, H_a L^{\Delta_1^{SB}/\nu}; c_b L^{\phi/\nu}, H_b L^{\Delta_1^{SB}/\nu} \right) \right\},$$

$$(2.3)$$

where the hyperscaling relation Eq.(1.6) and Eq.(1.13) has been used. In the case of O and E surfaces the corresponding surface enhancements do not contribute to the list of relevant scaling fields and the number of scaling arguments reduces accordingly (see Eqs.(1.10) and (1.11) and Ref.[20]). In the following we will only be concerned with the *leading* singular behavior of the scaling functions defined in Eq.(2.3) for a given combination (a, b) of surface universality classes. Finite or nonzero values of c_a, H_a, c_b, and H_b only generate *corrections to scaling* (see Sec.1.2 and Ref.[20]) so that the *leading singular* behavior of Eq.(2.3) can be expressed as

$$\lim_{A\to\infty} \frac{\mathcal{F}^{sing}(t, H; c_a, H_a; c_b, H_b; L)}{k_B T_c A} = L^{-(d-1)} \left\{ (|t|^\nu L)^d f_{bulk\pm} \left(H L^{\Delta/\nu} \right) \right.$$

$$\left. + (|t|^\nu L)^{d-1} \left[f_{s\pm,a} \left(H L^{\Delta/\nu} \right) + f_{s\pm,b} \left(H L^{\Delta/\nu} \right) \right] + \delta f_{\pm a,b} \left(|t| L^{1/\nu}, H L^{\Delta/\nu} \right) \right\}. \quad (2.4)$$

The dependence of the scaling functions in Eq.(2.4) on the surface universality classes characterized by $c_a, c_b \in \{-\infty, 0, \infty\}$ (see Sec.1.2) is captured by the subscripts a and b. Note, that the combination $(a, b) = (E, E)$ is not unique. The realization of the extraordinary surface universality class by surface fields in $O(N)$–symmetric systems allows the formation of *interfaces* $(N = 1)$ or *helical* magnetization profiles $(N \geq 2)$ [44] (see below).

The finite–size scaling functions $\delta f_\pm$ and $\delta f_{a,b\pm}$ defined in Eq.(2.3) and (2.4) have to meet two requirements. First, the finite–size scaling functions must *compensate* the critical singularities of F_{bulk}^{sing}, $F_{s,a}^{sing}$, and $F_{s,b}^{sing}$ in Eq.(2.1). Second, the finite–size scaling function has to account for the critical behavior of the film which in the idealized picture of the partial thermodynamic limit $A \to \infty$ is $d - 1$–dimensional. The precise meaning of this second requirement depends on the bulk universality class represented by the confined system, the range of the interaction, and the spatial dimension d. For brevity we restrict the following discussion to short–ranged interactions in $d = 3$. If the confined system belongs to the Ising universality class, critical singularities of the *film* are located at a critical temperature $T_c(L) \neq T_{c,b} \equiv T_c(\infty)$. In presence of surface fields the symmetry is already broken at $T = T_c(\infty)$ and an additional shift of the critical field to a finite value $H_c(L)$ must be taken into account [40, 41, 42]. The critical singularities at this shifted critical point are governed by the critical exponents of the Ising universality class in *two* dimensions (see Table 1.2). If the confined system belongs to the XY universality class a phase transition to a state with long–ranged order will not occur [45]. Instead, we encounter the Kosterlitz–Thouless transition at $T = T_{KT}(L)$, below which the correlation length remains infinite and the exponent η governing the pair correlation function becomes a *function* $\eta(T)$ of the temperature [46]. If, finally, the confined system belongs to the Heisenberg universality class no phase transition occurs at all [45]. To present knowledge the Kosterlitz–Thouless transition has no analogue in the Heisenberg universality class. Additional aspects and theoretical achievements in this *dimensional crossover* problem will be discussed in Secs.2.2 and 5.5.

These different kinds of behavior are illustrated and summarized in Fig.2.1, where the specific heat of an Ising–like film (a) and a XY–like film (b) is schematically shown as a function of temperature in comparison with the bulk specific heat in the absence of external fields. The singularity in the specific heat of the Ising film at $T = T_c(L)$ in Fig.2.1(a) is governed by a logarithm $(\alpha(d = 2) = 0)$, whereas the specific heat

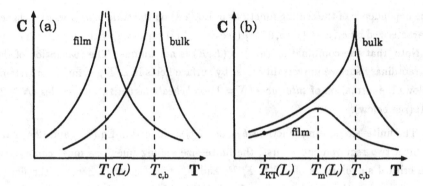

Fig. 2.1:   Specific heat of a film in the Ising universality class (a) and in the
XY universality class (b) in $d = 3$ without external fields. The bulk specific heat
is shown for comparison. In the Ising universality class two–dimensional critical
behavior occurs at a finite temperature $T_c(L)$, whereas in the XY universality
class a rounded maximum occurs at a temperature $T_m(L)$. In the Heisenberg
universality class the behavior of the specific heat is similar to (b), where the
Kosterlitz–Thouless transition is absent.

of the XY film in Fig.2.1(b) displays a *rounded maximum* at $T = T_m(L)$ and the
Kosterlitz–Thouless transition at $T = T_{KT}(L)$ which remains invisible in the specific
heat curve. The singularities of the bulk specific heats are governed by the exponent
α in $d = 3$ (see Table 1.5). The specific heat of a Heisenberg film behaves as shown
in Fig.2.1(b) except that a Kosterlitz–Thouless transition does not occur.

The position of the shifted critical point $(T_c(L), H_c(L))$ in a (T, H) phase diagram
(see Fig.1.1) is determined by the finite–size scaling function $\delta f_{\pm a,b}(x, y)$ to which we
will refer exclusively in the following (see Eq.(2.4)). The shifted critical point is
given by a singular point $(x, y) = (x_c, y_c)$ of $\delta f_{a,b\pm}(x, y)$, which immediately implies
to leading order [40]

$$t_c(L) \equiv \frac{T_{c,b} - T_c(L)}{T_{c,b}} = x_c L^{-1/\nu}, \text{and} \quad H_c(L) = y_c L^{-\Delta/\nu}, \tag{2.5}$$

where $t_c(L)$ is known as the *fractional shift*. Note, that $y_c \neq 0$ only in presence
of surface fields. The exponent of the fractional shift is often denoted by Λ which
Eq.(2.5) predicts to be given by $\Lambda = 1/\nu$. A schematic phase diagram of an Ising
system in $d = 3$ confined to a film geometry in presence of surface fields is shown in

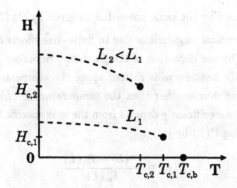

Fig. 2.2: Schematic phase diagram for an Ising
system confined to a film geometry of thicknesses L_1 and $L_2 < L_1$. The shifted critical
point is located at $(T_{c,i}, H_{c,i}) = (T_c(L_i), H_c(L_i))$
for $i = 1, 2$, respectively. The shifted coexistence lines are shown as dashed curves (taken
from Ref.[47]).

Fig.2.2. For a film of thickness L the shifted critical point is located at $T = T_{c,i} \equiv$
$T_c(L_i)$ and $H = H_{c,i} \equiv H_c(L_i)$ for $i = 1, 2$. Upon reduction of the film thickness the
coexistence line shown as dashed lines in Fig.2.2 becomes shorter and more strongly
curved (see Ref.[47]). Note, that the coexistence line of the bulk Ising system is given
by the temperature interval $(0, T_{c,b})$ on the T axis (see Fig.1.1(b)).

When the fractional shift was first considered the general value $\Lambda = 1$ had been
obtained from a mean–field like argument [48]. Exact calculations for Ising model
strips in two dimensions, where the specific heat becomes *rounded* (see Fig.2.1(b)),
cannot make a decision, because $\Lambda = 1 = 1/\nu$ in $d = 2$ and, moreover, $x_c = 0$ in
this case so that $t_c(L)$ is governed by the first correction term $L^{-2} \ln(L/a)$, where
a is the lattice constant [49, 50]. The value $\Lambda = 1$ for the shift exponent was also
obtained for the spherical model and the ideal Bose fluid in a film geometry with free
boundary conditions in $d \geq 3$ [51, 52]. However, in retrospective the result $\Lambda = 1$
in the latter two cases seems to be due to the *implementation* of the mean spherical
and the density constraint, respectively [42]. The *concept* of finite–size scaling itself
remains valid in either case [52, 53]. Finally, we note that the position of $T_m(L)$ in

Fig.2.1(b) is determined by the same power law as given by Eq.(2.5) for $T_c(L)$.

The rounding of critical singularities due to finite–size effects as shown in Fig.2.1
(b) can be captured by the definition of the *fractional rounding* [42, 48, 49, 51, 52].
If the rounded specific heat curve is shifted along the temperature axis such that
$T_m(L)$ and $T_{c,b}$ lie one over another then the temperature $T^*(L)$ where the shifted
film specific heat curve significantly departs from the bulk specific heat curve [3] defines
the fractional rounding $t^*(L)$ by [51]

$$t^*(L) \equiv \frac{T^*(L) - T_m(L)}{T_m(L)}. \tag{2.6}$$

The physical origin of the fractional rounding is the growth of the correlation length
ξ_+ as $T_{c,b}$ is approached from above. At $T = T^*(L) > T_m(L)$ we have $\xi_+(t^*) \simeq L$
which leads to [51, 52]

$$t^*(L) \simeq \left(L/\xi_0^+\right)^{-1/\nu}. \tag{2.7}$$

As for the fractional shift a new exponent θ has been introduced for the fractional
rounding which is identified as $\theta = 1/\nu$ by Eq.(2.7) [42, 51].

The singular part $C_{a,b}^{sing}(t, H; L)$ of the specific heat of a film per unit volume can
be immediately inferred from Eq.(2.4) as

$$\lim_{A \to \infty} \frac{C_{a,b}^{sing}(t, H; L)}{k_B A L} = L^{\alpha/\nu} C_{\pm a,b}\left(|t| L^{1/\nu}, H L^{\Delta/\nu}\right), \tag{2.8}$$

where the hyperscaling relation Eq.(1.6) has been used. At the critical point $(t, H) =
(0, 0)$ of the bulk system (see Fig.2.1) the specific heat scales as $L^{\alpha/\nu}$ which is an
experimentally testable prediction (see Sec.6.1). For an Ising strip in $d = 2$ $L^{\alpha/\nu}$ is
replaced by $\ln L$ [49, 50].

The scaling forms of the thermodynamical quantities discussed above are only
valid if the dimension d is smaller than the upper critical dimension d_c. In $d = d_c$
logarithmic corrections to the usual scaling behavior in $d < d_c$ must be expected
due to the presence of logarithmic corrections to the pure Landau critical behavior
(see below). In $d > d_c$ the breakdown of hyperscaling even modifies the form of the
scaling arguments. For periodic boundary conditions applied to a critical system with
short–ranged interactions confined to a cube L^d the singular part of the free energy

---

[3]There is some arbitrariness in the definition of $T^*(L)$ (see Ref.[51]).

can be written in the form [54, 55, 56]

$$\frac{\mathcal{F}^{sing}(T,H;L)}{k_B T_{c,b} L^d} = L^{-d} f_{\pm}\left(|t|L^{d/2}, HL^{3d/4}\right).$$ (2.9)

If, however, the surfaces of the cube represent the surface universality classes described in Sec.1.2, a *third* scaling variable $|t|L^2$ may be neccessary in a scaling relation like Eq.(2.9) [55].

Long–ranged interactions, too, require a separate finite–size scaling analysis, because the decay of the bulk correlation functions can no longer be characterized by a finite correlation length $\xi_{\pm}$ [57]. In this case a length scale $\lambda_L(t)$ in the finite system can be defined such that $\lambda_\infty(t) \equiv \lim_{L\to\infty} \lambda_L(t) \propto t^{-\nu}$ for $t \searrow 0$, so that $L/\lambda_\infty(t)$ forms a proper scaling argument. However, $\lambda_\infty(t)$ does not neccessarily provide the *only* length by which L can be scaled (see Eq.(2.42) in Sec.2.2) [57].

2.1.1 Logarithmic Corrections

The decomposition of the free energy of a finite system according to Eq.(2.1) follows a geometrical hierarchy. If L is a typical length of the system each contribution is weighted by a factor $L^{d'}$, where d' is the spatial dimension of the geometrical constituent of the configuration. Within the framework of finite–size scaling the zero–dimensional corners attract special attention if there is a finite extension L in the geometry, because $d' = 0$ includes *logarithmic* growth with L. From a phenomenological point of view such logarithmic contributions originate from an interference between the corner contributions to the *singular* and the *regular* part of the free energy [58]. For a cube in $d = 3$ the decomposition of the total free energy reads for short–ranged interactions [56, 58]

$$\frac{\mathcal{F}(T,H;L)}{k_B T_{c,b} L^3} = L^{-3} f_{\pm}^{sing}\left(|t|L^{1/\nu}, HL^{\Delta/\nu}\right) - L^{-3} f_{c\pm} \ln(L/l) + F_b^{reg}(T,H)$$

$$+ L^{-1} F_s^{reg}(T,H) + L^{-2} F_e^{reg}(T,H)$$

$$+ L^{-3} F_c^{reg}(T,H) + L^{-3} \delta F^{reg}(T,H;L),$$ (2.10)

where l is an arbitrary length scale, $f_{c\pm}$ is a constant amplitude, $F_x^{reg}(T,H)$ denotes a *regular* bulk ($x = b$), surface ($x = s$), edge ($x = e$), or corner ($x = c$) contribution, and δF^{reg} is the *regular* finite–size contribution, which tends to zero as L grows. The finite–size scaling function $f_{\pm}^{sing}(x, y)$ of the *singular* part of the total free energy $\mathcal{F}$ can be decomposed into bulk, surface, edge, corner, and finite–size contributions so that we have for $x \neq 0$ [56, 58]

$$f_{\pm}^{sing}(x, y) = |x|^{3\nu} f_{b\pm}^{sing}(y) + |x|^{2\nu} f_{s\pm}^{sing}(y) + |x|^{\nu} f_{e\pm}^{sing}(y) + f_{c\pm} \ln(|x|^{\nu}/l) + \delta f_{\pm}^{sing}(x, y),$$
(2.11)

where $f_{c\pm}$ is the same amplitude as in Eq.(2.10) and the scaling functions $f_{b\pm}^{sing}$, $f_{s\pm}^{sing}$, $f_{e\pm}^{sing}$, and $\delta f_{\pm}^{sing}$ correspond to the regular contributions in Eq.(2.10). Moreover, these scaling functions and the amplitude $f_{c\pm}$ are *universal* [56, 58]. Both Eqs.(2.10) and (2.11) involve a second length scale l which is needed in order to provide a dimension-less argument in the logarithmic terms. At $T = T_{c,b}$ only L remains as a macroscopic length so that l is of *microscopic*, i.e., *nonuniversal* origin. In continuum theory (see Sec.1.4) one can only account for such microscopic length scales in an indirect manner by imposing *normalization conditions* on the free energy at a reference temperature $T_1 > T_{c,b}$ ($t_1 > 0$), where the correlation length ξ_+ takes on a finite nonuniversal value $\xi_{1+} \equiv \xi_+(t_1)$ for short–ranged interactions. This procedure provides a microscopic length scale $l \equiv \xi_{1+}$ and gives rise to logarithmic corrections especially in $d = d_c$ (see Chap.4 and Chap.5). However, for $T_1 < T_{c,b}$ the correlations remain long–ranged in systems with continuous symmetries and the above normalization procedure does not yield the desired microscopic length l.

In the limit of very large systems $L \gg \xi_\pm$ or sufficiently far from bulk criticality the $\ln(L/l)$ corner contribution in Eq.(2.10) disappears in favor of a temperature–dependent contribution. Inserting Eq.(2.11) into Eq.(2.10) one obtains [56, 58]

$$F_{c\pm}^{sing}(t; L) = \nu f_{c\pm} \ln |t| L^{-3}$$
(2.12)

for the singular part $F_{c\pm}^{sing}$ of the corner free energy in the limit $L \to \infty$ at finite reduced temperature t. The above considerations hold as well in $d = 2$, where "sur-faces" are edges and "edges" are corners. In fact, one of the first explicit calculations has been performed in $d = 2$ for a critical model using conformal invariance [59] (see Sec.2.3).

From the phenomenological point of view a purely geometrical evidence for corners is not sufficient to generate logarithmic size dependent terms in the free energy of a critical system. If, for example, one *periodic* boundary condition is applied across the cubic geometry behind Eqs.(2.10) and (2.11) the topology of the cube changes to the topology of a torus with a quadratic cross section. Such a geometry has no corners and logarithmic corrections as in Eqs.(2.10) and (2.11) are absent. In a *film* geometry therefore such logarithmic corner contributions do not occur either.

Various logarithmic corrections appear as modifications of power laws with *integer* critical exponents which typically show up at the upper critical dimension $d = d_c$. For example one has $\alpha = 0$ for critical systems in $d = 4$ (logarithmic specific heat) and $\alpha_s = 1$ for tricritical semiinfinite systems in $d = 3$ (logarithmic excess energy density). A corresponding logarithmic term appears in the finite–size scaling function in order to compensate the bulk and surface singularities as will be demonstrated in Sec.5.4 for critical and tricritical films. Further logarithmic corrections can appear in $d = d_c$ as modifications of the Landau critical behavior even for non–integer critical exponents. An important example is the bulk correlation length for a critical $O(N)$–symmetric system for short–ranged interactions in $d = 4$ which reads (see, e.g., Ref.[34])

$$\xi_+(t) \simeq \xi_0^+ |t|^{-\frac{1}{2}} |\ln |t||^{\frac{N+2}{2(N+8)}} \qquad (2.13)$$

for $t > 0$ and general N. For $t < 0$ Eq.(2.13) only holds, if $N = 1$. Logarithmic size–dependent terms must therefore be expected in addition to the usual scaling contributions to the singular free energy [56].

2.1.2   Interfacial Tensions and the Helicity Modulus

Under the influence of opposing surface fields an Ising–like system confined to a film geometry forms an *interface* between the outer surfaces which separates the differently magnetized regions. The free energy contribution associated with the formation of the interface is called interfacial free energy or *interfacial tension*. If the system is confined to a box of height L and cross section $A = L'^{d-1}$, the interface is confined to a finite area and finite–size effects on the interfacial tension especially at bulk criticality will occur. The finite–size effects of course depend on the boundary conditions applied at the surrounding walls. In order to *pin* the interface the upper and the lower half of the box can be exposed to opposing surface fields [56, 60]. The interfacial energy grows with the area A, but similar to Eq.(2.10) there are edge, corner, and finite–size contributions to the interfacial tension and logarithmic corrections must be expected as well. In terms of the free energy *densities* $f_{+-}(t; L, L')$ of the system for opposing surface fields and $f_{++}(t; L, L')$ for a uniform surface field the interfacial tension can be defined as [56]

$$\sigma_{+-}(t; L, L') = \frac{L}{k_B T_{c,b}} \left(f_{+-}(t; L, L') - f_{++}(t; L, L') \right), \qquad (2.14)$$

where $t = (T - T_{c,b})/T_{c,b}$. If *antiperiodic* (*aper*) boundary conditions are applied across L and *periodic* (*per*) boundary conditions a applied otherwise, a *floating* interface forms with an interfacial tension $\sigma_{aper}(t; L, L')$ defined analogous to Eq.(2.14), where f_{+-} is replaced with f_{aper} and f_{++} is replaced with f_{per} [56]. For $L' = L$ and in $d = 3$ one obtains in analogy with Eq.(2.10) for $t \leq 0$ [56]

$$\sigma_{+-}(t; L, L') = L^{-2} \left[\Sigma \left(t L^{1/\nu} \right) + s \ln(L/l) \right] + L^{-1} \sigma_e^{reg}(t) + L^{-2} \sigma_c^{reg}(t) + \ldots, \quad (2.15)$$

where the scaling function $\Sigma(x)$ and the amplitude s are universal and $\sigma_x^{reg}(t)$ denotes the regular edge ($x = e$) and corner ($x = c$) contribution to the interfacial tension $\sigma_{+-}(t; L, L')$. Note, that $\Sigma(x) \sim x^{2\nu}$ for $x \to \infty$. The logarithmic term in Eq.(2.15) has been explicitly found for interfaces in circular and rectangular geometries in $d = 2$ using conformal invariance [61] (see Sec.2.3).

For $T < T_{c,b}$ but still above the roughening temperature extensive calculations have been performed on a Gaussian *capillary wave* model in rectangular geometries for various boundary conditions in general dimension [60]. For a single antiperiodic and $d - 1$ periodic boundary conditions, where the center of mass of the interface is kept fixed, the interfacial tension reads [60]

$$\sigma_{aper}(t; L, L) = \sigma_s(t) + L^{-(d-1)} \ln(L/l) + \ldots, \qquad (2.16)$$

whereas for an interface pinned by surface fields the expression [60]

$$\sigma_{+-}(t; L, L) = \sigma_s(t) + L^{-1}\sigma_e(t) + L^{-2}\sigma_c(t) - (2L)^{-(d-1)} \ln(L/l) + \ldots \qquad (2.17)$$

has been obtained. Note, that the coefficients of the logarithmic terms are simple *universal numbers*.

Surface fields acting on a confined $O(N)$–symmetric system may enclose any angle θ between $\theta = 0$ (parallel) and $\theta = \pi$ (antiparallel), if $N \geq 2$. As a response the system forms a helical magnetization profile which smoothly interpolates between the different orientations at the boundaries. If the distance between these boundaries is L, the relative angle $\delta\theta$ between neighboring spins in an XY or Heisenberg model is of the order $\delta\theta \sim \theta/L$. The amount of energy stored in this little twist between the spins can be estimated as $\delta\epsilon \sim 1 - \cos \delta\theta \sim \theta^2/2L^2$.

In analogy with Eq.(2.14) we confine the $O(N)$–symmetric system to a box of height L and cross section $A = L'^{d-1}$ so that the surface fields act on the top and bottom surface and periodic boundary conditions are applied to the surrounding walls. In terms of the free energy densities $f_\theta(t; L, L')$ of the twisted configuration and $f_0(t; L, L')$ of the configuration without twist the *rigidity* of the system under an imposed *orientational gradient* of the magnetization is captured by the *helicity modulus* $\Upsilon(t)$ which can be defined by [44]

$$\Upsilon(t) = \lim_{L\to\infty} \frac{2L^2}{k_B T_{c,b}\theta^2} \lim_{L'\to\infty} (f_\theta(t; L, L') - f_0(t; L, L')) \qquad (2.18)$$

in the thermodynamic limit. Note, that $f_\pi(t; L, L')$ corresponds to $f_{+-}(t; L, L')$ and that $f_0(t; L, L')$ corresponds to $f_{++}(t; L, L')$ (see Eq.(2.14)).

For short–ranged interactions critical behavior in $O(N)$–symmetric systems can be observed in $d > 2$ [45]. The singular behavior of $\Upsilon(t)$ for $t \nearrow 0$ is then governed by the power law [44]

$$\Upsilon(t) = u_b(-t)^\upsilon \qquad (2.19)$$

with $\upsilon = (d-2)\nu$ in $d < 4$ and $\upsilon = 2 - \alpha - 2\nu$ in general. The critical behavior of the helicity modulus $\Upsilon(t)$ for $2 < d < 4$ is of special relevance for liquid ^{4}He near the λ–transition, because $\Upsilon(t)$ is proportional to the *superfluid density* $\rho_s(t)$ for $N = 2$ [44] (see Sec.6.2).

A definition of the helicity modulus $\Upsilon(t; L, L')$ for a finite system seems to be straightforward by omitting the limits in Eq.(2.18). However, a helicitiy modulus defined this way and therefore its finite–size properties like scaling functions or logarithmic corrections will depend on the boundary conditions. Moreover, the orientational gradient in the system can also be imposed from outside by *helical boundary conditions* which replace the surface fields at the top and the bottom surface of the box. In a finite–size scaling analysis of the helicity modulus $\Upsilon(t; L, L')$ the influence of the boundary conditions must therefore be carefully considered.

The finite–size scaling analysis for $\Upsilon(t; L, L')$ can be performed along the lines of Eq.(2.10) and (2.11). For short–ranged interactions in $d = 3$ and for $L = L'$, and one has close to $t = 0$ [62]

$$\Upsilon(t; L, L) = L^{-1} \left[U \left(tL^{1/\nu} \right) - u \ln(L/l) \right] + L^{-1}\Upsilon_1^{reg}(t) + L^{-2}\Upsilon_2^{reg}(t) + \dots \qquad (2.20)$$

in analogy with Eq.(2.10). The functions $\Upsilon_1^{reg}(t)$ and $\Upsilon_2^{reg}(t)$ denote regular background contributions and $U(x)$ is the universal finite–size scaling function for the singular part of the helicity modulus. The scaling function $U(x)$ can be decomposed for $x < 0$ according to Eq.(2.11) so that [62]

$$U(x) = u_b(-x)^\nu + u\ln((-x)^\nu/l) + \ldots, \qquad (2.21)$$

where u_b and u are universal constants. Inserting Eq.(2.21) into Eq.(2.20) the asymptotic finite–size effect on the helicity modulus for $L \to \infty$ up to regular terms in t is found to be [62]

$$\Upsilon(t; L, L) = u_b(-t)^\nu + u\nu L^{-1}\ln(-t) + \ldots. \qquad (2.22)$$

Note, that Eq.(2.22) also proposes the leading singular finite–size correction to the superfluid density $\rho_s(t; L)$ ($N = 2$) which is an intensively studied quantitiy in experiments on confined ^{4}He (see Sec.6.2).

If the leading term in Eq.(2.21) is interpreted as the *bulk* contribution to the helicity modulus then the logarithmic correction should be a *surface* effect. In contrast to the free energy the helicity modulus can therefore be modified by logarithmic size–dependent terms even in a *film geometry*. Note, that the above statements on the helicity modulus depend on the boundary conditions chosen in the definition of $\Upsilon(t; L, L')$.

2.1.3 Curvature Effects

A system confined in a finite geometry will in general be bounded by *curved* surfaces. In three dimensions the local behavior of a curved surface can be described in terms of the *mean curvature* $\mathcal{H}$ and the *Gaussian curvature* $\mathcal{K}$. Assuming short–ranged interactions a curvature effect on the critical behavior becomes visible if the bulk correlation length $\xi_\pm$ has become comparable to the length scales associated with the curvatures of the surface. With respect to a decomposition of the free energy according to a geometrical hierarchy as in Eq.(2.10) dimensional analysis suggests that a shape contribution or *shape tension* due to a finite *mean curvature* $\mathcal{H}$ replaces the *edge* contribution in Eq.(2.10). Higher order shape tensions due to a finite *Gaussian curvature* $\mathcal{K}$ or due to a deviation $\mathcal{H}^2 - \mathcal{K}$ from the spherical shape may be suspected to replace the corner terms in Eq.(2.10). According to the phenomenological arguments given in Ref.[58] these higher order shape tensions can give rise to logarithmic size–dependent contributions to the free energy similar to those shown in Eq.(2.10). The simplest geometry which posesses both a finite mean and Gaussian curvature is given by a sphere. The singular part of the free energy of an $O(N)$–symmetric system confined to a sphere of radius R with a surface representing the *ordinary* surface universality class can be written as [63]

$$\frac{\mathcal{F}^{sing}(t, H; R)}{k_B T_{c,b}} = \frac{1}{d} S_d R^d |t|^{2-\alpha} f_{b\pm} \left(H|t|^{-\Delta} \right) + S_d R^{d-1} |t|^{2-\alpha_s} f_{s\pm} \left(H|t|^{-\Delta} \right)$$

$$+ \ S_d R^{d-2} |t|^{2-\alpha-2\nu} f_{\mathcal{H}\pm} \left(H|t|^{-\Delta} \right) + \cdots, \tag{2.23}$$

where S_d denotes the surface area of the unit sphere and $f_{b\pm}$, $f_{s\pm}$, and $f_{\mathcal{H}\pm}$ are the scaling functions of the bulk, surface, and the mean curvature contribution, respectively. The Gaussian curvature term and eventually associated logarithmic corrections have not yet been identified explicitly. Note, that short–ranged interactions and $2 < d < 4$ have been assumed in Eq.(2.23).

In two dimensions a single curvature κ characterizes the local behavior of the boundary. Likewise, curvature contributions to the free energy replace the corner contributions and lead to logarithmic size–dependent terms in the free energy of the confined system. For two–dimensional critical systems this has been rigorously shown by the principle of *conformal invariance* [59]. The same effect occurs in a system which is confined to a two–dimensional *curved* manifold [59].

2.2   Lattice Models in Finite Geometries

A widely used approach to finite–size scaling is based on the analysis of lattice models in finite geometries. The boundaries of these geometries always follow the symmetry axes of the lattice in order to obtain simple parameterizations of the surfaces in terms of lattice vectors. For an adequate lattice description of systems with curved boundaries very large lattices are needed so that a theoretical treatment by analytical methods or computer simulations is beyond the scope of the present techniques.

2.2.1 The Ising Model

The best studied system on finite lattices is the ferromagnetic nearest–neighbour spin–$\frac{1}{2}$ Ising model which can be exactly solved in two dimensions on finite square, honeycomb, and triangular lattices for periodic, free (missing bonds, *ordinary*) and fixed (surface fields, *extraordinary*) boundary conditions [48, 49]. The specific heat of the finite system displays a rounded maximum as a function of temperature (see Sec.2.1 and Fig.2.1). In a rectangle with $n_1 \times n_2$ lattice sites the specific heat C per spin evaluated *at* the critical temperature $T_{c,b}$ of the corresponding infinite Ising lattice model grows as $\ln n_1$ for $n_1 \to \infty$ with n_1/n_2 fixed [48]. The position T_m of the specific heat maximum approaches $T_{c,b}$ from *below* for *free* boundaries and from *above* for *periodic* boundary conditions. The shift $(T_m - T_{c,b})/T_{c,b}$ of the specific heat maximum with respect to $T_{c,b}$ decays as n^{-1} for $n_1 = n_2 = n \to \infty$ [49] which is in accordance with the general expectation in $d = 2$ (see Eq.(2.5)). The height of the specific heat *maximum* $C(T_m)$ per spin shows logarithmic growth with n as well [49].

The ability to find exact solutions for the Ising model in $d = 2$ gives the opportunity to study the full finite–size scaling behavior of the free energy in presence of surface fields and, in this special case, the influence of *logarithmic corrections* on the scaling form of the free energy in a rigorous way. This has been done for a square–lattice nearest neighbour Ising *strip* which is the two–dimensional analogue of a *film geometry* [50]. With a free boundary condition on one edge and a surface field H_1 on the other edge of the strip the singular part of the total free energy per spin can be written in the *modified* scaling form [50] (see also Ref.[51])

$$\lim_{m \to \infty} \frac{\mathcal{F}^{sing}(t, H_1; n, m)}{k_B T_{c,b} m n} = -t^2 \ln |t| f_{1\pm}\left(|t|n, H_1 n^{1/2}\right) + t^2 f_{2\pm}\left(|t|n, H_1 n^{1/2}\right)$$

$$= n^{-2}\left[\ln n\, g_{1\pm}\left(|t|n, H_1 n^{1/2}\right) + g_{2\pm}\left(|t|n, H_1 n^{1/2}\right)\right], \quad (2.24)$$

where n is the number of lattice rows, $\Delta_1 = \frac{1}{2}$, and the scaling functions $g_{1\pm}(x, y)$ and $g_{2\pm}(x, y)$ given by

$$g_{1\pm}(x, y) = x^2 f_{1\pm}(x, y) \quad \text{and} \quad g_{2\pm}(x, y) = -x^2 \ln x\, f_{1\pm}(x, y) + x^2 f_{2\pm}(x, y) \quad (2.25)$$

are *analytic* functions of x and y (see Ref.[50] for explicit expressions). The free energy given by Eq.(2.24) can be decomposed into bulk, surface, and finite–size contributions (see Eq.(2.1)). The singular bulk free energy density follows directly from Eq.(2.24) in the limit $n \to \infty$ and the singular part of the total surface free energy per spin has the representation [50]

$$\lim_{m \to \infty} \frac{\mathcal{F}_s^{sing}(t, H_1; m)}{k_B T_{c,b} m} = -t \ln |t| f_{s,1\pm} \left(H_1 |t|^{-1/2} \right) + t f_{s,2\pm} \left(H_1 |t|^{-1/2} \right)$$

$$= n^{-1} \left[\ln n \, g_{s,1\pm} \left(H_1 |t|^{-1/2} \right) + g_{s,2\pm} \left(H_1 |t|^{-1/2} \right) \right] \quad (2.26)$$

in analogy with Eq.(2.24). Note, that Eq.(2.26) is in accordance with the value $\alpha_s = 1$ for the exponent of the excess specific heat for a semiinfinite Ising system in $d = 2$ (see Table 1.2 and Eq.(1.13)). The logarithmic size–dependent corrections to the finite–size scaling behavior of the free energy shown in Eqs.(2.24) and (2.26) are due to the logarithmic singularities in the bulk free energy ($\alpha = 0$) and the surface free energy ($\alpha_s = 1$) at $t = 0$. The *simultaneous* occurrence of these logarithmic contributions is a special feature of the Ising universality class in $d = 2$.

The advantage of exactly solvable model systems is that besides the scaling functions $f_{1\pm}$ and $f_{2\pm}$ in Eq.(2.24) for the *singular* part of the free energy the *total* free energy $\mathcal{F}$ of the system is obtained as well. It is particularly interesting to calculate the total free energy of an Ising strip at $t = 0$ which has been done in Ref.[50] for $H_1 = 0$. One obtains

$$\lim_{m \to \infty} \frac{\mathcal{F}(0, 0; n, m)}{k_B T_{c,b} m n} = f_b^0 + f_s^0 n^{-1} - \frac{\pi}{48} n^{-2} + \ldots, \quad (2.27)$$

where f_b^0 and f_s^0 are the (nonuniversal) values of the bulk and surface free energies per spin for $T = T_{c,b}$ and $H_1 = 0$, respectively. The dots indicate terms which decay *faster* than n^{-2}. The amplitude of the n^{-2}–contribution to Eq.(2.27) is *universal*, i.e., it only depends on the bulk universality class, the two surface universality classes (here ordinary) and the spatial dimension d. This simple expression incorporates the critical analogue of the well known *Casimir effect* which will be discussed in Chap.3 in detail.

Local information on an Ising system confined to a strip in $d = 2$ can be obtained from, e.g., the *magnetization profile* for which rigorous statements for various combinations of boundary conditions or surface fields can be made [64]. Profiles and wall effects in critical films with special reference to two–dimensional systems are studied in Chap.4 so that we refrain from giving any details here.

The specific heat of certain *layered* systems (e.g., liquid crystals) exhibits two peaks as a function of temperature which, coming from high temperatures, can be associated with the onset of surface ordering and subsequent bulk ordering upon lowering the temperature. In order to mimic the main aspects of this behavior an *anisotropic* nearest–neighbour Ising system with different horizontal and vertical bulk coupling constants J_h and J_v, respectively, which is confined to a strip in $d = 2$ with *enhanced* surface couplings has been studied in Ref.[65]. It turns out that the specific heat of such a strip exhibits two rounded maxima reminiscent of the two peaks in the three–dimensional system. The surface coupling J_s sets a length scale ξ_s on which the surface magnetization decays into the bulk. The asymptotic scaling limit is only reached if the width L of the strip is much larger than ξ_s which itself can be much larger than the lattice constant [65]. The position of the specific heat maximum reminiscent of the singularity at $T_{c,b}$ follows a *nonmonotonic* scaling function of the argument L/ξ_s [65].

The near–critical layered Ising model in $d = 2$ which is characterized by two coupling constants $J_\parallel(z)$ and $J_\perp(z)$ varying arbitrarily along the z–direction of the coordinate system has recently been investigated by an exact local free energy functional [66, 67]. The construction of this functional generalizes a rigorous implementation of an entropy like functional in $d = 1$ [68]. The energy density profiles of a two–dimensional semi–infinite Ising system in presence of a surface field have been obtained from the corresponding variational principle. The implementation of this variational approach in $d = 4 - \varepsilon$ may give new approximations for critical Ising systems in $d = 3$ [66].

Spin–$\frac{1}{2}$ Ising models in finite geometries have also been intensively studied by *Monte–Carlo simulations* in $d \geq 2$. An early test of the finite–size behavior of the energy density, the specific heat, the magnetization, the magnetic susceptibility, and the spin–spin correlation function against the predictions of finite–size scaling theory has been performed in Refs.[69] and [70] by means of a Monte–Carlo simulation. A spin–$\frac{1}{2}$ Ising model with nearest–neighbour interactions on a square lattice in $d = 2$ with $L \times L$ ($L \leq 100$) lattice sites [69] and on a cubic lattice in $d = 3$ with $L \times L \times L$ ($L \leq 20$) lattice sites [70] for periodic and free boundary conditions served as model systems. The simulation data are consistent with the finite–size scaling theory in the sense that *data collapse* is obtained in a scaling plot using the theoretically expected scaling variables. The shape of the finite–size scaling functions is therefore displayed by these scaling plots. For the three–dimensional $L \times L \times L$ lattice with free boundary conditions the *leading* size corrections to the specific heat, the magnetization, and the susceptibility are governed by the corresponding *excess* quantities (see Table 1.3) [70].

The violation of hyperscaling for critical systems in $d > 4$ leads to a modification of the scaling variables (see Eq.(2.9)). A Monte–Carlo simulation of a five–dimensional spin–$\frac{1}{2}$ Ising model in a hypercube with periodic boundary conditions supports the new choice of the scaling variables by showing data collapse in scaling plots of the simulation data [55].

Under the influence of surface fields the critical point of an Ising model confined to a film geometry in $d = 3$ should be shifted according to Eq.(2.5). The data of a Monte–Carlo simulation of a spin–$\frac{1}{2}$ Ising model with nearest–neighbour interactions on a $L \times L \times D$ cubic lattice with surface fields acting on the spins in the top and bottom $L \times L$–layer and otherwise periodic boundary conditions are consisitent with the shift exponents in Eq.(2.5) [47]. The sizes L and D were chosen as $L = 128$ and $8 \leq D \leq 40$ lattice sites. For temperatures sufficiently below $T_{c,b}$ the shift of the *coexistence line* is found to be in accordance with *Kelvin's law* $H_{coex}(D) \propto D^{-1}$ for $D \geq 16$ [47] (see also Fig.2.2). The Ising model is equivalent to a lattice gas model so that the Ising film studied in Ref.[47] simultaneously provides a reasonable model for the phenomenon of *capillary condensation* of a fluid in a slit [71] (see also Ref.[28]). In the lattice gas picture the chemical potential $\mu_{coex}(D)$ at liquid–vapor coexistence in a slit of width D replaces the magnetic field $H_{coex}(D)$ so that $\mu_{coex}(D) - \mu_{coex}(\infty)$ is governed by Kelvin's law away from the critical point. For temperatures between the *wetting temperature* T_w of the corresponding semiinfinite lattice gas and the bulk critical temperature $T_{c,b}$ the surfaces of the slit are covered by thick (quasi–) wetting films which causes a correction to Kelvin's law [71]. For nearest–neighbour interactions the corrected version of Kelvin's law reads

$$\mu_{coex}(D) - \mu_{coex}(\infty) = \frac{\Delta\Omega_s}{\Delta\rho}D^{-1}\left(1 + \frac{a}{\Delta\rho}\frac{\ln D}{D} + \ldots\right), \qquad (2.28)$$

where $\Delta\Omega_s = \Omega_{ls} - \Omega_{vs}$ denotes the difference of the surface contributions to the grand canonical potential Ω for the liquid and the vapor at *bulk* coexistence $\mu = \mu_{coex}(\infty)$ and $\Delta\rho = \rho_l - \rho_v$ is the corresponding density difference. The amplitude a in Eq.(2.28) depends on the microscopic interaction parameters. Note, that the logarithm in Eq.(2.28) is due to the *finite* range of the interactions. For van–der–Waals forces $\ln D$ must be replaced with $D^{1/3}$ so that the correction becomes $D^{-2/3}$

instead of $\ln D/D$. The second–order *wetting transition* (see Ref.[25]) which occurs in a semiinfinite lattice gas (Ising) model for a finite surface field becomes *rounded* in a slit of finite width D. The local density at the surface increases steeply but in a smooth fashion as $T \nearrow T_w$ and leads to a large value of the excess density for $T > T_w$ [71].

A three–dimensional Ising model confined to a film geometry has been investigated near the surface–bulk (SB) transition by an improved single–cluster Monte–Carlo method [72]. The ratio J_1/J of the surface and bulk coupling constants at the SB transition (see Fig.1.11) is to great accuracy given by $J_1/J = K_1^c/K^c = 1.5$. Special attention has been paid to the crossover exponent ϕ (see Eq.(1.11)) and the exponent β_1^{SB} of the surface magnetization (see Table 1.4). The values found for ϕ and β_1^{SB} in Ref.[72] deviate substantially from the results of an earlier Monte–Carlo analysis [73] which may be due to a slightly different value for T_c obtained in Ref.[72].

For opposing surface fields applied to the boundaries of an Ising model confined to a film an *interface* forms separating the differently magnetized regions for $T < T_{c,b}$ (see also Sec.2.3). A corresponding spin–$\frac{1}{2}$ Ising model in a strip in two dimensions can be solved exactly [74]. Beyond this exactly solvable model it is particularly interesting to study the parallel and transverse spin–spin correlation functions in an Ising strip under the influence of an interface between two *defect lines* of weakened bonds which modify the fluctuation spectrum of the interface [74]. The interface itself is imposed on the system by fixing all spins on one edge to, say, $+\frac{1}{2}$ and all spins on the opposite edge to $-\frac{1}{2}$ which is equivalent to the application of infinitely strong opposing surface fields. Without the defect lines the parallel correlation length $\xi_{\parallel}$ is a smooth function of the temperature and the width of the strip. In presence of defect lines, however, a weak singularity may occur in $\xi_{\parallel}$ as a function of temperature due to transverse fluctuations of the interface which span the distance between the defect lines [74]. Unfortunately the model with defect lines can no longer be solved exactly so that one has to resort to a Monte–Carlo simulation. The simulation data indicate, that the singularity in $\xi_{\parallel}$, if present at all, should be very weak [74].

Thin films of FeF_2 near their antiferromagnetic Ising–like phase transition can be studied in superlattices of $(FeF_2)_n(ZnF_2)_m$ with $3 < m < 8$ by means of high resolution x–ray diffraction and Monte–Carlo simulations [75]. The specific heat of these films, which can be inferred from the thermal expansion coefficient, is found to be in accordance with the finite–size scaling prediction given by Eq.(2.8) with $H = 0$, and the shift of the critical temperature is given by Eq.(2.5). These findings have been confirmed independently by corresponding Monte–Carlo simulations [75]. Therefore the complete scenario of the *dimensional crossover* in Ising–like films from $d = 3$ to $d = 2$ can be observed experimentally in the above $(FeF_2)_n(ZnF_2)_m$ superlattices [75].

2.2.2 *The XY – and the Heisenberg Model*

Lattice models with continuous $O(N)$ symmetries are of great importance for the description of the superfluid–like phase transitions (see Fig.1.6) and the ferromagnetic phase transition in isotropic ferromagnets (see Fig.1.7). The most prominent examples for such lattice models are given by the XY model (^{4}He, $N = 2$) and the Heisenberg model (isotropic ferromagnet, $N = 3$, see Eq.(1.18)). Unfortunately exact solutions of these lattice models do not even exist in $d = 2$ so that one has to resort to approximate descriptions or computer simulations in order to obtain the finite–size behavior of these systems, where the *dimensional crossover* from $d = 3$ to $d = 2$ in film geometries is particularly interesting (see also Fig.2.1 and Sec.5.5).

A dimensional crossover from $d = 3$ to $d = 2$ can be enforced by the introduction of a *spatial anisotropy* to the coupling constants of the lattice model. If $\{e_1, e_2, e_3\}$ denotes the basis of a lattice in $d = 3$, in such models the couplings along two of these vectors are chosen to be equal, e.g., $J_1 = J_2 = J$, whereas the third coupling J_3 differs from J by an amount given by an anisotropy parameter α according to $J_3 = \alpha J$ with $0 \leq \alpha \leq 1$. In the case $\alpha = 0$ the lattice planes spanned by $\{e_1, e_2\}$ are completely decoupled ($d = 2$) and for $\alpha = 1$ full isotropy is restored ($d = 3$). The partition function of a nearest neighbour XY model with a spatial anisotropy along the e_3–direction in the periodic Gaussian (Villain) form has the representation [76]

$$Z(T, \alpha) = \prod_R \int_{-\pi}^{\pi} \frac{d\theta(R)}{2\pi} \sum_{\{n(R)\}} \exp\left(-S\{\theta(R), n(R)\}\right),$$

$$S\{\theta(R), n(R)\} = \frac{J}{2k_B T} \sum_R \left[\sum_{i=1}^{2}(\Delta_i\theta(R) - 2\pi n_i(R))^2 + \alpha(\Delta_3\theta(R) - 2\pi n_3(R))^2\right],$$

$$(2.29)$$

where R denotes lattice vectors, $\theta(R)$ is the angle of the classical spin vector at site R with respect to, say, the e_1–direction, $n(R) = (n_1(R), n_2(R), n_3(R))$ is a triple of integers at site $\mathbf{R}$, and $\Delta_i \theta(\mathbf{R}) = \theta(\mathbf{R} + \mathbf{e}_i) - \theta(\mathbf{R})$. The sum over all configurations $\{\mathbf{n(R)}\}$ of integers on the lattice guarantees the 2π–periodicity of the expression under the integral in Eq.(2.29) in each angle variable $\theta(\mathbf{R})$ (see Eq.(1.18) for $N = 2$). A Migdal renormalization group analysis of the partition function in Eq.(2.29) maps any anisotropic case with $\alpha > 0$ onto the isotropic case $\alpha = 1$, i.e., critical behavior as in $d = 3$ is obtained. The case $\alpha = 0$ ($d = 2$) is mapped onto itself, but the Migdal renormalization procedure fails to reproduce the Kosterlitz–Thouless transition [76]. However, the Gaussian form of the partition function given by Eq.(2.29) allows a direct Monte–Carlo analysis which has also been performed in Ref.[76] on a $L \times L \times L$ simple cubic lattice with periodic boundary conditions and $L \leq 24$ lattice sites. The simulation data show that the specific heat curves up to $\alpha \simeq 0.2$ resemble the curve for $\alpha = 0$ ($d = 2$) very much, whereas a sharp peak in the specific heat occurs for $\alpha \geq 0.4$ which is typical for $d = 3$ (see Fig.2.1(b)). The size dependence of the simulation data increases rapidly as $\alpha \to 0$ which can be inferred from the finite–size scaling behavior of the internal energy, the specific heat, the magnetization, or the susceptibility [76]. In summary, the shapes of these observables show a smooth crossover from three–dimensional to two–dimensional behavior at $\alpha \simeq 0.1 \ldots 0.2$ which may be due to a change in the vortex–loop configuration taking place at about the same value of α. However, the typical Kosterlitz–Thouless singularities are *only* observed for $\alpha = 0$, whereas for *any* $\alpha > 0$ the singularities are *algebraic* which is typical for a second–order phase transition. The locations of the algebraic singularity and the maximum of the specific heat which coincide in $d = 3$ ($\alpha = 1$) follow *different* paths as α decreases such that the algebraic singularity is finally converted to an exponential singularity in $d = 2$ ($\alpha = 0$) [76]. In a recent Monte–Carlo study of Eq.(2.29) for $\alpha = 1$ on $L \times L \times N$ lattices with $N \leq 16$, $L \gg N$, free boundary conditions on the top and bottom $L \times L$ layer, and otherwise periodic boundary conditions the exponent Λ of the fractional shift has been determined [77]. The onset of two–dimensional Kosterlitz–Thouless behavior for a given number of layers N defines a critical temperature $T_c(N)$ which is shifted with respect to the critical temperature $T_{c,b} \equiv T_c(\infty)$ in $d = 3$ by an amount $\propto N^{-\Lambda}$. The value for Λ found in Ref.[77] seems to be *inconsistent* with the expectation $\nu = 1/\Lambda = 0.67$ for the XY universality class in $d = 3$ (see Table 1.5 and Eq.(2.5)). A corresponding analysis has been performed on the XY model in the form given by Eq.(1.18) on $L \times L \times N$ lattices with $L \leq 64$ and $N \leq 16$ [78], where the boundary conditions are chosen as in Ref.[77]. The analysis of the fractional shift exponent Λ in this case is found to be in accordance with $\Lambda = 1/\nu$ contrary to the findings in Ref.[77]. This contradiction has not yet been resolved. The transition temperature approaches the position $T_m(N)$ of the maximum in the specific heat curve from below as the number of layers N is increased which is in qualitative accordance with the behavior of the transition temperature and the specific heat maximum found in Ref.[76] as the anisotropy parameter α is increased. Moreover, the class of vortex structures responsible for the low temperature behavior of the layered XY system can be identified [78].

In a further Monte–Carlo analysis of a layered XY model in the form given by Eq.(1.18) on a $L \times L \times N$ lattice with $L \leq 64$ and $N \leq 8$ and boundary conditions as in Refs.[77, 78] special attention has been paid to the helicity modulus (see Eq.(2.18)) for phase twists in different spatial directions [79]. The helicity modulus, which is proportional to the superfluid density, can be studied as a function of temperature for different layer thicknesses N. For $N = 1$ the superfluid density displays the universal jump expected from the XY model in $d = 2$ at $T = T_{KT}$. As N grows the jump diminishes and the transition temperature $T_{KT}(N)$ grows towards the three–dimensional transition temperature according to Eq.(2.5) with $\nu = 2/3 \simeq 0.67$, where the jump has finally vanished and the three dimensional bulk singularity in the superfluid density occurs (see Eq.(2.19)) [79].

A special variant of the Heisenberg model, the spin-$\frac{1}{2}$ XXZ model, has been studied between two and three dimensions by the cluster variation method which is an extension of the mean–field approximation towards the treatment of correlations between neighbouring spins [80]. The nearest–neighbour spin-$\frac{1}{2}$ XXZ model is defined by the Hamiltonian

$$H = - \sum_{<ij>} \left[J_{ij} \left(\sigma_i^x \sigma_j^x + \sigma_i^y \sigma_j^y \right) + J_{z,ij} \sigma_i^z \sigma_j^z \right] - h \sum_i \sigma_i^z, \qquad (2.30)$$

where $< ij >$ denotes nearest–neighbour pairs and σ_i^α is the α–component of the Pauli matrix at site i. The model Hamiltonian in Eq.(2.30) incorporates the Ising model ($J_{ij} = 0$), the XY model without external fields ($J_{z,ij} = 0, h = 0$) and the Heisenberg model ($J_{ij} = J_{z,ij}$) for spin-$\frac{1}{2}$ particles. In order to obtain a crossover from three to two dimensions an anisotropy as in Eq.(2.29) has been introduced so that the interaction *between* neighbouring layers of the lattice can be weakend compared to the interaction *within* a layer. In addition a finite stack of n layers has been considered for spatially isotropic nearest–neighbour couplings and free boundary conditions on the top and bottom layer [80]. For anisotropic interactions the transition temperature $T_c(\alpha)$ decreases monotonically for a decreasing anisotropy parameter α and reaches a *finite* value at $\alpha = 0$ ($d = 2$) in the Ising and XY case of Eq.(2.30). In the Heisenberg case the critical temperature vanishes for $\alpha = 0$ which is in accordance with the general expectation [45]. In the XY case of Eq.(2.30), however, the cluster variation method does not reproduce the Kosterlitz–Thouless transition for $\alpha = 0$. Instead, a transition to a state with a *finite* spontaneous magnetization is predicted [80]. In the slab geometry with n layers the transition temperature $T_c(n)$ for decreasing n shows the same behavior as $T_c(\alpha)$ for decreasing α. For the XY case the cluster variation method again predicts a transition to a state with long–ranged order instead of the Kosterlitz–Thouless transition. Moreover, one expects $T_c(n) = 0$ in the Heisenberg case for any finite n [45] which in Ref.[80] is only found for $n = 1$. These discrepancies between the above results and the expected behavior are due to the mean–field origin of the cluster variation method. The phase diagram of the XXZ model calculated in Ref.[80] for different anisotropies α or layer thicknesses n, respectively, therefore always displays transitions from a disordered state to Ising or XY type states with *long–ranged* order.

Finite–size scaling has also become a standard method to extract bulk or surface critical behavior from Monte–Carlo simulation data of Heisenberg models on finite lattices. For example, the *helicity modulus* of the classical Heisenberg model for combined periodic and antiperiodic boundary conditions on simple cubic and face–centered cubic lattices yields an estimate for the (nonuniversal) critical couplings [81]. Using this information the critical exponents β_1 of the surface magnetization and γ_{11} of the local susceptibility (see Tables 1.3 and 1.4) can be determined from finite–size scaling for free boundaries in combination with periodic or helical boundary conditions for the remaining surfaces [81]. High–precision estimates of the bulk critical coupling and the bulk critical exponents have beeen obtained from a single–cluster Monte–Carlo simulation of the classical Heisenberg model on a simple cubic $L \times L \times L$ lattice with $L \leq 48$ and periodic boundary conditions [82]. A corresponding study can be found in Ref.[83] for $L \leq 40$ on simple cubic and body–centered cubic lattices. Finally, we note that a generalized form of the finite–size scaling analysis can be applied to *crossover phenomena* between different bulk universality classes or near multicritical points so that the corresponding *crossover exponents* can be obtained from Monte–Carlo data on finite lattices [84].

2.2.3   The Potts Model

A particularly interesting candidate for a finite–size scaling analysis on lattices is the q–state Potts model (see Eq.(1.27)), because it represents various types of critical behavior in terms of the parameter q (see Table 1.6). For $0 < q \leq 4$ the two–dimensional Potts model without external fields displays a second–order phase transition at a finite critcal temperature, whereas in $d = 3$ the phase transition for $q \geq 3$ is first–order (see Fig.1.12). Interesting crossover phenomena in finite geometries can therefore be expected for $q = 3$ and $q = 4$. A similar crossover phenomenon occurs for an anti-ferromagnetic Ising model on a face–centered cubic lattice in an external magnetic field confined to a slab geometry in $d = 3$, where the phase transition is first order (4–state Potts like) for thick slabs (more than 11 layers) and second order (Ising like) for thin slabs (less than 11 layers) [85].

An extended finite–size scaling analysis of the q–state Potts model in strips of width n and $n \times n$ squares in two dimensions with periodic boundary conditions has been performed in Ref.[86]. For $0 < q \leq 4$ the critical finite–size scaling behavior of the free energy density in a $n \times \infty$ quadratic lattice strip is described by

$$f(T = T_{c,b}; n, q) = f(T = T_{c,b}; \infty, q) + a_1(q)n^{-2} + a_2(q)n^{-4} + \dots, \qquad (2.31)$$

where $a_1(q)$ is the *Casimir amplitude* for the q–state Potts universality class in $d = 2$ for periodic boundary conditions (see Eq.(2.27) and Chap.3). The finite–size corrections in Eq.(2.31) have been obtained from a fit to the data of a numerical evaluation of transfer matrix methods. Rather accurate values for the critical exponents $\alpha(q)$ of the bulk specific heat and $\delta(q)$ of the critical isotherm can be inferred from an extrapolation of the *effective exponents* $\alpha(q, n)$ and $\delta(q, n)$ of the strip for $0 < q < 4$ to $n \to \infty$ [86]. The convergence of the effective exponents for $q = 4$, which forms the borderline to the first–order regime in $d = 2$, appears to be significantly worse than for $q < 4$. A corresponding procedure for $n \times n$ squares yields the same results but with different convergence properties of $\alpha(q, n)$ and $\delta(q, n)$ for $n \to \infty$. The results for $\alpha(q)$ and $\delta(q)$ obtained this way confirm conjectured Padé approximants for these exponents, where the deviations are largest for $q = 4$.

In the case $q > 4$ the Potts model on $n \times \infty$ and $n \times n$ lattices shows a crossover to a first–order phase transition for $n \to \infty$ which leads to *diverging* effective exponents $\alpha(q, n)$ and $\delta(q, n)$. Moreover, the specific heat and the magnetic susceptibility per lattice row on $n \times \infty$ strips develop a sharp peak at some finite temperature $T_m(n)$, where the peak height grows *linearly* with n as $n \to \infty$. This is a clear indication for the *discontinuity* which the internal energy and the magnetization display at the transition temperature of the infinite system.

Interfaces can be introduced to the Potts model on a finite $L_1 \times L_2$ lattice with periodic boundary conditions by placing *permutation operators* $G_{r,r'}$ acting on a bond between two Potts spins p_r and $p_{r'}$ on the lattice along certain lines [87]. The $L_1 \times L_2$ rectangle with periodic boundary conditions can be identified with a *torus*, where L_1 is the toroidal and L_2 is the poloidal perimeter. If a permutation operator $G_{r,r'}$ is placed along the toroidal perimeter and a second permutation operator $H_{r,r'}$ is placed along the poloidal perimeter transverse to $G_{r,r'}$, the torus becomes equivalent to a $L_1 \times L_2$ rectangle with *boundary conditions* induced by G and H on opposite edges, where the *identity operator* corresponds to a *periodic boundary condition*. At $T = T_{c,b}$ the interfacial free energy σ then scales as [87]

$$\sigma = A_q^{GH}(s)/L_2 = s^{-1} A_q^{GH}(s)/L_1, \tag{2.32}$$

where $s = L_2/L_1$ is the aspect ratio of the torus and $A_q^{GH}(s)$ is the universal finite–size scaling amplitude of the interfacial free energy induced by the permutation operators G and H. [39]. The case $s = 0$ corresponds to a cylinder of perimeter L_2 with an interface induced by G or a strip of width L_2 with the appropiate boundary condition. The universal amplitude $A_q^{GH}(s)$ is defined as (see also Eq.(2.14)) [87]

$$A_q^{GH}(s) = s \lim_{L_1, L_2 \to \infty} \frac{F_{GH}(T_{c,b}; L_1, L_2, q) - F_{per}(T_{c,b}; L_1, L_2, q)}{k_B T_{c,b}}, \tag{2.33}$$

where the aspect ratio s is kept fixed and F_{GH} and F_{per} denote the free energies of the torus with and without the interfaces, respectively. The universal amplitudes $A_q^{GH}(s)$ are caluclated in Ref.[87] for $q = 2, 3, 4$ and a few values of the aspect ratio s by evaluating the leading finite–size contribution to the partition function of the q–state Potts model with the appropiate boundary conditions. Note, that the case $q = 2$ corresponds to the Ising model.

The finite–size behavior of the 3–state Potts model in $d = 3$ has been studied by a finite–size scaling analysis of high statistics Monte–Carlo data on simple cubic $L \times L \times L$ lattices with $16 \leq L \leq 64$ [88]. The specific heat and the susceptibility develop sharp peaks at a finite temperature and the peak heights scale linearly with the volume L^3 of the lattice. Moreover, the correlation length shows a finite jump at the same temperature which is reminiscent of a first–order phase transition occuring in the infinite–volume limit. However, the jumps of the internal energy and the Potts magnetization are rather small so that the phase transition of the three–state Potts model in $d = 3$ is only weakly first–order [88]. A corresponding Monte–Carlo study of the 3–state Potts model on a diamond lattice with $20^3 \ldots 64^3$ lattice sites supports the above findings [89].

If a critical system is confined to a finite geometry in d dimensions, the shift of the specific heat maximum $T_m(L)$ with respect to the critical temperature $T_{c,b}$ of the infinite system in terms of the system size L is governed by the shift exponent $\Lambda = 1/\nu$ (see Eq.(2.5)). Near a first–order phase transition the correlation length ξ for short–ranged interactions remains finite, and the corresponding shift of $T_m(L)$ with respect to the transition temperature of the bulk system is expected to be governed by $\Lambda = d$ once the system size L exceeds ξ. A rigorous proof of this statement for the q–state Potts model confined to a cylindrical $L^{d-1} \times L'$ geometry in $d \geq 2$ has been given in Ref.[90] for large q.

2.2.4   The Spherical Model

Rigorous statements on the finite–size scaling properties of a model system representing one of the most interesting universality classes are usually confined to two dimensions if any exact results are available at all. In this respect the spherical model has the invaluable advantage that exact solutions can be obtained in any spatial dimension of interest for short–ranged and long–ranged interactions. This gives the opportunity to address special issues in finite–size scaling such as the influence of long–ranged interactions or logarithmic corrections rigorously.

The geometry which is usually considered as a confinement of the spherical model in d dimensions is infinite in $d' < d$ directions and finite in the remaining $d - d'$ directions. In short notation this geometry is often characterized as $L^{d-d'} \times \infty^{d'}$. The boundary conditions are taken to be periodic, antiperiodic, free (ordinary), or fixed (extraordinary). If the pair interaction potential $J_{ij} = J(|\mathbf{r}_i - \mathbf{r}_j|) > 0$ is short–ranged (see Eq.(1.21) and Ref.[13]) and $2 < d' < d < 4$, the singular part of the specific heat, the spontaneous magnetization, and the magnetic susceptibility are governed by the usual scaling relations (see Eqs.(2.4), (2.8), and (1.21)) [91, 92]. The condition

$d' > 2$ guarantees the existence of a phase transition in the finite system at a shifted critical temperature $T_c(L)$ which confirms Eq.(2.5) [91]. The critical behavior of the confined spherical model near $T_c(L)$ is governed by the d'–dimensional counterparts of the d–dimensional critical exponents (see Eq.(1.21) for $\sigma^* = 2$). A corresponding statement holds for the *amplitudes* of the singular thermodynamic quantities and the correlation length [91, 92].

At $T = T_{c,b}$ the correlation length ξ_L of the spin–spin correlation function in a *finite* system is governed by a *universal* amplitude A. This amplitude can be calculated exactly for a nearest–neighbour spherical model with periodic or antiperiodic boundary conditions in a $L^2 \times \infty$ geometry [93]. Specifically, one has in the large L limit

$$\xi_L/L = A/x, \qquad (2.34)$$

where x is the scaling dimension of the spin density in the spherical model and $A = 0.1361\ldots$ for antiperiodic boundary conditions is the universal amplitude. In fact, the aforementioned value for A comes very close to the corresponding value $A \simeq 0.12$ for the Ising model in $d = 3$ [93]. In two dimensions the amplitude A for periodic boundary conditions is given by $A = 1/(2\pi)$ which is a consequence of *conformal invariance* [39].

Above the upper critical dimension d_c finite–size scaling still holds, but the scaling arguments must be modified (see Eq.(2.9)). For short–ranged interactions d_c is usually so large that $d > d_c$ cannot be realized by physically attainable dimensions d. However, for *long–ranged* interactions $J(r \to \infty) \sim r^{-(d+\sigma)}$ with $\sigma < 2$ the upper critical dimension is given by $d_c = 2\sigma$, and if σ is small enough, the case $d > d_c$ becomes physically relevant. The spherical model with long–ranged interactions can be treated by analytical methods which have been generalized to $L^{d-d'} \times \infty^{d'}$ geometries with periodic boundary conditions [94]. The results show that Eq.(2.9), which holds for hypercubes ($d' = 0$), must be modified for $d' > 0$ according to [94]

$$\lim_{L' \to \infty} \frac{\mathcal{F}^{sing}(t, H; L, L')}{k_B T_{c,b} L^{d-d'} L'^{d'}} = L^{-d^*} f_\pm \left(|t| L^{d^*/(2-\alpha)}, H L^{d^* \Delta/(2-\alpha)} \right), \qquad (2.35)$$

where the reduced temperature t contains a finite–size shift of the critical temperature and

$$d^* = d - d' \frac{d\nu + \alpha - 2}{d'\nu + \alpha - 2}. \qquad (2.36)$$

For $d < d_c$ (hyperscaling) or $d' = 0$ one has $d = d^*$ (see Eq.(2.9)). The factor L^{-d^*} in Eq.(2.35) measures the inverse *correlated volume* [94].

In bulk systems with long–ranged interactions the bulk correlation length $\xi_{\pm}$ in the usual sense does not exist (see Eq.(5.10) in Sec.5.1). For a spherical model confined to a $L^{d-d'} \times \infty^{d'}$ geometry, however, an *effective correlation length* $\xi_L(t, H)$ can be introduced such that the pair correlation function $G_L(\mathbf{r}; t, H)$ takes the scaling form [95]

$$G_L(\mathbf{r}; t, H) = a(t) r^{-(d-2+\eta)} g\left(r/\xi_L, L/\xi_L\right), \tag{2.37}$$

where $a(t)$ is analytic in the reduced temperature t and $g(x, y)$ is a *universal* finite–size scaling function. The dependence of $G_L(\mathbf{r}; t, H)$ on the magnetic field H is completely absorbed by the effective correlation length $\xi_L(t, H)$. This is a special feature of the spherical model which does not hold in general. The *effective bulk correlation length* $\xi_{\infty}(t, H)$ defined by [95]

$$\xi_{\infty}(t, H) = \lim_{L \to \infty} \xi_L(t, H) \tag{2.38}$$

provides a finite macroscopic length scale at least for $T > T_{c,b}$ although the usual bulk correlation length is always infinite. In the limit $L \to \infty$ and for $H = 0$ one has for the bulk susceptibility

$$\chi_{\infty}(t, 0) = \int d^d r\, G_{\infty}(\mathbf{r}; t, 0) \propto \left(\xi_{\infty}(t, 0)\right)^{2-\eta}.$$

From the critical singularity of the susceptibility $\chi_{\infty}(t, 0) \sim |t|^{-\gamma}$ near $t = 0$ we conclude that $\xi_{\infty}(t, 0) \sim |t|^{-\nu}$ with $\nu = \gamma/(2 - \eta)$. Therefore $\xi_{\infty}(t, 0)$ displays the same singularity as the usual bulk correlation length. Provided, that the spatial dimension d of the confined sperical model is in the range $\sigma < d < 2\sigma = d_c$, the effective correlation length $\xi_L(t, 0)$ has the *scaling form* [57, 95]

$$\xi_L(t, 0) = L\,\Xi(L/\xi_{\infty}(t, 0)) \tag{2.39}$$

so that in view of the property $\xi_{\infty}(t, 0) \sim |t|^{-\nu}$ Eq.(2.39) is a prescription which allows one to replace L/ξ_L in Eq.(2.37) by the usual scaling argument $|t|L^{1/\nu}$. For dimensions d *above* the upper critical dimension $d_c = 2\sigma$ the effective correlation length attains a different scaling form, namely [57]

$$\xi_L(t, 0) = l(L/l)^{\omega}\,\tilde{\Xi}\left((L/l)^{\omega}l/\xi_{\infty}(t, 0)\right) \quad \text{with} \quad \omega = (d - d')/(2\sigma - d'), \tag{2.40}$$

where l is a *microscopic* length scale which cannot be eliminated due to $\omega > 1$. Therefore Eq.(2.40) requires the introduction of a *second* macroscopic bulk length λ_∞ according to [57]

$$\lambda_\infty(t,0) = l\left(\xi_\infty(t,0)/l\right)^{1/\omega} \tag{2.41}$$

which can be used to rewrite the finite–size scaling form of the pair correlation function in Eq.(2.37) for $H = 0$ and $d > d_c$ as [57]

$$G_L(\mathbf{r}; t, 0) = a(t) r^{-(d-2+\eta)} h\left(r/\xi_\infty, L/\xi_\infty, L/\lambda_\infty\right) \tag{2.42}$$

Usual critical finite–size scaling is recovered from Eq.(2.42) for $d > d_c$, but here *two* macroscopic bulk length scales are involved.

Concerning logarithmic corrections to finite–size scaling explicit results from the spherical model in finite geometries are available as well. First, we note that *at the* upper critical dimension $d_c = 2\sigma$ the effective correlation length at bulk criticality $\xi_L(0,0)$ in the $L^{d-d'} \times \infty^{d'}$ geometry with periodic boundary conditions differs from the form shown in Eqs.(2.39) and (2.40) by a logarithmic factor. One finds [95]

$$\xi_L(0,0) = \xi_0(d',\sigma) L \left(\ln(L/l)\right)^{1/(2\sigma-d')}, \tag{2.43}$$

where $0 < \sigma \leq 2$ characterizes the range of the interactions. For the precise form of the amplitude $\xi_0(d',\sigma)$ see Ref.[95]. Second, the free energy in the critical region of a nearest neighbour ferromagnetic mean spherical model confined to a hypercube L^d in d dimensions with free (ordinary) or fixed (extraordinary) boundary conditions can be studied analytically and compared to Eq.(2.10) [96]. In order to control the presence of edges and corners *periodic* boundary conditions are applied along $0 \leq d_p \leq d$ directions. In the vicinity of $T_{c,b}$ and for $L \to \infty$ the logarithmic corrections ΔF_O to the free energy in $d = 3$ in the case of $d - d_p$ free (O) boundary conditions read [96]

$$\frac{\Delta F_O(t \simeq 0; L)}{k_B T_{c,b}} = \begin{cases} \ln(L/l) + 2^{-3}\ln\ln(L/l) & : \quad d_p = 0 \\ 2^{-2}\ln(L/l) & : \quad d_p = 1 \\ 0 & : \quad d_p \geq 2, \end{cases} \tag{2.44}$$

where l is the lattice constant. Although there are no real corners in the L^d geometry in the case $d_p = 1$ a logarithmic contribution occurs and, moreover, in presence of corners an additional double logarithmic term appears. The logarithmic corrections to the free energy displayed in Eq.(2.44) may be specific for the spherical model, but one should note that the structure of Eq.(2.44) has not been anticipated by the phenomenological arguments which lead to Eq.(2.10). For $d - d_p$ *fixed* (E) boundary conditions the corresponding corrections ΔF_E read in $d = 3$ [96]

$$\frac{\Delta F_E(t \simeq 0; L)}{k_B T_{c,b}} = \begin{cases} 2^{-3}(1 + 9\pi)\ln(L/l) - 2^{-1}\ln\ln(L/l) & : \quad d_p = 0 \\ 2^{-1}\pi\ln(L/l) - 2^{-1}\ln\ln(L/l) & : \quad d_p = 1 \\ 2^{-3}\pi\ln(L/l) - 2^{-1}\ln\ln(L/l) & : \quad d_p = 2 \\ 0 & : \quad d_p = 3, \end{cases} \quad (2.45)$$

where the logarithms are already generated by the *surfaces*. In summary, the spherical model with nearest–neighbour interactions in finite geometries generates more and other logarithmic corrections to the free energy than expected from Eq.(2.10).

The spin variables in the spherical model are allowed to vary *continuously* and therefore a *helicity modulus* can be defined analogous to Eq.(2.18) using antiperiodic versus periodic boundary conditions. In a finite system there are various ways to define a helicity modulus such that the bulk limit given by Eq.(2.18) is reproduced for an infinite system. In the hypercube L^d one possible choice is [97]

$$\Upsilon(t; L) = \frac{2L^2}{k_B T \pi^2} \left[f_{\tau_a}(t; L) - f_{\tau_p}(t; L) \right], \quad (2.46)$$

where $f_\tau(t; L)$ is the free energy density of the spherical model confined to a hypercube with the set $\tau = \{\tau_1, \ldots, \tau_d\}$ of boundary conditions applied to the directions $x_1, \ldots, x_d$. The boundary conditions τ_a and τ_p differ in the boundary condition τ_1 which is *periodic* in τ_p and *antiperiodic* in τ_a. Note, that the helicity modulus defined by Eq.(2.46) depends on the set $\{\tau_2, \ldots, \tau_d\}$ of boundary conditions which τ_p and τ_a have in common. Altogether d_p periodic, d_a antiperiodic, d_0 free, and d_1 fixed boundary conditions are applied, where $d_p + d_a + d_0 + d_1 = d$ [97]. The explicit calculation shows that a logarithmic term as predicted by Eq.(2.20) in $d = 3$ only occurs sufficiently *below* $T_{c,b}$. In the *vicinity* of $T_{c,b}$ a double logarithmic correction [97]

$$\Delta\Upsilon(t \simeq 0; L) = -\pi^{-2}2^{d_a-1}L^{-1}\ln\left[(d_1 - d_0)\ln(L/l)\right] \quad (2.47)$$

appears for $d_1 > d_0$ in $d = 3$ which has not been predicted by Eq.(2.20). We note again that Eq.(2.46) is not the only possible definition of the helicity modulus in a finite system. Other definitions may lead to a different finite–size behavior [97]. Finally, we mention that for $L^{3-d'} \times \infty^{d'}$ geometries with $d' > 0$ and periodic and antiperiodic boundary conditions logarithmic corrections to the helicity modulus defined according to Eq.(2.46) do *not* occur in the critical region [97]. Due to these significant discrepancies between the behavior of the spherical model and the phenomenological expectation expressed in Eqs.(2.10) and (2.20) logarithmic corrections to finite–size scaling deserve a further investigation especially by field–theoretical methods.

2.3   Continuum Theory in Finite Geometries

2.3.1   The Ginzburg – Landau Hamiltonian

The standard continuum model for critical or tricritical $O(N)$–symmetric systems confined to finite geometries is based on the Ginzburg–Landau Hamiltonian given by Eq.(1.28) or Eq.(1.30), respectively, where the domain of integration is restricted to the volume of the geometry. For a film geometry two surface contributions according to Eq.(1.29) or Eq.(1.31), respectively, complete the Hamiltonian which reads for a critical film

$$\mathcal{H}\{\Phi(\mathbf{r})\} = \int_0^L \int d^{d-1}r_\| \left[\tfrac{1}{2}(\nabla\Phi(\mathbf{r}))^2 + \tfrac{\tau}{2}\Phi^2(\mathbf{r}) + \tfrac{g}{4!}\left(\Phi^2(\mathbf{r})\right)^2 - \mathbf{h}(\mathbf{r})\cdot\Phi(\mathbf{r}) \right]$$

$$+ \int d^{d-1}r_\| \left[\tfrac{c_a}{2}\Phi^2(\mathbf{r}_\|,0) - \mathbf{h}_a(\mathbf{r}_\|)\cdot\Phi(\mathbf{r}_\|,0) \right]$$

$$+ \int d^{d-1}r_\| \left[\tfrac{c_b}{2}\Phi^2(\mathbf{r}_\|,L) - \mathbf{h}_b(\mathbf{r}_\|)\cdot\Phi(\mathbf{r}_\|,L) \right], \qquad (2.48)$$

where $\mathbf{r} = (\mathbf{r}_\|,z)$ and L denotes the film thickness. According to the theoretical description of surface critical phenomena (see Sec.1.2) the two walls a and b of the film are characterized by surface enhancements c_a and c_b and surface fields $\mathbf{h}_a$ and $\mathbf{h}_b$. Besides the usual bulk scaling fields all possible combinations (a,b) of the three surface universality classes O, SB, and E determine the leading critical behavior of the film. Note, that for $(a,b) = (E,E)$ the relative orientation of the surface fields is important (see Sec.2.1). The parameter τ in Eq.(2.48) is proportional to the reduced temperature (see Eq.(1.28)).

For the description of confinements with edges and corners it seems natural to add further edge and corner contributions of the type given by Eq.(1.29) to the Ginzburg–Landau Hamiltonian. Introducing an "edge enhancement" c_e and an "edge field" $\mathbf{h}_e$ one can write in complete analogy with Eq.(1.29) [20]

$$\mathcal{H}_{\mathcal{E}}\{\Phi(\mathbf{r})\} = \int_{\mathcal{E}} d^{d-2}r_\| \left[\tfrac{c_e}{2}\Phi^2(\mathbf{r}_\|,0) - \mathbf{h}_e(\mathbf{r}_\|)\cdot\Phi(\mathbf{r}_\|,0) \right], \qquad (2.49)$$

where the domain $\mathcal{E}$ of integration is $d-2$–dimensional. To which extent an edge contribution as given by Eq.(2.49) must be taken into account in the Ginzburg–Landau Hamiltonian depends on the *surface* universality class one is considering. If two surfaces which share a common edge belong to the *ordinary* surface universality class, the influence of c_e can be disregarded for the leading singular behavior of the free energy [20]. The edge field $h_e(\mathbf{r})$ in this case gives rise to a new scaling argument in the edge contribution $F_{e,O}^{sing}$ to the singular part of the free energy which reads (see Eqs.(1.10) and (2.11)) [20]

$$F_{e,O}^{sing}(t, h; h_1; h_e) = |t|^{(d-2)\nu} f_{e\pm}^{sing}\left(h|t|^{-\nu}, h_1|t|^{-\Delta_1^O}, h_e|t|^{-\Delta_e^O}\right), \qquad (2.50)$$

where Δ_e^O in analogy with Δ_1^O denotes the edge gap exponent for the ordinary surface universality class. Note, that Δ_e^O depends on the opening angle of the edge [20]. For other types of surfaces the full edge contribution according to Eq.(2.49) must be retained. In the renormalization group sense the scaling field c_e is *marginal* in any spatial dimension and h_e is *relevant* above two dimensions. In the same sense *corner* contributions to the Ginzburg–Landau Hamiltonian of the type given by Eq.(2.49) are *irrelevant*, if $d < 4$.

For confinements with curved surfaces curvature contributions to the Ginzburg–Landau Hamiltonian replace the edge and corner terms. The mean curvature $H(\mathbf{r})$ of the surface generates a contribution of the form [20]

$$\mathcal{H}_H\{\Phi(\mathbf{r})\} = \int_S dS H(\mathbf{r})\left[\tfrac{c_H}{2}\Phi^2(\mathbf{r}_{\parallel}, 0) - h_H(\mathbf{r}_{\parallel})\cdot\Phi(\mathbf{r}_{\parallel}, 0)\right], \qquad (2.51)$$

where the integration is carried out over the boundary S of the geometry. The scaling fields c_H and h_H in Eq.(2.51) play the same role as c_e and h_e in Eq.(2.49). This means in particular that for a surface S belonging to the ordinary surface universality class c_H has no influence on the leading singular behavior of the free energy [20]. In the renormalization group sense c_H is always marginal and h_H is relevant, if $d > 2$. Higher order curvature contributions due to, e.g., the Gaussian curvature $K(\mathbf{r})$ are irrelevant in the same sense as corner contributions.

Continuum models are also often used to describe fluctuations of interfaces *away* from critical points. The extended study on finite–size effects in fluid interfaces discussed in Ref.[60] (see Sec.2.1) for example relies on an *effective capillary wave Hamiltonian* which is of the Gaussian form

$$\mathcal{H}_{cw}\{z(\mathbf{r}_{\parallel})\} = \tfrac{1}{2}\sigma_{cw}\int d^{d-1}r_{\parallel}\left[\left(\nabla z(\mathbf{r}_{\parallel})\right)^2 + \left(z(\mathbf{r}_{\parallel})/\xi_{\parallel,cw}\right)^2\right]. \qquad (2.52)$$

The field $z(\mathbf{r}_{\|})$ denotes the height of the interface between two fluids above an ar-
bitrary flat zero level, and $\xi_{\|,cw}$ is the interfacial correlation length given by $\xi_{\|,cw}^2 =$
$\sigma_{cw}/(\Delta\rho g)$, where σ_{cw} is the interfacial tension, $\Delta\rho$ is the density difference between
the two fluids, and g is the gravitational acceleration. Due to the truncation of the
gradient expansion in Eq.(2.52) after the second–order term Eq.(2.52) describes a set
of decoupled modes in fourier space. Higher order gradient contributions are sometimes subsumed in a momentum dependent surface tension $\sigma_{cw}(\mathbf{q})$ which modifies
the capillary wave Hamiltonian according to [60]

$$\mathcal{H}_{cw}\{z(\mathbf{q})\} = \tfrac{1}{2} \int d^{d-1}q \left(\sigma_{cw}(\mathbf{q})\mathbf{q}^2 + \Delta\rho g\right) |z(\mathbf{q})|^2, \qquad (2.53)$$

where $\sigma(\mathbf{q})$ reduces to the usual interfacial tension σ in the limit $\mathbf{q} \to 0$. A large–
momentum cutoff q_{max} for liquid–vapor interfaces ($T < T_{c,b}$) is set by the bulk correlation length ξ_- in a natural way by $q_{max} \simeq \pi/\xi_-$. However, depending on $\sigma_{cw}(\mathbf{q})$
different choices for q_{max} are possible [60].

The principal structure of effective interface Hamiltonians can be unveiled from
the microscopic point of view using *density functional theory* [98]. This procedure
has the advantage that gradient terms of higher order than displayed in Eq.(2.52)
can be constructed systematically and, moreover, the expansion coefficients can be
unambiguously related to the microscopic interaction potential [98]. It turns out that
the expansion coefficients are governed by *moments* of the interaction potential which
for long–ranged interactions only exist up to a *finite* order. These deficiencies of the
gradient expansion for long–ranged interactions are also reflected by a *nonanalyticity*
in the $\mathbf{q}$–dependence of $\sigma_{cw}(\mathbf{q})$ for $\mathbf{q} \to 0$ as shown in Ref.[98] for nonretarded van-
der–Waals forces in $d = 3$.

2.3.2 Mean Field Theory

The simplest possible approximation to the free energy of a finite system in the critical
region described by a Ginzburg–Landau Hamiltonian like in Eq.(2.48) can be obtained
by interpreting $\mathcal{H}\{\Phi(\mathbf{r})\}$ as a *functional* for the free energy which upon minimization
with respect to $\Phi(\mathbf{r})$ yields the order parameter and the free energy in equilibrium.
The corresponding free energy is the leading term in a *saddle–point evaluation* of the
partition function for the Ginzburg–Landau Hamiltonian. The order parameter $\mathbf{M}$,
which exhibits a spatial variation in presence of boundaries, is the solution of the

Euler–Lagrange equation

$$\Delta M(r) = \tau M(r) + \tfrac{g}{6} M^2(r) M(r) - h(r) \tag{2.54}$$

subject to the *boundary conditions*

$$\frac{\partial M}{\partial z}(r_\parallel, 0) = c_a M(r_\parallel, 0) - h_a(r_\parallel)$$

$$\frac{\partial M}{\partial z}(r_\parallel, L) = -c_b M(r_\parallel, L) + h_b(r_\parallel) \tag{2.55}$$

generated by the *surface contributions* to $\mathcal{H}\{\Phi(r)\}$ for a film geometry (see Eq.(2.48)).
In lattice models the surface universality classes are realized by boundary conditions,
where free boundaries represent the ordinary (O) surface universality class ($c = +\infty$),
fixed boundaries represent the extraordinary (E) surface universality class ($c = -\infty$)
(see Sec.2.2), and the special value K_1^c/K^c of the surface to bulk coupling ratio
$J_1/J = K_1/K$ represents the surface–bulk (SB) or special surface universality class
($c = 0$) (see Fig.1.11). On the mean–field level Eq.(2.55) provides the direct link between the surface enhancement c and the surface field h_1 on one hand and *boundary
conditions* for the *order parameter* on the other hand. According to Eq.(2.55) the O
surface universality class corresponds to a *Dirichlet* boundary condition $M(r_\parallel, 0) = 0$
and the SB surface universality class corresponds to a *Neumann* boundary condition
$\partial M(r_\parallel, 0)/\partial z = 0$. In the E surface universality class, where $c \to -\infty$ or, equivalently, $h_1 \to \pm\infty$, the order parameter displays a z^{-1}-singularity at the surface
$z = 0$ (see, e.g., [20]). This singularity in the order parameter profile is an artefact of
the continuum description given by the Ginzburg–Landau Hamiltonian. In real systems or microscopic models the molecular diameter or the lattice constant provides
a microscopic cutoff (see Chap.4).

The framework of mean–field or Landau theory provided by Eq.(2.48), (2.54),
and (2.55) gives the opportunity to address a variety of questions in finite–size scaling
within a continuum model [41]. However, Landau theory disregards fluctuations
so that nontrivial results can only be obtained for situations in which the order
parameter does not vanish. This is only the case for nonvanishing bulk or surface fields
or for sufficiently negative reduced temperature or surface enhancements. A system
for which nonvanishing surface fields are important is provided by a binary liquid
mixture. If such a binary mixture is in contact with an external wall, the composition

of the mixture usually deviates from the bulk value in the vicinity of the surface due to a preferential affinity of the wall for one of the components. The chemical affinity of a wall is theoretically described by a surface field and the composition profile of the mixture basically represents the order parameter profile [41]. The scaling behavior of the free energy of a binary mixture confined between parallel plates according to Landau theory is governed by Eq.(2.3), where the bulk and surface critical exponents are given by their Landau values (see Ref.[19, p.37] for a complete list). The shift of the critical demixing transition of a binary mixture confined between plates behaves as shown in Eq.(2.5) [41] (see also Ref.[40]).

The nonzero order parameter profile in a film geometry with surface fields gives rise to a mean–field finite–size contribution to the free energy per unit area at the bulk critical point which is governed by the universal *Casimir amplitudes* Δ (see Chap.3) [99]. In presence of surface fields the combinations $(a, b) = (O, E)$, (SB, E), $(+, +)$, and $(+, -)$ of surface universality classes are possible for Ising–like systems, where $(+, +)$ refers to parallel and $(+, -)$ refers to antiparallel surface fields. Note, that for (O, E) and (SB, E) the E surface can be represented by a surface field of any orientation. Landau theory predicts nonzero values for the Casimir amplitudes $\Delta_{O,E}$, $\Delta_{SB,E}$, $\Delta_{+,+}$, and $\Delta_{+,-}$. The effect of the critical finite–size contribution to the free energy on the thickness of a *critical wetting layer* [99], e.g., a binary liquid at the bulk demixing point is discussed in Sec.6.3 in detail.

In a fluid confined to a film geometry true wetting transitions, which are characterized by a *diverging* coverage, are prevented by the finite separation L of the walls. However, a sharp *quasi–wetting* transition to a state characterized by wetting layers of finite thickness which diverges for $L \rightarrow \infty$ may occur in a fluid between walls. *Opposing* surface fields furthermore suppress *capillary condensation* which usually intervenes if the distance L between the walls is too small or the walls are chemically too similar (parallel surface fields). Landau theory for a film with opposing surface fields then reveals a richly structured phase diagram showing first–order, critical, and tricritical quasi–wetting transitions which can be treated as *shifted* wetting transitions [100]. According to Landau theory the full variety of quasi–wetting transitions persists in the vicinity of the bulk critical point.

If a fluid is confined between parallel plates characterized by opposing surface fields which are chosen such that above a certain *critical* wetting temperature T_w the fluid wets one wall and dries the other, coexistence between liquid and vapor occurs

for temperatures T below a certain critical temperature $T_c(L)$ [101]. Landau theory
for Eq.(2.48) in this case shows that the critical temperature $T_c(L)$ is located *below*
the critical wetting temperature T_w. The shift of $T_c(L)$ with respect to T_w is found to
be exponentially small as function of the ratio L/ξ_-, where ξ_- is the *bulk* correlation
length at $T = T_c(L)$. For the general shift law $T_w - T_c(L) \sim L^{-1/\beta_s}$, where β_s is
the critical wetting exponent governing the growth of the wetting layer thickness as
$T \nearrow T_w$, a heuristic scaling argument can be given [101]. The density profile of the
fluid for large plate separations L has scaling properties in the temperatures range
$T_w \leq T < T_{c,b}$. The *distant wall effect* of one wall on the profile near the other wall
is governed by wetting exponents, if $T < T_{c,b}$. For example, one has at $T = T_w$ for
$z \ll L$ [102]

$$\frac{M(z) - M_0}{M_0} \sim \left(\frac{z}{L}\right)^{\tau-1/\beta_s}, \tag{2.56}$$

where $\tau = 2(d-1)/(3-d)$ and the density profile is given by $M(z) = M_0 m(\tilde{\varepsilon}^{\beta_s} z, \tilde{\varepsilon}^{\beta_s} L)$
with $\tilde{\varepsilon} = (T_w - T)/T_w$. In $d = 2$ the exponent in Eq.(2.56) is unity and in $d = 3$ the
power law is converted to an exponential, i.e., the distant wall correction becomes
exponentially small [102]. Distant wall corrections at $T = T_{c,b}$ will be discussed in
Chap.4 in detail. The full spectrum of phase behavior of fluids confined between walls
with opposing surface fields is summarized in Ref.[103].

Mean–field theory for fluids confined to finite geometries is of course not restricted
to the Ginzburg–Landau Hamiltonian. A widely used and very powerful mean–field
description of fluids is based on density functional theory which in contrast to the
Ginzburg–Landau Hamiltonian incorporates the detailed microscopic pair interaction
potential between the particles. Especially the phenomenon of *capillary condensation*, which we have already encountered several times, has been extensively studied
by means of a density functional theory based on *short–ranged* pair interactions and
wall potentials with *equal* decay lengths [28]. During capillary condensation an undersaturated vapor condenses to a liquid inside sufficiently narrow capillaries or small
pores. The simplest model for a capillary is given by a slab (film) of width L. For
small undersaturations $\Delta\mu$ of the vapor and a width L *below* a certain threshold L_{min}
the fluid inside the capillary is always in the liquid state. For large undersaturations
$\Delta\mu$ *above* a certain threshold the fluid inside the capillary is always in the vapor state,
where two different vapor configurations coexist at a certain capillary width L_t [28].
In the intermediate regime for $\Delta\mu$ and L the total amount $\Gamma(T, L, \Delta\mu)$ of adsorbed

liquid in general jumps discontinuously to some finite value at a certain undersaturation $\Delta\mu(T, L)$ for given width L and temperature T. Interestingly, metastable
states exist in this case which give rise to hysteresis effects in Γ. These hysteresis
effects disappear at a certain *capillary critical point* $(\Delta\mu_c, L_c)$ for each temperature T
[28]. Very recently capillary condensation of liquid ^{4}He inside narrow slabs has been
studied with a density functional model [104]. It turns out that for widths $L \leq 100\text{Å}$
the capillary condensation condition obtained from the Kelvin equation, which has
been compared to the corresponding prediction of density functional theory, is quantitatively unreliable [104].

One issue which is often raised in the context of capillary condensation is the effect
of thick (quasi–) wetting layers on the inner walls of a capillary on the Kelvin equation,
i.e., on the shift of the liquid–vapor coexistence curve due to the finite capillary width
L [105]. It turns out that the thickness $l(\Delta\mu)$ of the wetting layer effectively *reduces*
the capillary width so that the leading correction to the Kelvin equation is governed
by the ratio $l(\Delta\mu)/L$. For short–ranged interactions one has $l(\Delta\mu) \sim -\ln\Delta\mu \sim \ln L$
(see Eq.(2.28)) which has been derived in Ref.[71] independently.

Mean–field theory either in the Landau version based on the Ginzburg–Landau
Hamiltonian or based on density functional theory reveals a large spectrum of phenomena in confined fluids which can be observed directly in experiments. However, as
far as the vicinity of critical points is concerned the theoretical description of finite–
size effects by mean–field theory suffers from well known shortcomings. In order to
account for the contributions of critical fluctuations to finite–size effects properly
one has to turn to the renormalization group treatment of the Ginzburg–Landau
Hamiltonian.

2.3.3   The Renormalization Group

The field–theoretical analysis of a critical system confined to a finite geometry which is
in general described by a Ginzburg–Landau Hamiltonian with surface, edge, and curvature contributions requires the renormalization of the surface, edge, and curvature
scaling fields (see Eqs.(2.48), (2.49), and (2.51)) besides the usual bulk renormalization procedure [20]. However, as far as the field–theoretical analysis of a critical *film*
is concerned the renormalization of all scaling fields occurring in the film Hamiltonian given by Eq.(2.48) has already been performed in the field–theoretical analysis of

semiinfinite systems [20] so that the renomalization Z–factors and Wilson functions needed for a film are known beforehand.

The special values $c = -\infty, 0, \infty$ which correspond to the E, SB, and O surface universality class, respectively, are the *fixed point values* of the surface enhancement c under the renormalization group. On the mean–field level these fixed point values (surface universality classes) can be translated to *boundary conditions* for the order parameter profile $M(r_{\|}, z)$ (see Eq.(2.55)). However, beyond the mean–field approximation $M(r_{\|}, z)$, now given by the thermal average $M(r_{\|}, z) = \langle \Phi(r_{\|}, z) \rangle$, does no longer fulfill these boundary conditions in general. The behavior of the order parameter profile $M(r_{\|}, z)$ near a plane wall at, say, $z = 0$ sufficiently far away from any edges or corners is governed by the power law

$$\left| M(r_{\|}, z \to 0) \right| \sim z^{(\beta_1^a - \beta)/\nu}, \qquad (2.57)$$

where the exponent β_1^a of the surface magnetization (see Table 1.4) depends on the surface universality class a to which the wall under consideration belongs. For $O(N)$–symmetric systems with short–ranged interactions in $2 < d < 4$ one has $\beta_1^O > \beta$, $\beta_1^{SB} < \beta$, and, formally, $\beta_1^E = 0$ which means that $M(r_{\|}, z \to 0) \sim z^{-\beta/\nu}$ for the E surface universality class [20]. The order parameter profile therefore still fulfills a *Dirichlet* boundary condition at an O surface, but critical fluctuations destroy the Neumann boundary condition at a SB surface. However, the Ginzburg–Landau Hamiltonian for a geometry which is exclusively bounded by surfaces representing the O surface universality class greatly simplifies, because such a surface is still represented by a Dirichlet boundary condition and surface, edge, or curvature contributions to the Hamiltonian become irrelevant [20]. For example, the Hamiltonian for a critical system confined to a sphere with an O surface reads [63]

$$\mathcal{H}\{\Phi(r)\} = \int_{|r| \leq R} d^d r \left[\frac{1}{2} (\nabla \Phi(r))^2 + \frac{\tau}{2} \Phi^2(r) + \frac{g}{4!} \left(\Phi^2(r) \right)^2 - h(r) \cdot \Phi(r) \right]. \qquad (2.58)$$

The decomposition of the free energy for Eq.(2.58) is shown in Eq.(2.23).

Finite–size scaling studies within the framework of the field–theoretical renormalization group have been carried out for hypercubic L^d or cylindrical $L^{d-1} \times \infty$ geometries with periodic boundary conditions, where surface, edge, and corner terms do neither appear in the Hamiltonian nor in the free energy. In these geometries phase transitions are always suppressed (for short–ranged interactions), i.e., the critical singularities are replaced with rounded maxima (see Sec.2.1 and Fig.2.1). In order to obtain a field–theoretical description of this rounding effect the *zero mode* of the theory must be treated nonperturbatively [106]. We will discuss this special procedure in Sec.5.5 in some detail. The nonperturbative treatment of the zero mode leads to a *rounded* specific heat for a $O(N)$–symmetric system confined to a hypercube L^d with periodic boundary conditions [107]. At the upper critical dimension $d_c = 4$ the specific heat per unit volume of an Ising system behaves as $\left(\ln(L/\xi_0^+)\right)^{1/3}$ for $T = T_{c,b}$ which is due to the usual logarithmic corrections to Landau critical behavior in $d = d_c$ [107]. Below the upper critical dimension in $d = 4 - \varepsilon$ the usual finite–size scaling behavior according to Eq.(2.8) sets in. Field–theoretical studies of the free energy in a critical film for various boundary conditions are presented in Chap.5.

The ordinary surface universality class (Dirichlet boundary condition) provides an appropiate description for confined ^{4}He (XY universality class, $N = 2$) near the bulk λ–transition (see Fig.(1.6)) [108, 109]. Again the proper treatment of the zero–mode of the theory is essential for reliable field–theoretical results for the specific heat near $T = T_\lambda$ (see Sec.5.5). Field theory in this case even yields *quantitative* agreement with experimental data on the specific heat down to temperatures slightly below T_λ [108] (see Sec.6.1).

The minimal subtraction scheme, which in dimensional regularization is used to absorb dimensional poles in the field–theoretical perturbation series, is often combined with the $\varepsilon = 4 - d$–expansion (see, e.g., Ref.[34]). However, there is no general necessity to follow this procedure, the minimal subtraction scheme can be formulated in *fixed* spatial dimension $d < 4$ as well [110, 111]. Moreover, there is evidence that better agreement of field–theoretical predictions with experimental data in $d = 3$ can be obtained this way [112] (see Sec.6.1).

The interfacial tension $\sigma_{aper}(t; L, L)$ according to Eq.(2.14) but with antiperiodic versus periodic boundary conditions in a L^d–geometry has been studied in Ref.[113] on the basis of a field–theoretical treatment of an $O(N = 1)$–symmetric system below bulk criticality. A logarithmic contribution to the scaling function $\Sigma\left(|t|L^{1/\nu}\right)$ of the interfacial tension (see Eq.(2.15)) has been found for $t < 0$ and large L in $d < 4$. Specifically, the scaling function Σ reads [113]

$$\Sigma(x) = \sigma_0 x^\mu - \tfrac{\mu}{2}\ln\left(\sigma_0^{1/\mu}x\right) + C + \mathcal{O}\left(\exp\left(-x^\nu/\sqrt{2}\right)\right), \qquad (2.59)$$

where $x = |t|L^{1/\nu}$, $\mu = (d - 1)\nu$, and σ_0 and C are constants. In $d = 4$ additional logarithmic size–dependent terms contribute to the interfacial tension which are not captured by Eq.(2.15). Due to logarithmic corrections to Landau critical behavior the scaling argument x in Eq.(2.59) is modified by a factor $|\ln |t||^{1/3}$ (see also Eq.(2.13)) [113].

A detailed description of the field–theoretical renormalization group can be found in Refs.[34, 35, 36, 37], and the extension to semiinfinite systems with some reference to other geometries is reviewed in Ref.[20] to which we refer the reader for further details.

2.3.4 Conformal Field Theory in Two Dimensions

Conformal coordinate transformations can be interpreted as local scale transformations with a position dependent scale factor. It seems therefore rather natural to extend the principle of scale invariance at critical points to conformal mappings, where the transformation laws for cumulants under scale transformations are generalized accordingly [114] (see Sec.3.3). Conformal invariance drastically restricts the allowed functional forms of cumulants and scaling density profiles. In *two dimensions* this restriction is particularly strong due to the exceptionally large conformal group which contains all holomorphic functions and their antiholomorphic counterparts on two–dimensional space, i.e., the complex plane [115, 116]. The conformal invariance principle can be generalized to two–dimensional *semiinfinite* systems, where the functional form of cumulants especially near the boundary is restricted as strongly as in unbounded space [117]. The general transformation law for cumulants (conformal Ward identity, see Eq.(3.39) in Sec.3.3) can be generalized to semiinfinite geometries (the upper halfplane) accordingly [117]. In the framework of finite–size scaling the behavior of cumulants, scaling density profiles, and the free energy of critical two–dimensional systems under conformal transformations which relate infinite and finite geometries is of special interest [118].

Within the framework of conformal field theory in two dimensions the universality class of a critical system is characterized by the value of a single parameter, the central charge or conformal anomaly number c which cannot be determined from general conformal invariance considerations. However, the requirement that the critical statistical system which one would like to describe by a conformal field theory has a *unitary transfer matrix* leads to a "quantization" of the possible values $c < 1$ of the central charge c according to [119]

$$c = 1 - \frac{6}{m(m+1)}, \tag{2.60}$$

where $m \in \{3, 4, 5, \ldots\}$. The allowed values for m can be translated to critical and tricritical models in the following way [119]

$$
\begin{array}{lll}
m = 3 & : & \text{critical Ising} \quad (c = \frac{1}{2}) \\
m = 4 & : & \text{tricritical Ising} \quad (c = \frac{7}{10}) \\
m = 5 & : & \text{critical 3-state Potts} \quad (c = \frac{4}{5}) \\
m = 6 & : & \text{tricritical 3-state Potts} \quad (c = \frac{6}{7}) \\
m = \infty & : & \text{critical Gaussian} \quad (c = 1).
\end{array}
\tag{2.61}
$$

The spectrum of critical exponents for these models and for other models in the same universality classes is uniquely determined by the value of m [119].

The knowledge of the central charge c given by Eq.(2.60) gives direct access to the *Casimir amplitudes* in strips with periodic, free (O, O), and fixed $(+, +)$ boundary conditions for the models, i.e., the universality classes listed in Eq.(2.61) [118, 120, 121] (see Sec.3.4). For other boundary conditions the Casimir amplitudes have been determined in Ref.[118] for $m = 3$ (Ising) and $m = 5$ (3-state Potts).

The correlation length in a strip at $T = T_{c,b}$ with periodic boundary conditions can be determined by mapping a bulk pair correlation function into the strip geometry using a branch of the logarithm on the complex plane as the conformal transformation. The result is displayed in Eq.(2.34), where $A = 1/(2\pi)$ [122]. For a strip with free boundaries (O, O) a corresponding analysis yields $\xi_L = L/(\pi x_s)$, where x_s is the *surface* scaling dimension of the scaling operator in the pair correlation function [122] (see also Ref.[39]). In this case the amplitude A takes the value $A = 1/\pi$.

Correlation functions or cumulants at the critical point can be found as solutions of differential equations in terms of the complex coordinates [39]. Bulk six-point functions and three-point functions in a semiinfinite geometry for the critical Ising model [123] and energy density correlation functions for $O(N)$ models in semiinfinite geometries with various boundary conditions [124] and in an Ising model with a defect line [125], only to give a few examples, have been calculated this way. The finite-size behavior of correlation functions for the tricritical Ising model as a paradigm for a two–dimensional multicritical system in finite geometries is particularly interesting. An extended study of two–point correlation functions in a halfplane, a strip, a circle, and a rectangle with free boundary conditions shows that the static structure factor for the energy–energy correlation function on, e.g., a circular disk displays a surprisingly slow crossover to the usual bulk power law for growing momentum transfers [126]. A fit of experimental or simulation data, which are usually taken at intermediate momenta, to the *bulk* power law for the finite system might then yield grossly flawed results due to strong finite–size effects [126]. These findings do not seem to be specific for the tricritical Ising model. The static structure factor for the spin–spin correlation function of a *critical* Ising model on a circular disk displays the same slow crossover behavior [126]. The specific heat of a tricritical Ising $L \times L'$ rectangle as a function of the aspect ratio $s = L'/L$ inreases steeply from $s = 1$ to $s \simeq 4$ and then bends over to a much slower growth with s. The correlation length for the energy–energy correlation function in a $L \times L'$ rectangle displays a corresponding behavior which seems to be a purely geometrical effect [126].

Much effort has been spent on the calculation of scaling density profiles for the magnetization and the energy density for the two–dimensional critical Ising and 3–state Potts model [123, 127, 128, 129], where special attention has been paid to the short–distance behavior of these profiles in strips [127, 128, 129]. The short–distance behavior of such profiles is examined in Chap.4 so that we refrain from giving further details here.

Edges in general dimension d correspond to *corners* in $d = 2$. The conformal transformation $w = z^{\alpha/\pi}$ maps the upper halfplane onto a wedge with the opening angle α, and therefore scaling density profiles and correlation functions can be easily obtained from the corresponding halfplane quantities [127]. The corner exponents $x^{(c)}$ are related to the corresponding surface (edge) exponents $x^{(s)}$ by [39, 117]

$$x^{(c)} = (\pi/\alpha) \, x^{(s)}. \tag{2.62}$$

Domains with curved boundaries can be considered along the same line of argument [130]. If the geometry is characterized by a finite macroscopic length scale L and the boundary is curved or has corners, *universal* logarithmically size–dependent terms contribute to the free energy of the confined critical system [59]. The universal size–dependent contribution $\Delta F(t = 0; L, L')$ to the free energy of a critical system confined to a $L \times L'$ rectangle with a uniform (free or fixed) boundary condition reads for large (or small) aspect ratios $s = L'/L$ [131]

$$\frac{\Delta F(t = 0; L, L')}{k_B T_{c,b}} \simeq -\frac{c}{8}\ln(LL') - \frac{c\pi}{24}\left(s + \frac{1}{s}\right), \qquad (2.63)$$

where the lengths L and L' are measured in dimensionless units and c is the central charge of the system (see Eqs.(2.60) and (2.61)). Note, that the form of Eq.(2.63) is in accordance with Eq.(2.10) (see also Eqs.(2.44) and (2.45)).

The free energy associated with *domain boundaries* (interfacial tension) can be obtained by mapping a simple configuration which includes a straight domain wall conformally onto other configurations, where the domain wall may have a more complicated shape [132]. In the simplest possible case the upper halfplane is subdivided into two differently magnetized (ordered) regions which form a domain wall along the imaginary axis. Such a configuration can then be mapped conformally onto a halfplane with a magnetic droplet at the boundary or onto a strip with either a magnetic droplet at one boundary or a domain wall [132]. Other conformal transformations map these droplet or domain wall configurations onto circles or rectangles with internal interfaces [61]. The associated interfacial tensions contain logarithmic size–dependent contributions depending on the width of the strip, the radius of the circle, or the size of the rectangle [61] which is in qualitative accordance with Eq.(2.15).

Opposing surface fields on randomly chosen sections of the boundary generate domain walls and magnetic droplets of various sizes in a critical Ising halfplane [133]. Magnetization profiles and n–point cumulants between the local magnetization and the energy density can be determined exactly from conformal invariance considerations. Depending on the relative sizes of neighbouring boundary sections the coagulation of droplets can be observed [133].

Finally, we would like to point out that the theoretical background of conformal field theory, many applications of conformal invariance, and the extension to superconformal invariance are reviewed in Ref.[39].

3. The Casimir Effect

3.1 Introduction

One of the most important discoveries in the theory of electromagnetism dates back to the year 1948, when the Dutch physicist H. B. G. Casimir realized that two perfectly conducting plates standing face to face in vacuum at a distance L much smaller than their lateral extensions attract each other with a force per unit area (pressure) F given by [134, 135]

$$F = -\frac{\pi^2}{240} \frac{\hbar c}{L^4}. \tag{3.1}$$

At first sight one might have the impression that the force in Eq.(3.1) is due to a superposition of the van–der–Waals attractions which two neutral atoms like those in the plates experience in vacuum. A careful analysis reveals a retardation effect if the separation of the atoms is larger than the wavelength λ corresponding to a transition of the atom from its ground state to the first excited states [136]. In this case the well known r^{-7}–dependence of the van–der–Waals force on the separation r of the two atoms is modified to a r^{-8}–law. (This crossover has been observed experimentally, e.g., for very smooth muscovite mica surfaces facing one another at distances well below 1000Å [137, 138].) If the aforementioned impression were correct then Eq.(3.1) could be obtained by integrating the retarded van–der–Waals force over the two plates. The retardation would then account for those distances L between the surfaces realized in the first experiments which had been performed at $L > 1000$Å for quartz plates [139] or aluminum samples [140] (see Refs.[141] and [142] for more examples). However, a simple integration over the mutual van–der–Waals forces between the atoms clearly disregards macroscopic properties of the plates like being a conductor or a dielectric. Therefore the above superposition procedure for van–der–Waals forces is only justified for a sufficiently diluted arrangement of neutral atoms, but this is neither in the spirit of Ref.[134] nor is it realized in the experiments. The modern view of Eq.(3.1), which has already been raised by Casimir himself, is a *macroscopic* one in contrast to the *microscopic* reasoning by means of interatomic forces. Following Ref.[134] the interaction between the plates is mediated by the *zero–point fluctuations* of the electromagnetic field in vacuum. The plates themselves act as *boundary conditions* for the electromagnetic field which modify the frequency spectrum of the fluctuations in such a way that the *zero–point energy* of the electromagnetic field exhibits an L–dependence. Then the force in Eq.(3.1) is just the response of the zero–point energy to a change in L. The above argument can immediately be generalized to finite temperatures by replacing the zero–point or *gound state energy* with the *free energy* and it can be applied to quantum field theories in general as well. Thus the consequences of the Casimir effect are reaching far beyond Eq.(3.1) and even beyond the concepts of electrodynamics.

As a first extension of the arrangement leading to Eq.(3.1) one can think of replacing the metallic plates with two plane dielectrics at a distance L separated by vacuum. Two dielectric functions $\varepsilon_1(\omega)$ and $\varepsilon_2(\omega)$ characterize these two media, respectively [143]. The original argument leading to Eq.(3.1), however, cannot be applied because it does not account for absorption described by the imaginary parts of $\varepsilon_1(\omega)$ and $\varepsilon_2(\omega)$. Instead, the Maxwell equations in the dielectric media are modified by introducing a random field $\mathbf{K}(\mathbf{r})$ which is characterized by the correlation function [143]

$$\langle K_i(\mathbf{r})K_j(\mathbf{r}')\rangle = 2\hbar \coth \frac{\hbar\omega}{2k_BT}\text{Im}\varepsilon(\omega)\delta_{ij}\delta(\mathbf{r}-\mathbf{r}'). \qquad (3.2)$$

For very diluted media, i.e., $\varepsilon_1,\varepsilon_2 \to 1$, the retarded van–der–Waals force can be recovered in terms of the static polarizabilities of two atoms, one in each medium [143]. Thus the van–der–Waals force, which fails to account for the Casimir force between dense macroscopic bodies by mere superposition, is in turn included in the macroscopic theory of the Casimir effect between dielectric media as a limiting case.

The considerations in Ref.[143] have been further generalized to three dielectrics, where the third replaces the vacuum between the other two. Moreover, these considerations have been rigorously justified using quantum field theory and the Feynman diagram technique [144]. The influence of finite temperatures T on the Casimir force, which has already been studied in Ref.[143], is rederived in Ref.[144]. The relation for the Casimir force between two perfectly conducting plates in vacuum for $T > 0$ follows from the general expression by taking the limit $\varepsilon_1,\varepsilon_2 \to \infty$. For $k_BT \ll \hbar c/L$ one finds a correction term $\propto T^4$ [145] [4]

$$F = -\frac{\pi^2}{240}\frac{\hbar c}{L^4}\left[1 + \frac{16}{3}\left(\frac{k_BTL}{\hbar c}\right)^4\right]. \qquad (3.3)$$

---

[4]The relation for the Casimir force between conducting plates for $T > 0$ given in Refs.[143] and [144] is in error [145]. However, the *general* expression for the Casimir force between dielectric media in Refs.[143] and [144] is correct.

In the other extreme case $k_B T \gg \hbar c / L$ the force between metallic plates displays a L^{-3}-behavior with an additive correction which vanishes exponentially as L increases. One obtains [145]

$$F = -\frac{k_B T}{4\pi L^3} \zeta(3) - \frac{k_B T}{2\pi L^3} \left(1 + t + \frac{t^2}{2} \right) e^{-t}, \tag{3.4}$$

where $\zeta(3) \simeq 1.202$ is a special value of the Riemann zeta function and the dimensionless absolute temperature t is defined by $t = 4\pi k_B T L / (\hbar c)$.

A very common example of a system which consists of three parallel layers of dielectrics is represented by a liquid film of thickness L in contact with the vapor of the liquid on one side and with a substrate on the other side. The thickness L of the film is an equilibrium quantity, which adjusts itself such, that the chemical potentials in the liquid and the vapor are equal. The chemical potential in the film deviates from its value in the bulk liquid at coexistence with the vapor by an amount $\Delta\mu$ representing the energy per atom in the film due to the presence of two boundaries at a finite distance L. For thin films the usual van–der–Waals forces lead to $\Delta\mu \propto L^{-3}$, for thick films one has $\Delta\mu \propto L^{-4}$ due to retardation [144]. The inversion of these relations allows a study of the liquid film thickness L as a function of thermodynamic variables like temperature and pressure. The theory of the interaction between dielectric media is therefore intimately linked to the phenomenon of *wetting* (see Sec.6.3 and Ref.[25]).

The forces between the dielectrics or the conducting plates are in general determined from one spatial diagonal component of the Maxwell stress tensor $T_{\mu\nu}$ [143, 144]. For the same arrangement of two perfectly conducting plates in vacuum considered in Ref.[134] the ground state expectation value of $T_{\mu\nu}$ as a whole has been studied in Ref.[146]. The result is

$$\langle T_{\mu\nu} \rangle = \frac{\pi^2 \hbar c}{720 L^4} \begin{pmatrix} -1 & & & \\ & 1 & 0 & \\ & 0 & 1 & \\ & & & -3 \end{pmatrix}, \tag{3.5}$$

where the coordinates in 3+1-dimensional space–time are arranged as (t, x, y, z) and the Minkowski metric is chosen to be $g_{\mu\nu} = \mathrm{diag}(-1, 1, 1, 1)$. The z–axis of the coordinate system underlying Eq.(3.5) is perpendicular to the plates so that

$\langle T_{zz} \rangle$ corresponds to F in Eq.(3.1) (see Sec.3.4). Analogous to the considerations in Refs.[143, 144] Eq.(3.1) has been generalized to finite temperatures in Ref.[146]. Some of these results have been recently used in order to estimate the Casimir energy of light and electrons in cavities inside *disordered media* [147].

In the decades that passed since the discovery of the Casimir effect a large number of other examples and applications in many different fields have emerged from the basic considerations we briefly scetched out above. Today the richness of this field ranges from studies of nonplanar geometries of conducting or dielectric macroscopic bodies immersed in fluids over the bag model for the description of quark confinement in hadrons according to quantum chromodynamics to the Casimir effect in general curved space–time in cosmology. To account for all these topics is certainly beyond the scope of this article. Therefore we would like to refer the reader to Ref.[141] for a larger collection and review of examples in the aforementioned fields and to the references given there for further reading. Supplementary theoretical and experimental examples can be found in Ref.[142]. An extended review on the Casimir effect is given in Ref.[148]. A concise summary of basic physical ideas and results concerning the Casimir effect can be found in Ref.[149]. In order to account for the research activity in this field we give a brief overview over some of the more recent achievements concerning the Casimir effect.

3.1.1   The Casimir Effect in Dielectric Media

In the spirit of the plane geometry considered in Ref.[144] a spherical shell filled with dielectric material surrounded by vacuum has attracted special attention [150] after a strong attractive contribution to the Casimir force on the surface of a rigid dielectric sphere has been found [151]. The material is characterized by a dielectric function or permittivity $\varepsilon(\omega)$ and a permeability $\mu(\omega)$ subject to the constraint $\varepsilon(\omega)\mu(\omega) = 1$. The total surface force acting on the shell is evaluated for the step–shaped permeability $\mu(\omega) \to 0$ or ∞ for frequencies ω below a certain cutoff frequency ω_0 and $\mu(\omega) = 1$ for $\omega > \omega_0$. Moreover, only thin shells are considered in the sense that the thickness of the shell is supposed to be much smaller than its inner or outer radius. The result is strongly cutoff–dependent and indicates attraction for sufficiently large cutoff frequencies ω_0. The analytic expression for the Casimir force contains a divergent series, which needs to be truncated by another cutoff. This peculiarity seems to be due to the small shell–thickness limit [150].

The interaction of two dielectric *slabs* of *finite* thickness D separated by a distance L has been considered in Ref.[152]. The permittivities in the two slabs are equal and account for the possibility of absorption, the region between the slabs and outside the whole arrangement is assumed to be vacuum. The matter in the slabs is modelled by a continuous set of quantum harmonic oscillators driven by δ–correlated Langevin forces which cause a polarisation of the matter. The Casimir force then consists of a vacuum contribution originating from the electromagnetic field and a contribution due to the Langevin forces [152] each of which is represented in terms of the reflection and transmission coefficients of a single slab. In the limit of infinitely thick slabs $(D \rightarrow \infty)$ the results of Ref.[143] are recovered. The energy dissipation per unit area in this case is found to be less than in the case of a homogeneous medium [153]. However, there is some freedom in the interpretation of the Casimir force. It is not neccessary to decompose the Casimir force into a vacuum and a source (Langevin) contribution as in Ref.[152]. Instead, the Casimir force can be attributed *entirely* to source fields if *normal ordering* is applied to the field operators describing the electromagnetic field [154]. The appearance of a vacuum contribution is due to a different ordering scheme. This picture is in accordance with Schwinger's derivation of the Casimir force, which is solely based on source fields [145]. Here we would like to add that the free energy of bosonic and fermionic systems confined between two parallel plates at distance L exhibit simple transformation laws under the so–called temperature inversion mapping $k_B T L/(\hbar c) = x \rightarrow 1/x$. This symmetry relates the Casimir energy for scalar fields at $T = 0$ to the Stefan–Boltzmann law of a Bose gas at high temperatures [155].

Atoms confined between two parallel conducting plates exhibit modified transition probabilities to excited states due to the presence of the boundaries [156]. The calculation presented in Ref.[156] applies to experiments on atoms trapped in micro-cavities [157]. Moreover, the interaction effect between a microcavity and a beam of ground–state sodium atoms has recently been exploited to probe the retarded van–der–Waals force derived in Ref.[136] experimentally in vacuum [158]. A recent theoretical study of the Casimir energy inside waveguides and rectangular cavities based on the basis of quantized Hertz potentials can be found in Ref.[159]. If the confined particle is charged, the Casimir effect generates a correction to the electro-static potential which the particle experiences due to its image charges. Specifically, the potential of a charged particle at distance z from a single, plane, and conducting wall is modified by a z^{-2}-contribution at $T = 0$ [160]. The van–der–Waals energy between an atom and a metallic cylindrical surface has been studied in Ref.[161]. In the limit of vanishing cylinder radius the energy is significantly enhanced over the corresponding energy of an atom facing a plane metallic surface. This finding serves as an explanation for the observed increased reactivity of the *edges* in comparison with the *terraces* on *stepped surfaces* in chemisorption experiments (e.g., hydrogen on stepped Ni(100) surfaces) or adsorption experiments (e.g., Xe on stepped Pd(100) surfaces) [161].

In an experiment the Casimir force F (see Eqs.(3.1), (3.3), and (3.4)) is measured as an *average* of one stress tensor component, e.g., $\langle T_{zz} \rangle$ (see Eq.(3.5)) over a certain probe area A and a duration τ of the measurement. In general neither the stress tensor nor its integrals commute with the Hamiltonian of the quantized electromagnetic field and therefore one finds a *finite* mean square deviation of the time and area averaged stress tensor [162]. For parallel conducting plates at distance L the standard deviation ΔF of the Casimir force F in Eq.(3.1) is given by $\Delta F/F \propto (L/(c\tau))^3$ for $L/(c\tau) \ll 1$, where c is the velocity of light. In an experiment usually $L/(c\tau) \sim 10^{-4}$ so that ΔF is minute in comparison with F. The analysis of the mean square deviation can be performed on the general basis of the stress tensor – stress tensor correlation function, which allows a generalization to conducting bodies with curved surfaces [163]. These considerations have been applied to conducting spheres and hemispheres [164] and a conducting circular disk [165], each of radius R. For $R \ll c\tau$ the one–sided disk and the hemisphere give $\Delta F \sim R^2/(c\tau)^4$, whereas the two–sided disk and the sphere exhibit fluctuations $\Delta F \sim R^3/(c\tau)^5$. Thus the two–sided objects cause fluctuations which are suppressed by a factor $R/(c\tau)$ compared to the fluctuations on one–sided objects.

Moving boundaries introduce an explicit time dependence into the Casimir effect. If one of the two plates in a parallel conducting plate setup moves according to a given law $z = z(t)$ with $|\dot{z}|/c \ll 1$, the Casimir force is modified by a $(\dot{z}/c)^2$ and a $z\ddot{z}/c^2$ contribution [166], where the latter gives rise to a mass renormalization. The moving plate fulfills the equation of motion $m\ddot{z} + m\gamma\dot{z}^2 + V'(z) = 0$, where the potential $V(z)$ contains both the static Casimir potential and an external one. If $V(z)$ is a harmonic potential, the $\dot{z}^2$-force in the equation of motion leads to a small shift of the resulting oscillator levels [166]. Moreover, the above continuous change in the boundary conditions leads to an emission of radiation [167]. However, the emission rate is governed by a coefficient $(\dot{z}/c)^2$, and it is therefore extremely small for macroscopic moving bodies.

The behavior of the vacuum energy of a field in the presence of dielectric surfaces has been studied on quite general grounds in field theory for, e.g., quantum electrodynamics or quantum chromodynamics [168]. Special emphasis has been put on the discussion of the cutoff dependence of the vacuum energy, where the cutoff is related to the microscopic structure of the boundary [168]. In general a boundary is defined as a surface, where the position dependent permittivity $\varepsilon(\omega, \mathbf{r})$ crosses over from $\varepsilon_1(\omega)$ for one medium (e.g., a dielectric) to $\varepsilon_2(\omega)$ for another medium (e.g., vacuum). If this crossover is smooth, a finite interface width can be defined which gives rise to a surface specific cutoff [168, 169]. The vacuum energy $\mathcal{E}$ associated with the boundary can then be represented by surface tensions $\mathcal{E}^S$ and curvature contributions (shape tensions) $\mathcal{E}^C$ in a geometrical expansion in powers of the principal curvatures κ_1 and κ_2 of a surface embedded in three dimensions [168]

$$\mathcal{E} = \mathcal{E}^S \int dS + \mathcal{E}^C \int dS(\kappa_1 + \kappa_2) + \mathcal{E}_1^C \int dS(\kappa_1 - \kappa_2)^2 + \mathcal{E}_2^C \int dS \kappa_1 \kappa_2 + \dots. \quad (3.6)$$

These considerations have been applied to the vacuum energies of, e.g., a dielectric sphere and conducting spherical and cylindrical shells. Moreover, the restriction $\varepsilon(\omega)\mu(\omega) = 1$ [150] on the permittivity ε and the permeability μ can be relaxed to a certain extent [169].

Pointlike masses attached to a straight string of uniform energy density and tension experience an attractive force due to the transverse quantum mechanical fluctuations of the string [170]. Since all forces are attractive the system exhibits an instability which is a one–dimensional analogue of the instability in gravity. For large distances r the force decays as r^{-3} and for small distances as r^{-1}. For small inertial masses m_1 and m_2 attached to a straight string the force between them is proportional to the product $m_1 m_2$ so that the equivalence principle holds in this limit [170].

Further applications are provided by confined statistical systems to which we will now turn.

3.1.2   The Casimir Effect in Statistical Systems

Boundaries modify the spectrum of electromagnetic fluctuations between them in such a way that the vacuum energy of the electromagnetic field becomes size and shape dependent. This is the basic idea behind the Casimir effect leading to Eq.(3.1) in the simplest case. This idea has been used in Ref.[171] in order to determine the attractive forces between a *large* number of nonabsorbing dielectric planes and, moreover, between the (111)–planes of an fcc crystal of spheres. Both arrangements are spatially periodic and the force displays the usual L^{-4}-behavior, where L is the separation between *adjacent* planes. The planes themselves are characterized by a refractive index $n = n_0 + 2n_1$, where n_0 is the refractive index of the nonabsorbing surrounding medium (n_1 just measures the *difference* in the refractive index between the planes and the medium). For a sufficiently large number of planes the Casimir pressure

$$P = -\frac{\pi^2 \hbar c n_1}{4 n_0^4 L^4} \tag{3.7}$$

is obtained [171], which is in close analogy with Eq.(3.1). A realization of such systems is provided by a colloidal suspension of charged polystyrene spheres in water forming an fcc structure. In this case n_1 in Eq.(3.7) is given by $n_1 = \frac{3}{8\sqrt{2}} \frac{m^2-1}{m^2+1} \Phi$, where $m = 1.19$ is the ratio of the refractive indices of polystyrene and water and Φ denotes the volume fraction of the colloidal crystal [171]. Then Eq.(3.7) gives the Casimir pressure between the (111)–planes of the colloidal crystal. The separation between the planes is of the order of the wavelength of visible light so that these systems can be probed by light scattering.

Near a critical point of a fluid a fluctuating field of a completely different origin becomes important, the *order parameter*. The modern view of critical phenomena is based on field theory and therefore boundaries in a critical system act as they do in electrodynamics, they modify the spectrum of the fluctuating order parameter so that the *singular part* of the *free energy* becomes size and shape dependent. Consequently, there is an *additional* force present in the fluid near a critical point, namely the Casimir force generated by *critical fluctuations*. The first field–theoretical calculation of this force at the bulk critical temperature can be found in Ref.[172]. If a colloidal suspension like the one decribed in Ref.[171] still exists at a critical point of the medium, the Casimir pressure in Eq.(3.7) will be modified by an additional term coming from the *critical* fluctuations. The Casimir forces in critical systems have been studied for various examples in a film geometry in d dimensions [173] (see also Secs.3.2, 3.4, and 6.4).

A correlated fluid like superfluid ^{4}He sufficiently far *below* the λ–transition induces forces on surfaces immersed in the fluid due to the presence of *Goldstone modes* [174]. In particular, such fluctuations induce many–body forces so that the sum over all interactions between pairs of surfaces immersed in a correlated fluid does not account for the whole interaction between three or more surfaces [174]. The model Hamiltonians considered in Ref.[174] are of the Gaussian Ginzburg–Landau type and therefore they do not apply to critical fluids. Special attention has been paid to anisotropic liquids like nematic liquid crystals [175]. If two rigid parallel plates are immersed in a nematic liquid, fluctuations of the nematic director around its average orientation generate a long ranged force between the plates. The average orientation of the nematic director itself is prescribed to be perpendicular to the walls as a boundary condition ("strong homeotropic anchoring") [175, 176], which imposes a size dependence on the fluctuation spectrum and therefore on the free energy. Here it is interesting to note that the Casimir force is *attractive* for *equal* boundary conditions on the walls, whereas it is *repulsive* for *different* ones [174, 176]. Furthermore, there is a one to one correspondence between the vacuum fluctuations of the electromagnetic field and the orientational fluctuations of the nematic director in the so called isotropic case [176].

From the more technical point of view the Casimir effect in various geometries has provided many applications for the zeta–function regularization technique [177, 178, 179, 180, 181, 182, 183] sometimes in combination with the so–called heat–kernel approach [180, 182]. The zeta–function regularization scheme can be reinterpreted in terms of an algebraic cutoff scheme [184] which, as combined with an exponential cutoff procedure, can be demonstrated to be equivalent to the pure cutoff procedure. This has been explicitly shown for the Casimir energy between two perfectly conducting plates in two– and three–dimensional space–time [184, 185]. A physically motivated cutoff frequency in the mode sum for the Casimir energy yields finite results for very small cavities, which may serve to formulate a new Casimir model of the electron [186]. Certain aspects of these approaches will show up in the following sections (see especially Secs.3.2 and 5.3) which are devoted to the field–theoretical treatment of critical fluctuations in $O(N)$–symmetric systems between rigid parallel walls with various boundary conditions. The field–theoretical approach is designed for the treatment of the critical region and contrasts another approach based on the Ornstein–Zernicke equation [187].

According to present knowledge the Casimir effect is a phenomenon common to all systems characterized by fluctuating quantities with external constraints. The nature of that quantity as well as the nature of the fluctuations seem to be irrelevant for the basic functional form of the Casimir force compared to rather unspecific properties like the spatial dimension or certain symmetries. From both the point of view of *finite–size scaling* for the free energy and the *stress tensor* in critical systems we will now see that critical fluctuations of the order parameter for various examples fit into this scheme.

3.2   The Casimir Effect in Critical Films

If a fluid is confined between two parallel plates each of area A separated by a distance L, the total free energy $\mathcal{F}$ of the fluid can be decomposed into four distinct contributions as shown in Eq.(2.1) for the singular part of $\mathcal{F}$. In absence of an external bulk field H the dominant term in this decomposition is given by the bulk free energy $F_{bulk}(T)$ per unit volume of the unconfined fluid. The next to leading order terms are the two surface free energies $F_{s,a}(T)$ and $F_{s,b}(T)$ per unit surface area A generated by the two plates a and b, respectively. The finite separation L of the plates finally gives rise to a finite–size contribution $\delta F_{a,b}(T, L)$ per unit surface area. In the following we will use $k_B T_{c,b}$ as a natural energy scale for the free energy contributions involved in the above decomposition. We then have in the limit $A \to \infty$ [188]

$$\lim_{A \to \infty} \frac{\mathcal{F}(T, L)}{k_B T_{c,b} A} = L F_{bulk}(T) + F_{s,a}(T) + F_{s,b}(T) + \delta F_{a,b}(T, L). \qquad (3.8)$$

In the vicinity of the bulk critical temperature of the fluid the total free energy of the film given by Eq.(3.8) can be further decomposed into a *regular* and a *singular* part [58, 188], where the latter is responsible for the *critical singularities* of the free energy at the bulk critical point. The decomposition of the *leading* singular part of the free energy into bulk, surface, and finite–size contributions is displayed in Eq.(2.4). At $T = T_{c,b}$ ($t = 0$) and $H = 0$ the finite–size scaling function $\delta f_{\pm a,b} \left(|t| L^{1/\nu}, H L^{\Delta/\nu} \right)$ reduces to a *universal* finite–size amplitude $\Delta_{a,b}$, and we have [188] (see Eqs.(2.1) and (2.4))

$$\delta F_{a,b}^{sing}(t = 0, L) = \delta f_{\pm a,b}(0, 0) L^{-(d-1)} = \Delta_{a,b} L^{-(d-1)}. \qquad (3.9)$$

One option to obtain field–theoretical results for the amplitude $\Delta_{a,b}$ defined by Eq.(3.9) is provided by the calculation of the *renormalized free energy* f^R of the film. The basic ingredient of the renormalized free energy is the *singular* part of the total free energy, but furthermore f^R satisfies a certain *normalization condition* (see Eq.(5.1) in Sec.5.1). In terms of f^R the total free energy $\mathcal{F}(T, L)$ can be written in the form [58, 188]

$$\lim_{A \to \infty} \frac{\mathcal{F}(T, L)}{k_B T_{c,b} A} = f^R(t, L) + f^{reg}(t, L), \tag{3.10}$$

where $f^{reg}(t, L)$ collects all *regular* contributions to $\mathcal{F}(T, L)$. The decomposition of f^R into bulk, surface, and finite–size contributions yields a renormalized finite–size part $\delta f^R_{a,b}$ of f^R which is related to the scaling function $\delta f_{\pm a,b}$ in Eq.(2.4) by (see Eq.(5.20) in Sec.5.1)

$$\delta f^R_{a,b} = L^{-(d-1)} \left(\delta f_{\pm a,b} + \mathcal{O}\left(e^{-L/\xi_1}\right) \right), \tag{3.11}$$

where the reference length ξ_1 is of the order of several Å up to a few tens of Å . For short–ranged interactions and $H = 0$ we have $\xi_1 = \xi_{\pm}(t_1) = \xi_0^{\pm}|t_1|^{-\nu}$, where $t_1 \neq 0$ is a *reference* reduced temperature on which the renormalized free energy depends in addition to t (see Sec.5.1). In experimental realizations of critical films one usually has $L \gg \xi_1$ so that the exponential in Eq.(3.11) can be neglected, and we obtain $\delta f^R_{a,b} = \Delta_{a,b} L^{-(d-1)}$ at the bulk critical point to a very high accuracy. We will return to the renomalization procedure for the free energy in detail in Chap.5.

The L–dependence of $\delta F^{sing}_{a,b}(t = 0, L)$ given by Eq.(3.9) gives rise to a *force* $F = (d - 1)\Delta_{a,b} L^{-d}$ between the plates which is a close analogue of the *Casimir force* in electromagnetism given by Eq.(3.1). The universal finte–size amplitude $\Delta_{a,b}$ defined by Eq.(3.9) is therefore commonly denoted as the *Casimir amplitude* [38, 172, 188, 189].

If one of the surfaces a or b belongs to the extraordinary surface universality class, mean field theory for the film Hamiltonian given by Eq.(2.48) already yields a finite result for $\Delta_{a,b}$. However, a simple analysis shows that $\Delta_{a,b} = \mathcal{O}(\varepsilon^{-1})$ in these cases, where $\varepsilon = 4 - d$. In $d = 4$ Eq.(3.9) is therefore modified by an L–dependent *logarithmic correction* if a surface field is present at the wall a or b. For any other combination (a, b) of surface universality classes such a logarithmic correction does not occur (see below).

In order to identify $\Delta_{a,b}$ for a specific lattice model in a film geometry one has to find its free energy evaluated at the critical temperature of this lattice model in unbounded space. The free energy decomposes according to Eq.(3.8), and the finite size contribution $L^{-(d-1)}\delta f_{a,b\pm}$ according to Eq.(2.4) yields the Casimir amplitude $\Delta_{a,b}$. For the Ising model this has been done rigorously in $d = 2$ [50] (see Eq.(2.27)) and by real space renormalization in $d = 3$ [190]. Moreover, Eqs.(3.9) and (3.11) give direct access to the Casimir amplitude using the field–theoretical renormalization group [34, 35, 36, 37] for the calculation of the renormalized free energy f^R. This has been done for the Ising universality class [172, 188, 189], for the XY – and Heisenberg universality classes [188, 189] in $d = 4 - \varepsilon$, and for the Baxter model, the q–state Potts model, and the N–component cubic model [86, 120, 191] for various values of q and N in $d = 2$.

3.2.1   The Critical Casimir Amplitudes

The field–theoretical analysis of the Casimir effect in $d = 4 - \varepsilon$ at bulk criticality which has been performed so far is based on the Ginzburg–Landau Hamiltonian for $O(N)$ symmetric systems given by Eq.(2.48) [172, 188]. The critical point we have in mind for the bulk system may both be a usual critical point or a critical end point in the bulk phase diagram of, e.g., ^{4}He (see Fig.1.6). The boundary conditions (surface universality classes) we now consider are the symmetry conserving ones introduced in Sec.1.2 [20], which can be combined to $(a, b) = (O, O)$, (O, SB), and (SB, SB) for a film geometry. The combination $(a, b) = (SB, O)$ does not provide new informa-tion, because the Casimir amplitude is an *integrated* film quantity and it is therefore symmetric with respect to a and b: $\Delta_{a,b} = \Delta_{b,a}$. Also periodic (*per*) and antiperiodic (*aper*) boundary conditions can be imposed in our film geometry without breaking the symmetry. Note, that for these two cases the surface free energies in Eq.(3.8) vanish identically [56, 58, 188].

For an $O(N)$ symmetric critical (massless) Gaussian model in d dimensions a simple one–loop calculation yields the following exact results [172, 188]

$$\Delta_{O,O}^{(1)} = -2^{-d}\pi^{-d/2}N\Gamma\left(\tfrac{d}{2}\right)\zeta(d)$$
$$\Delta_{SB,SB}^{(1)} = \Delta_{O,O}^{(1)}$$
$$\Delta_{O,SB}^{(1)} = \left(2^{1-d} - 1\right)\Delta_{O,O}^{(1)}$$
$$\Delta_{per}^{(1)} = 2^d\Delta_{O,O}^{(1)}$$

$$\Delta_{aper}^{(1)} \;\; = \;\; 2^d \Delta_{O,SB}^{(1)}, \tag{3.12}$$

where $\Gamma(z)$ denotes the Gamma function and $\zeta(z)$ is the Riemann zeta function. The Casimir amplitudes in Eq.(3.12) can be regarded to be exact at the upper critical dimension $d_c = 4$ at critical points or critical end points. The above results can be obtained using dimensional regularization [34, 35], which in this case basically amounts to the analytic continuation of the Riemann zeta function $\zeta(z)$ to arguments z with Re $z < 0$. This procedure is equivalent to the so called zeta–function regularization, which has been applied to a massless bosonic field in $d+1$–dimensional spacetime in Refs.[177, 178, 180, 181]. The Casimir amplitude for periodic boundary conditions $\Delta_{per}^{(1)}$ in Eq.(3.12) can be recovered from Ref.[177] if $d+1$ there is replaced by our d.

In order to extrapolate the results given by Eq.(3.12) to the experimentally relevant case $d = 3$ we employ a two–loop approximation using dimensional regularization techniques and the ε–expansion (see Ref.[34] and Sec.2.3). To first order in ε we obtain for the critical $(\mathbf{\Phi}^2)^2$–model in $d = 4 - \varepsilon$ [172, 188, 189]:

$$\Delta_{O,O} \;\; = \;\; -\frac{\pi^2}{1440} N \left[1 + \varepsilon \left(\ln(2\sqrt{\pi}) + \frac{\gamma-1}{2} - \frac{\zeta'(4)}{\zeta(4)} - \frac{5\,N+2}{4\,N+8} \right) \right] = \Delta_{SB,SB}$$

$$\Delta_{O,SB} \;\; = \;\; \frac{7}{8}\frac{\pi^2}{1440} N \left[1 + \varepsilon \left(\ln(2\sqrt{\pi}) + \frac{\gamma-1}{2} - \frac{\zeta'(4)}{\zeta(4)} - \frac{\ln 2}{7} + \frac{5\,N+2}{14\,N+8} \right) \right]$$

$$\Delta_{per} \;\; = \;\; -\frac{\pi^2}{90} N \left[1 + \varepsilon \left(\ln\sqrt{\pi} + \frac{\gamma-1}{2} - \frac{\zeta'(4)}{\zeta(4)} - \frac{5\,N+2}{4\,N+8} \right) \right]$$

$$\Delta_{aper} \;\; = \;\; \frac{7}{8}\frac{\pi^2}{90} N \left[1 + \varepsilon \left(\ln\sqrt{\pi} + \frac{\gamma-1}{2} - \frac{\zeta'(4)}{\zeta(4)} - \frac{\ln 2}{7} + \frac{5\,N+2}{14\,N+8} \right) \right], \tag{3.13}$$

where $\gamma \simeq 0.577216$ denotes Euler's constant and $\zeta'(4) \simeq -0.068911$ is a first derivative of the Riemann zeta function with $\zeta(4) = \pi^4/90$. The result for $\Delta_{O,O}$ in Eq.(3.13) can also be found in Ref.[172] for $N = 1$. The Casimir amplitudes are no longer proportional to N due to the presence of non–Gaussian fluctuations in the two–loop contribution. Eq.(3.12) suggests several relations among the Casimir amplitudes which can be tested using the two–loop results in Eq.(3.13). We find [188]

$$\Delta_{O,O} = \Delta_{SB,SB} \quad \text{and} \quad \frac{\Delta_{per}}{\Delta_{O,O}} = \frac{\Delta_{aper}}{\Delta_{O,SB}} = 2^d \tag{3.14}$$

up to and including $\mathcal{O}(\varepsilon)$. Concerning the validity of these relations beyond the ε-expansion there is some information from exact results for $N = 1$ and $d = 2$. Conformal invariance (see Sec.3.3) implies that the Casimir amplitudes in two–dimensional strips are equal for equal boundary conditions: $\Delta_{a,a} = \Delta_{b,b}$ [118, 192]. Note, that $\Delta_{SB,SB}$ for $N \geq 1$ does not exist in $d = 2$. Furthermore, $\Delta_{per}/\Delta_{O,O} = 4 = 2^d$ holds in $d = 2$ for $N = 1$ [120]. Finally, the relations in Eq.(3.14) hold at the upper critical dimension $d_c = 4$ and for a Gaussian model in arbitrary dimension (see Eq.(3.12)) so that there is some reason to expect that they hold for $d = 3$ and arbitrary N.

In view of experimental tests (see Secs.6.3 and 6.4) the numerical precision of the extrapolation formulas in Eq.(3.13) for $d = 3$, i.e., $\varepsilon = 1$ is important, but from a first order ε–expansion we cannot expect high accuracy. Unfortunately higher order contributions to the Casimir amplitudes are unavailable at the moment so that Borel resummation techniques [193] cannot be applied to improve quantitative predictions. One remaining option is to include information from exact results in $d = 2$ in an interpolation scheme. Such results are known in the case $N = 1$ for $\Delta_{O,O}, \Delta_{per}$, and Δ_{aper}. The exact values known for $d = 2$ and $d = 4$ compared to real space Migdal–Kadanoff renormalization results in $d = 3$ [190] suggest, that $\Delta(d)$ is a monotonous function of the dimension d. Therefore the interpolation formula

$$\Delta(d) = \Delta_0 + \frac{(4 - d)\Delta_1}{1 + (4 - d)\delta} \tag{3.15}$$

has been used in Ref.[188] in order to achieve improved numerical values for the Casimir amplitudes in $d = 3$. The coefficients Δ_0 and Δ_1 can be read off from Eq.(3.13) writing $\Delta = \Delta_0 + \varepsilon\Delta_1$ for $\Delta = \Delta_{O,O}, \Delta_{per}$, and Δ_{aper}. The third coefficient δ is determined by equating $\Delta(d = 2) = \Delta_0 + 2\Delta_1/(1 + 2\delta)$ to the exact two–dimensional values for $N = 1$ [50, 56, 118, 120]. The results are shown in Table 3.1 [188, 189] along with the Migdal–Kadanoff estimates [190] for comparison. The entries in Table 3.1 without superscript are obtained from the evaluation of Eq.(3.13) for $\varepsilon = 1$. Especially for $d = 3$ and $N \geq 2$ no other information is available in the literature. The case $d = 3$ and $N = 2$ for $\Delta_{O,O}$ is of special importance for films of ^{4}He at one of the λ–points (see Fig.1.6 and Sec.6.3). An inspection of Table 3.1 shows that for equal boundary conditions a and b $\Delta_{a,b}$ is negative, whereas different boundary conditions generate a positive $\Delta_{a,b}$. The former case corresponds to an attractive Casimir force and the latter corresponds to a repulsive Casimir force between the boundaries of the film due to critical fluctuations of the order parameter (see Sec.6.4

Table 3.1: Casimir amplitudes Δ. The results for $d = 4$ can be regarded to be exact. For $N \geq 1$ the surface–bulk transition is absent in $d = 2$.

Amplitude	$d = 2$ $N = 1$	$d = 3$ $N = 1$	$N = 1$	$N = 2$	$N = 3$	$d = 4$ $N = 1, 2, 3$
$\Delta_{O,O}$	-0.065[a]	-0.012[b]	(-0.015[c,d])	-0.022	-0.032	-0.0069 N
$\Delta_{O,SB}$		0.013	(0.017[d])	0.026	0.039	0.0060 N
$\Delta_{SB,SB}$		-0.012	(0.019[d])	-0.022	-0.032	-0.0069 N
Δ_{per}	-0.26[a]	-0.11	(-0.15[c])	-0.20	-0.28	-0.11 N
Δ_{aper}	0.52[a]	0.14	(0.17[c])	0.28	0.43	0.096 N

[a] Exact [56, 50, 120, 118].
[b] Ref.[172].
[c] Interpolated values obtained from Eq.(3.15).
[d] Migdal–Kadanoff estimates [190].

and Refs.[174] and [176]). The only deviation from this picture is provided by the Migdal–Kadanoff result for $\Delta_{SB,SB}$ for $N = 1$ in $d = 3$ [190], which carries a sign opposite to the one expected from the field–theoretical calculation (see Eqs.(3.13) and (3.14)). There is still no explanation for this discrepancy. Note, that the Casimir force in our case forms an *additive critical* contribution to the force balance in the film besides other *noncritical* contributions, which are basically given by van–der–Waals forces (see Secs.3.1 and 6.3).

A second option for the improvement of the numerical estimates for the Casimir amplitudes is provided by a partial re–exponentialization of the ε–expansion results in Eq.(3.13). In order to maintain consistency with the ε–expansion up to and including $\mathcal{O}(\varepsilon)$ we can write for $d = 4 - \varepsilon$

$$\Delta_{O,O} = \Delta_{O,O}^{(1)}\left(1 - \frac{5}{4}\varepsilon\frac{N+2}{N+8}\right) = \Delta_{SB,SB}$$

$$\Delta_{O,SB} = \Delta_{O,SB}^{(1)}\left(1 + \frac{5}{14}\varepsilon\frac{N+2}{N+8}\right)$$

$$\Delta_{per} = \Delta_{per}^{(1)}\left(1 - \frac{5}{4}\varepsilon\frac{N+2}{N+8}\right)$$

$$\Delta_{aper} = \Delta_{aper}^{(1)}\left(1 + \frac{5}{14}\varepsilon\frac{N+2}{N+8}\right), \tag{3.16}$$

Table 3.2: Casimir amplitudes Δ in $d = 3$ for $N = 1, 2, 3$. The Gaussian amplitudes in the last column are given for comparison.

Amplitude	$N = 1^a$	$N = 2^a$	$N = 3^a$	Gauss[b]
$\Delta_{O,O}, \Delta_{SB,SB}$	-0.014	-0.024	-0.031	-0.024 N
$\Delta_{O,SB}$	0.020	0.041	0.063	0.018 N
Δ_{per}	-0.11	-0.19	-0.25	-0.19 N
Δ_{aper}	0.16	0.33	0.50	0.14 N

[a] according to Eq.(3.16).
[b] according to Eq.(3.12).

where the factors $\Delta_{a,b}^{(1)}$ are taken from Eq.(3.12). The numerical values for $d = 3$ are summarized in Table 3.2 together with the values for the Gaussian model from Eq.(3.12) for comparison. Apart from the values for $\Delta_{O,SB}$ these numbers are fairly close to the strict ε–expansions evaluated at $\varepsilon = 1$. In view of the accuracy of our estimates for the Casimir amplitudes $\Delta_{O,O}$, $\Delta_{SB,SB}$, Δ_{per}, and Δ_{aper} the direct comparison between Table 3.1 and Table 3.2 and in the case $N = 1$ the comparison between the ε–expansion and Eq.(3.15) (see Table 3.1) gives an impression of the size of the error bars attached to the above field–theoretical values for $\Delta_{a,b}$. As far as $\Delta_{O,SB}$ is concerned the large discrepancy between the estimates from the strict ε–expansion (see Table 3.1) and from Eq.(3.16) (see Table 3.2) shows that both estimates are not very reliable.

Concerning symmetry breaking boundary conditions there is substantially less information available. In this case at least one of the two plates belongs to the extraordinary surface universality class (see Sec.1.2) which has been shown to be equivalent to the presence of a surface field (see Sec.2.3). In the Ising universality class a surface field can either act along a prescribed direction or antiparallel to it. These two cases will be distinguished by assigning the symbol $+$ or $-$ to one or both of the indices a and b in $\Delta_{a,b}$, respectively. For the XY – and Heisenberg universality classes the surface fields may have any orientation and thus cause helical order parameter profiles. The only known results for the Casimir amplitudes for $O(N)$–systems are confined to the Ising universality class ($N = 1$), where $\Delta_{O,+}$, $\Delta_{SB,+}$, $\Delta_{+,+}$, and $\Delta_{+,-}$ can be studied. Exchanging $+$ with $-$ in these four amplitudes is equivalent to turning the system upside down and therefore leaves the amplitudes invariant.

Table 3.3: Exact Casimir amplitudes Δ in $d = 2$ for $N = 1$ (Ising universality class), taken from Refs.[120] and [118].

$\Delta_{0,0}$	Δ_{per}	Δ_{aper}	$\Delta_{0,+}$	$\Delta_{+,+}$	$\Delta_{+,-}$
$-\frac{\pi}{48}$ [a]	$-\frac{\pi}{12}$	$\frac{\pi}{6}$	$\frac{\pi}{24}$	$-\frac{\pi}{48}$	$\frac{23}{48}\pi$
-0.065 [b]	-0.262 [b]	0.524 [b]	0.131	-0.065	1.51

[a] see also Ref.[50]

[b] see Table 3.1

Table 3.4: Migdal–Kadanoff renormalization group estimates for the Casimir amplitudes Δ in $d = 2$, 3, and 4 for $N = 1$, taken from Ref.[190].

d	$\Delta_{0,0}$	$\Delta_{0,SB}$	$\Delta_{SB,SB}$	$\Delta_{0,+}$	$\Delta_{SB,+}$	$\Delta_{+,+}$	$\Delta_{+,-}$
2	-0.055			0.130		0	0.881
3	-0.015	0.017	0.019	0.051	0.017	0	0.279
4	-0.007	-0.003	$\simeq 10^{-5}$	0.017	0.012	0	0.100

Conformal field theory in $d = 2$ provides rigorous results for an Ising model on a strip [118, 120] which we have summarized in Table 3.3. The second row just displays the numerical evaluation of the first row. The amplitudes for symmetry–conserving boundary conditions are repeated here for comparison and completeness. Note, that $\Delta_{SB,+}$ does not exist in $d = 2$ for $N = 1$ and that the exact relations

$$\Delta_{0,0} = \Delta_{+,+} \quad \text{and} \quad \frac{\Delta_{per}}{\Delta_{0,0}} = 4 = 2^d \qquad (3.17)$$

hold (see Eq.(3.14)). From Migdal–Kadanoff renormalization group studies numerical predictions for many Casimir amplitudes realized in $d = 2$, 3, and 4 for $N = 1$ are available [190], they are summarized in Table 3.4. The discrepancy between $\Delta_{0,0}$ and $\Delta_{SB,SB}$ in $d = 3$ has already been mentioned in the discussion of Table 3.1. In addition we note that the entry for $\Delta_{+,+}$ in $d = 2$ contradicts the exact relation $\Delta_{0,0} = \Delta_{+,+}$ [118] (see Eq.(3.17) and Table 3.3), whereas the entries for $\Delta_{0,0}$ and $\Delta_{0,+}$ match the exact values quite well (see Table 3.3). In $d = 4$ the same discrepancy between $\Delta_{0,0}$ and $\Delta_{SB,SB}$ occurs, which we already mentioned in $d = 3$. Moreover

Table 3.5:   Tricritical Casimir amplitudes in $d = 3$ for
$N = 1, 2, 3$. The values can be regarded to be exact (see
Ref.[188]).

$\Delta_{o,o}$	$\Delta_{O,SB}$	$\Delta_{SB,SB}$	Δ_{per}	Δ_{aper}
$-\frac{N}{16\pi}\zeta(3)$	$\frac{3N}{64\pi}\zeta(3)$	$-\frac{N}{16\pi}\zeta(3)$	$-\frac{N}{2\pi}\zeta(3)$	$\frac{3N}{8\pi}\zeta(3)$
-0.0239 N	0.0179 N	-0.0239 N	-0.1913 N	0.1435 N

the $d = 4$–entry for $\Delta_{O,SB}$ differs even in sign from the exact field–theoretical result
in Table 3.1. In view of these problems with the Migdal–Kadanoff estimates for the
Casimir amplitudes in Table 3.4 an independent check by, e.g., computer simulations
would be desirable.

Field–theoretical predictions for the Casimir amplitudes in presence of symmetry
– breaking boundary conditions in $d = 4 - \varepsilon$ are still unavailable. The main obstacle
towards a field–theoretical calculation is the nonvanishing position–dependent order
parameter profile imposed by the surface field or fields. This modifies the propagator
in a nontrivial way resulting in rather complicated loop integrals. The simplest case
in this respect may be given by two parallel surface fields giving rise to the Casimir
amplitude $\Delta_{+,+}$. The knowledge of $\Delta_{+,+}$ or other Casimir amplitudes in presence of
surface fields is relevant for wetting experiments at the demixing transition in binary
liquid mixtures (see Sec.6.3).

3.2.2   The Tricritical Casimir Amplitudes

Tricritical points and tricritical end points naturally show up in systems of higher
complexity such as ^{3}He $-^4$He mixtures (see Fig.1.9) or ternary liquid mixtures. The
tricritical point itself is an intersection between three lines of critical points. The
finite size analysis for the free energy described before in this section can be directly
applied to the tricritical $(\Phi^2)^3$–Hamiltonian (see Eqs.(1.30) and (1.31)) in our film
geometry. Some analysis has been done in the literature on the mean field level
concerning the order parameter profile in films and in semiinfinite geometries below
the bulk critical temperature [194], but the free energy of a tricritical film has not
been discussed so far. However, Eq.(3.9) enables us to define the *universal* tricritical
Casimir amplitudes just like the critical ones giving direct access to these amplitudes

by a renormalization group analysis of the free energy. The upper critical dimension for tricritical phenomena is $d_c = 3$, which means that for symmetry–conserving boundary conditions and in $d = 3$ the Gaussian amplitudes given by Eq.(3.12) for $d = 3$ are *exact* [188]. For convenience of the reader we summarize the analytical results and their numerical evaluation in Table 3.5. The amplitudes in Table 3.5 are generated by purely Gaussian fluctuations at the bulk tricritical point and therefore they trivially have all the properties discussed for the Gaussian amplitudes given by Eq.(3.12). The Casimir amplitude $\Delta_{O,O}$ for $N = 2$ is of experimental relevance for films of tricritical ^{3}He–^{4}He mixtures (see Secs.6.3 and 6.4). Compared to the critical $\Delta_{O,O}$ the tricritical $\Delta_{O,O}$ is about twice as large for any $N = 1, 2, 3$. This is also true for all the other boundary conditions except the antiperiodic one. The tricritical Casimir amplitudes can be tested independently for a metamagnet (see Fig.1.8 and Eq.(1.19)) by, e.g., Monte–Carlo simulations of the corresponding lattice model in a film geometry. In the simplest case $N = 1$ the Hamiltonian given by Eq.(1.19) reduces to a spin–1 Ising model in a crystal field [195].

Tricritical Casimir amplitudes in two–dimensional strips can be determined for a tricritical (spin–1) Ising model and a tricritical 3–state Potts model (see Eq.(1.27)) from conformal invariance considerations (see Sec.3.4). One obtains

$$\Delta_{O,O} = -\frac{7\pi}{240} \simeq -0.0916 \quad \Delta_{per} = -\frac{7\pi}{60} \simeq -0.3665 \quad : \quad \text{tricritical Ising}$$
$$\Delta_{O,O} = -\frac{\pi}{28} \simeq -0.1122 \quad \Delta_{per} = -\frac{\pi}{7} \simeq -0.4488 \quad : \quad \text{tricritical 3–state Potts,}$$
$$(3.18)$$

where $\Delta_{+,+} = \Delta_{O,O}$ in the Ising case (aligned surface fields) and $\Delta_{1,1} = \Delta_{O,O}$ in the 3–state Potts case (equal surface Potts fields). With respect to Eq.(3.18) we note, that the tricritical spin–1 Ising model displays *superconformal invariance* [196].

3.3 Conformal Invariance

At a critical point a system can be regarded as *scale invariant* due to violent critical fluctuations of the order parameter which lack any characteristic length and time scale. Consequently thermodynamic quantities and correlation functions are predominantly governed by power laws as functions of their arguments (for examples see Sec.1.1). The modern theory of critical phenomena has provided a natural explanation for scale invariance within the framework of the renormalization group. Besides invariance under uniform scale transformations $\mathbf{r'} = \mathbf{r}/b$ invariance under rigid rotations and translations can also be included into the principle of scale invariance. Equipped with todays experience and knowledge in the theory of critical phenomena it seems rather natural to extend this invariance principle to mappings which only *locally* behave as a uniform scale transformation (combined with a rigid rotation and a translation) [114]. Such mappings are known as *conformal transformations* and we are therefore concerned with *conformal invariance* (for an introduction and reviews consult Refs.[39, 197]). Note, that the Ginzburg–Landau Hamiltonian (see Sec.2.3) remains invariant under conformal coordinate transformations at the renormalization group fixed point. In two dimensions conformal field theory has demonstrated an enormous predictive power due to the infinite group of conformal transformations, which contains the holomorphic functions and their antiholomorphic counterparts [39, 115, 116, 119, 197]. These functions interrelate semiinfinite and finite as well as different finite geometries and therefore they drastically restrict the possible functional forms of correlation functions or scaling density profiles [123, 128] (see Sec.2.3).

In general dimensions $d > 2$, however, the situation is remarkably different. Here the group of conformal transformations is a finite parameter group which is spanned by the uniform scale transformation, rigid translations and rotations, and the inversion at a sphere of radius R centered at the origin (see Refs.[39, 114]):

$$\begin{aligned}
\mathbf{r}' &= \mathbf{r}/b && \text{(scale transformation)} \\
\mathbf{r}' &= \mathcal{A}\mathbf{r} + \mathbf{c} && \text{(rotation and translation)} \\
\mathbf{r}' &= (R/r)^2 \mathbf{r} && \text{(inversion at a sphere).}
\end{aligned} \qquad (3.19)$$

Note, that $\mathcal{A} = \left(\mathcal{A}^T\right)^{-1}$ is the $d \times d$ rotation matrix and $\mathbf{c}$ is a d–dimensional constant vector. A general conformal transformation for $d > 2$ is given by any combination of the three mappings in Eq.(3.19). The conformal group in $d > 2$ is sometimes called the *small conformal group* [39]. The so–called special conformal transformation is given by the mapping [114]

$$\frac{\mathbf{r}'}{r'^2} = \frac{\mathbf{r}}{r^2} + \mathbf{a} \quad , \quad \mathbf{a} = \frac{\mathbf{c}}{R^2}, \qquad (3.20)$$

where R is the radius of the sphere and $\mathbf{c}$ is the translation vector. Equation (3.20) is of great practical importance, because it gives access to an *infinitesimal* conformal transformation by choosing $\mathbf{c}$ arbitrarily close to the origin. Expanding Eq.(3.20) to first order in $\mathbf{a}$ we obtain [39, 114]

$$\mathbf{r}' = \mathbf{r} + r^2 \mathbf{a} - 2\,(\mathbf{a} \cdot \mathbf{r})\,\mathbf{r}. \qquad (3.21)$$

The meaning of infinitesimal here is that $\delta \mathbf{r} = \mathbf{r}' - \mathbf{r} \to 0$ as $\mathbf{a} \to 0$, but we will nevertheless refer to Eq.(3.21) as an infinitesimal conformal transformation. To illustrate the loss of variety in the set of conformal transformations in $d > 2$ compared to $d = 2$ we quote the two–dimensional analogue of the small conformal group. Identifying the two–dimensional space with the complex plane the small conformal group is given by the mappings [39, 115, 116]

$$\zeta' = \frac{\alpha\zeta + \beta}{\gamma\zeta + \delta} \quad , \quad \alpha\delta - \beta\gamma \neq 0, \tag{3.22}$$

which are known as the fractional linear or *Möbius* transformations. The Möbius transformations are the only ones which map the *completed* or *entire* complex plane[5] *onto* itself in a conformal and invertible (sometimes called biholomorphic) way [39, 198]. Such mappings are called automorphisms. The *only* automorphisms of the complex plane itself (i.e., infinity is mapped to infinity) are given by the linear functions $\zeta' = \alpha\zeta + \beta$, $\alpha \neq 0$. Obviously, the linear functions correspond to the *subgroup* generated by the scale transformations in combination with rigid rotations and translations in arbitrary dimension (see Eq.(3.19)). To give an example we quote the Möbius transformation

$$\zeta' = \frac{i\zeta + 1}{\zeta + i},$$

which maps the upper half plane (i.e., the semiinfinite geometry) conformally onto the unit disk, where the real axis is mapped onto the unit circle. A Möbius transformation which maps the upper half plane onto a strip (i.e., the two–dimensional analogue of a film) does not exist (see Sec.2.3). It is thus *impossible* to map the semiinfinite geometry in $d > 2$ conformally onto a film geometry.

The finite geometries considered here are always embedded in the Euclidean space. Locally, any conformal transformation contains a uniform scale transformation characterized by a scale factor b (see Eq.(3.19)). The effect of a conformal mapping on the metric is then solely given by this scale factor b, which is now position–dependent. Writing the metric in terms of the coordinate differentials dx^μ we have [39]

$$dx'^\mu dx'_\mu = (b(\mathbf{r}))^{-2} dx^\mu dx_\mu \tag{3.23}$$

---

[5]The complex plane is completed with the point ∞ in such a way, that the completed plane becomes *compact* (Alexandrov compactification). The completed complex plane is topologically equivalent to the Riemann sphere [198].

for any conformal transformation. For example the inversion at a sphere of radius R leads to $b(\mathbf{r}) = r^2/R^2$ and for the infinitesimal conformal mapping in Eq.(3.21) we have [39, 114]

$$b(\mathbf{r}) = 1 + 2\mathbf{a} \cdot \mathbf{r}. \tag{3.24}$$

The conformal image of an Euclidean metric is therefore *locally* still an Euclidean metric. In order to probe the physical content of the conformal invariance principle the behavior of correlation functions under conformal transformations will now be studied.

3.3.1 Correlation Functions and Primary Scaling Operators

The Ginzburg–Landau Hamiltonian described in Secs.1.4 and 2.3 is a *local* functional of the fluctuating local magnetization $\mathbf{\Phi}(\mathbf{r})$. A conformal mapping induces a local real–space renormalization–group transformation. The locality of this conformal renormalization–group transformation suggests, that the scaling fields $h_i(\mathbf{r})$ and the corresponding scaling operators $\varphi_i(\mathbf{r})$ [6] obey the same transformation laws as for a usual scale transformation except that the scale factor b and the rotation matrix $\mathcal{A}$ are position dependent. However, in general a scaling operator will not transform in this simple fashion. If, for example, the operator contains derivatives with respect to its position, the transformation law will depend on *derivatives* of $b(\mathbf{r})$ as well. Those special scaling operators which actually *do* display the simple scaling behavior under a conformal transformation have been called *primary operators* [39, 115, 116, 199]. The n–point cumulant of n scalar primary scaling operators $\varphi_i(\mathbf{r}_i)$, $i = 1,\ldots,n$, located at pairwise different positions $\mathbf{r}_i$ *locally* fulfills the transformation law for a uniform scale transformation (see Eq.(3.19)) [39, 114]. The transformation law reads

$$\langle \varphi_1(\mathbf{r}_1)\varphi_2(\mathbf{r}_2)\ldots\varphi_n(\mathbf{r}_n)\rangle = (b(\mathbf{r}_1))^{-x_1}(b(\mathbf{r}_2))^{-x_2}\ldots(b(\mathbf{r}_n))^{-x_n}\langle \varphi_1(\mathbf{r}_1')\varphi_2(\mathbf{r}_2')\ldots\varphi_n(\mathbf{r}_n')\rangle, \tag{3.25}$$

where x_i, $i = 1,\ldots,n$, is the scaling dimension of the scaling operator $\varphi_i(\mathbf{r}_i)$ and $\langle\ldots\rangle$ denotes the cumulant. For nonscalar $\varphi_i(\mathbf{r}_i)$ the transformation law in Eq.(3.25) contains the local rotation matrices operating on the right hand side besides the scaling factors $b(\mathbf{r}_i)$ [39, 200]. We will be solely concerned with scalar operators in the following and we therefore refrain from giving the general law. (For a detailed

---

[6]The term "operator" is used in order to distinguish the *fluctuating* quantity $\varphi_i(\mathbf{r}_i)$ from its *nonfluctuating* thermal average $\langle\varphi_i(\mathbf{r}_i)\rangle$.

discussion of nonscalar operators we refer to Refs.[39, 115, 199, 200].) In order to write Eq.(3.25) as an *invariance* statement we reformulate it now for *infinitesimal* conformal mappings (see Eq.(3.21)) which we here more generally write as

$$\mathbf{r}' = \mathbf{r} + \mathbf{a}(\mathbf{r}). \tag{3.26}$$

Infinitesimal here means that $|\mathbf{a}(\mathbf{r})| \ll |\mathbf{r}|$ in a bounded subdomain $\mathcal{D}$ of our geometry so that $\mathcal{D}$ contains all the positions $\mathbf{r}_1, \ldots, \mathbf{r}_n$ of the scalar primary scaling operators $\varphi_1, \ldots, \varphi_n$ in Eq.(3.25). From Eq.(3.26) we immediately find

$$b(\mathbf{r}) = 1 - \frac{1}{d} \nabla \cdot \mathbf{a}(\mathbf{r}) \tag{3.27}$$

to first order in $\mathbf{a}(\mathbf{r})$. An expansion of Eq.(3.25) to first order in $\mathbf{a}(\mathbf{r})$ then yields [39, 199, 200]

$$\sum_{i=1}^{n} \left(\frac{x_i}{d} \nabla \cdot \mathbf{a}(\mathbf{r}_i) + \mathbf{a}(\mathbf{r}_i) \cdot \frac{\partial}{\partial \mathbf{r}_i} \right) \langle \varphi_1(\mathbf{r}_1) \varphi_2(\mathbf{r}_2) \ldots \varphi_n(\mathbf{r}_n) \rangle = 0. \tag{3.28}$$

Equation (3.28) expresses the desired invariance of the n–point cumulant under the infinitesimal conformal transformation given by Eq.(3.26). The explicit form of Eq.(3.28) again displays the fact that any of the $\varphi_i(\mathbf{r}_i)$ is a scalar *primary* operator. If $\delta_a\varphi(\mathbf{r})$ denotes the amount, by which the primary operator $\varphi_i(\mathbf{r}_i)$ changes under the infinitesimal conformal transformation in Eq.(3.26), then $\delta_a\varphi(\mathbf{r}) = \mathbf{a}(\mathbf{r}) \cdot \nabla\varphi(\mathbf{r}) + \frac{x}{d} \nabla \cdot \mathbf{a}(\mathbf{r}) \varphi(\mathbf{r})$ *inside* the cumulant in Eq.(3.28) [199]. This can be taken as a shorthand notation for the transformation law of scalar primary operators. For example, the local magnetization $\Phi(\mathbf{r})$ behaves this way. If φ is a scalar but not a primary operator like, e.g., $\Delta\Phi(\mathbf{r})$ then additional terms will contribute to $\delta_a\varphi(\mathbf{r})$ containing higher order derivatives of $\mathbf{a}(\mathbf{r})$ multiplied by operators other than $\varphi(\mathbf{r})$. For statements on the behavior of tensor operators we refer to Ref.[200].

In order to demonstrate the effect of Eq.(3.28) on correlation functions we consider a two–point cumulant of two scalar primary operators φ_1 and φ_2 with *different* scaling dimensions x_1 and x_2, respectively, in unbounded space. From homogeneity and isotropy of the unbounded space we conclude that $\langle \varphi_1(\mathbf{r}_1)\varphi_2(\mathbf{r}_2) \rangle = g(r_{12})$, where $r_{12} = |\mathbf{r}_1 - \mathbf{r}_2|$. Evaluating Eq.(3.28) for the special infinitesimal conformal transformation in Eq.(3.21) and for $n = 2$ we find the following differential equation for $g(r_{12})$:

$$\mathbf{a} \cdot (\mathbf{r}_1 + \mathbf{r}_2) r_{12} g'(r_{12}) + 2\mathbf{a} \cdot (x_1\mathbf{r}_1 + x_2\mathbf{r}_2) g(r_{12}) = 0. \tag{3.29}$$

The coordinates $\mathbf{r}_1 + \mathbf{r}_2$ and $\mathbf{r}_1 - \mathbf{r}_2$ are independent variables and therefore Eq.(3.29) is equivalent to the system

$$r_{12}g'(r_{12}) \quad + \quad (x_1 + x_2)g(r_{12}) = 0$$
$$(x_1 - x_2)g(r_{12}) = 0, \tag{3.30}$$

which implies $g(r_{12}) = 0$ because $x_1 \neq x_2$. Thus two–point cumulants of primary scaling operators with *different* scaling dimensions always vanish at a critical point [39]. For a detailed analysis see Ref.[200]. Note, that for $\varphi_1(\mathbf{r}_1) = \Phi(\mathbf{r}_1)$ and $\varphi_2(\mathbf{r}_2) = \Delta\Phi(\mathbf{r}_2)$ $\langle\varphi_1(\mathbf{r}_1)\varphi_2(\mathbf{r}_2)\rangle = \Delta_{\mathbf{r}_2}\langle\Phi(\mathbf{r}_1)\Phi(\mathbf{r}_2)\rangle \neq 0$, although $x_1 = x_\Phi \neq x_2$. In two dimensions each primary operator φ is an *ancestor* of a *conformal family* [115] which consists of the primary field φ itself and basically its spatial derivatives. The two–point cumulant between scaling operators from *different* conformal families in $d = 2$ vanishes, even if these operators are not primary. However, this does not hold for operators from the *same* conformal family. In this sense $\Delta\Phi$ is a descendant of the (primary) local magnetization Φ and therefore belongs to the same conformal family.

The n–point cumulant of n scalar primary scaling operators is fixed up to a function of $n(n - 3)/2$ invariants by the conformal invariance condition Eq.(3.28) [114]. Note, that for $n = 3$ the cumulant is fixed up to a multiplicative constant [114]. For a more general discussion we again refer to Ref.[200].

In a semiinfinite geometry the position vector $\mathbf{r}$ is split into a component $\mathbf{r}_{\parallel}$ parallel to the surface and a component $z\mathbf{e}_z$ perpendicular to it so that $\mathbf{r} = (\mathbf{r}_{\parallel}, z)$. We assume that the surface is located at $z = 0$ and that $z > 0$ characterizes the *interior* points of our semiinfinite system. Consider now the two–point cumulant of two scalar primary scaling operators $\varphi_1(\mathbf{r}_1)$ and $\varphi_2(\mathbf{r}_2)$ with bulk scaling dimensions x_1 and x_2, respectively, where $\mathbf{r}_2 = (0,0)$, i.e., φ_2 is located *in* the surface. The scaling dimension of $\varphi_2(0,0)$ is then given by the *surface* exponent $x_2^{(s)}$ [20], which is in general *different* from x_2. The two–point cumulant $\langle\varphi_1(\mathbf{r}_{\parallel}, z)\varphi_2(0,0)\rangle$ is a function of two macroscopic lengths, namely the distance $r = \sqrt{r_{\parallel}^2 + z^2}$ between φ_1 and φ_2 and the distance $z > 0$ between φ_1 and the surface. From *scale* invariance we obtain

$$\langle\varphi_1(\mathbf{r}_{\parallel}, z)\varphi_2(0,0)\rangle = r^{-x_1 - x_2^{(s)}} f(z/r), \tag{3.31}$$

where the scaling function $f(z/r)$ remains undetermined. *Conformal* invariance is more restrictive. The infinitesimal conformal mapping in Eq.(3.21) can be applied to Eq.(3.31), where the constant vector a is chosen to be *parallel* to the surface in order to map the surface onto itself. The evaluation of the conformal invariance condition Eq.(3.28) for the infinitesimal conformal transformation in Eq.(3.21) gives the differential equation $y f'(y) = (x_2^{(s)} - x_1) f(y)$ for the scaling function f, where $y = z/r$. The solution is given by $f(y) = A y^{x_2^{(s)} - x_1}$, A =const, and we therefore obtain from Eq.(3.31)

$$\langle \varphi_1(\mathbf{r}_{\parallel}, z) \varphi_2(\mathbf{0}, 0) \rangle = A r^{-x_1 - x_2^{(s)}} \left(\frac{z}{r} \right)^{x_2^{(s)} - x_1} \tag{3.32}$$

for two *primary* scaling operators φ_1 and φ_2. In contrast to scale invariance *conformal* invariance completely determines the functional form of the above two–point cumulant. Only the amplitude A is left undetermined. If $\mathbf{r}_1 = (\mathbf{r}_{\parallel}, z)$ in $\langle \varphi_1(\mathbf{r}_1) \varphi_2(\mathbf{r}_2) \rangle$ as before, but additionally $\mathbf{r}_2 = (\mathbf{0}, z')$ with $z' > 0$, scale invariance leads to

$$\langle \varphi_1(\mathbf{r}_{\parallel}, z) \varphi_2(\mathbf{0}, z') \rangle = r^{-x_1 - x_2} \tilde{f}(z/r, z'/r), \tag{3.33}$$

where $\tilde{f}$ is an unknown scaling function of *two* varables. Conformal invariance (see Eq.(3.28)) here determines the two–point cumulant in Eq.(3.33) up to an unknown function of a *single* variable (see Ref.[117] for the explicit result in $d = 2$). Hence conformal invariance is a valuable tool to study critical *semi–infinite* systems besides critical bulk systems, and it has been intensively used in $d = 2$ (see Sec.2.3).

3.3.2   The Stress Tensor

The transformation law in Eq.(3.25) displays the behavior of a n–point cumulant of n primary scaling operators under a conformal mapping. The influence of the conformal transformation on Eq.(3.25) is twofold. First, it gives rise to the local scaling factors $b(\mathbf{r}_i)$, $i = 1, \ldots, n$, where $\mathbf{r}_i$ is the position of the scaling operator φ_i. Second, the cumulant on either side of Eq.(3.25) is taken with respect to the *same* fixed–point Hamiltonian $\mathcal{H}^*$, because $\mathcal{H}^*$ is invariant under conformal transformations. Here we have assumed, that the geometry is conformally mapped onto itself in order to relate the n–point cumulant to itself in the same geometry.

As before, we now consider the general infinitesimal transformation $\mathbf{r}' = \mathbf{r} + \mathbf{a}(\mathbf{r})$ (see Eq.(3.26)), but this time we assume, that $|\mathbf{a}(\mathbf{r})| \ll |\mathbf{r}|$ *globally*. Therefore the infinitesimal transformation can no longer be globally conformal. Instead we introduce a bounded domain $\mathcal{D}$ which contains the position vectors $\mathbf{r}_1, \ldots, \mathbf{r}_n$ of the scaling operators in the n–point cumulant we are considering. Now a smooth $\mathbf{a(r)}$ can be designed such that $\mathbf{r}'$ is conformal *and* $|\mathbf{a(r)}| \ll |\mathbf{r}|$ for all $\mathbf{r} \in \mathcal{D}$ and that $|\mathbf{a(r)}| \ll |\mathbf{r}|$ in the complement of $\mathcal{D}$, where $\mathbf{r}'$ is still differentiable but no longer conformal [39]. The complement of $\mathcal{D}$ will be denoted by $\mathcal{R}$ in the following. The mapping $\mathbf{r}'(\mathbf{r})$ is still conformal in the vicinity of any of the positions $\mathbf{r}_1, \ldots, \mathbf{r}_n$, so that the transformation law in Eq.(3.25) still holds *except* that the cumulant on the right hand side of Eq.(3.25) is to be taken with respect to a Hamiltonian $\mathcal{H} = \mathcal{H}^* + \delta\mathcal{H}$ *different* from the fixed–point Hamiltonian $\mathcal{H}^*$. The change $\delta\mathcal{H}$ in the Hamiltonian originates from the region $\mathcal{R}$, where $\mathbf{r}'(\mathbf{r})$ is not conformal. Using Eq.(3.27) for $\mathbf{r}_i \in \mathcal{D}$ we find the new transformation law for the n–point cumulant of n *primary* scaling operators analogous to Eq.(3.28) [39, 199]

$$\sum_{i=1}^{n} \left(\frac{x_i}{d} \nabla \cdot \mathbf{a}(\mathbf{r}_i) + \mathbf{a}(\mathbf{r}_i) \cdot \frac{\partial}{\partial \mathbf{r}_i} \right) \langle \varphi_1(\mathbf{r}_1)\varphi_2(\mathbf{r}_2) \ldots \varphi_n(\mathbf{r}_n) \rangle = \langle \delta\mathcal{H}\varphi_1(\mathbf{r}_1)\varphi_2(\mathbf{r}_2) \ldots \varphi_n(\mathbf{r}_n) \rangle,$$

$$(3.34)$$

where the right hand side is the *cumulant* of $\delta\mathcal{H}, \varphi_1, \ldots, \varphi_n$. In Eq.(3.34) the cumulants are taken with respect to the fixed–point Hamiltonian $\mathcal{H}^*$. The change $\delta\mathcal{H}$ in the Hamiltonian is linearly related to the Jacobian $\frac{\partial a^\mu}{\partial x_\nu}\big|_{\mathbf{r} \in \mathcal{R}}$ via

$$\delta\mathcal{H} = \int_{\mathcal{R}} d^d r \, \frac{\partial a^\mu}{\partial x_\nu} T_{\mu\nu}(\mathbf{r}). \qquad (3.35)$$

Equation (3.35) defines the *stress energy tensor* or *stress tensor* $T_{\mu\nu}(\mathbf{r})$. This definition is in accordance with Refs.[199, 201, 202] and differs from the definition in Refs.[39, 192] by a factor $-S_d^{-1}$ in general dimension d, where S_d is the surface of d–dimensional unit sphere. The Jacobian $\frac{\partial a^\mu}{\partial x_\nu}$ can be decomposed into a sum of a diagonal contribution $d^{\mu\nu}$ describing the local scale transformation, an antisymmetric contribution $\rho^{\mu\nu}$ describing the local rotation, and a symmetric traceless contribution $\sigma^{\mu\nu}$ describing the local shear. For conformal transformations $\sigma^{\mu\nu}$ vanishes identically. On the other hand $\delta\mathcal{H} = 0$ for conformal mappings, and therefore both $d^{\mu\nu}$ and $\rho^{\mu\nu}$ must annihilate the right hand side of Eq.(3.34) independently. In other words, the stress tensor $T_{\mu\nu}(\mathbf{r})$ is symmetric and traceless

$$T_{\mu\nu}(\mathbf{r}) = T_{\nu\mu}(\mathbf{r}) \quad \text{and} \quad T^\mu_\mu(\mathbf{r}) = 0 \qquad (3.36)$$

for $\mathbf{r} \in \mathcal{R}$ *inside* the cumulant on the right hand side of Eq.(3.34) [39, 199]. Because $\mathcal{D}$ can be chosen as an arbitrary small neighborhood of the points $\mathbf{r}_1, \ldots, \mathbf{r}_n$, $\mathbf{r}$ may be any position in the system *except* $\mathbf{r}_1, \ldots, \mathbf{r}_n$ (see Eq.(3.34)). Inserting Eq.(3.35) in Eq.(3.34) gives the so called *conformal Ward identity* [39] in d dimensions. Using the Gauss theorem we can rewrite the right hand side of Eq.(3.34) as

$$\int_{\mathcal{R}} d^d r \, \frac{\partial a^\mu(\mathbf{r})}{\partial x_\nu} \langle T_{\mu\nu}(\mathbf{r})\varphi_1(\mathbf{r}_1)\dots\varphi_n(\mathbf{r}_n)\rangle = \int_{\partial\mathcal{R}} dS \, n^\nu(\mathbf{r}) a^\mu(\mathbf{r}) \langle T_{\mu\nu}(\mathbf{r})\varphi_1(\mathbf{r}_1)\dots\varphi_n(\mathbf{r}_n)\rangle$$

$$- \int_{\mathcal{R}} d^d r \, a^\mu(\mathbf{r}) \left\langle \frac{\partial}{\partial x_\nu} T_{\mu\nu}(\mathbf{r})\varphi_1(\mathbf{r}_1)\dots\varphi_n(\mathbf{r}_n) \right\rangle, \tag{3.37}$$

where $\mathbf{n}$ is the surface normal vector on the boundary $\partial\mathcal{R}$ of $\mathcal{R}$. Note, that for geometries with surfaces $\partial\mathcal{R} \neq \partial\mathcal{D}$. For $\mathbf{r} \epsilon \mathcal{R}$ the functions $a^\mu(\mathbf{r})$ are arbitrary and influence neither the surface integral on the right hand side of Eq.(3.37) nor the left hand side of Eq.(3.34) which depends only on $a^\mu(\mathbf{r}_i)$, $i = 1,\dots,n$. Therefore one has [39]

$$\frac{\partial}{\partial x_\nu} T_{\mu\nu}(\mathbf{r}) = 0 \tag{3.38}$$

for $\mathbf{r} \in \mathcal{R}$ or, more general, $\mathbf{r} \neq \mathbf{r}_i$, $i = 1,\dots,n$ *inside* the cumulant on the right hand side of Eq.(3.37). Equation (3.38) expresses a conservation law for the stress tensor $T_{\mu\nu}(\mathbf{r})$. The conformal Ward identity then takes the final form [39, 199]

$$\sum_{i=1}^{n} \left(\frac{x_i}{d} \nabla \cdot \mathbf{a}(\mathbf{r}_i) + \mathbf{a}(\mathbf{r}_i) \cdot \frac{\partial}{\partial \mathbf{r}_i} \right) \langle \varphi_1(\mathbf{r}_1)\varphi_2(\mathbf{r}_2)\dots\varphi_n(\mathbf{r}_n)\rangle$$

$$= \int_{\partial\mathcal{R}} dS \, n^\nu(\mathbf{r}) a^\mu(\mathbf{r}) \langle T_{\mu\nu}(\mathbf{r})\varphi_1(\mathbf{r}_1)\dots\varphi_n(\mathbf{r}_n)\rangle. \tag{3.39}$$

In two dimensions Eq.(3.39) can be transformed to a *local* relation using Cauchy's theorem [39, 117, 199] giving rise to short distance expansions of scaling operators φ with $T_{\mu\nu}$. In a film geometry Eqs.(3.34) and (3.39) put the Casimir amplitudes (see Sec.3.2) into a much broader perspective (see Sec.3.4).

The construction of the stress tensor $T_{\mu\nu}$ in general is not an easy task [201, 202]. In order to give at least an example we resrict ourselves to the case of a critical Gaussian theory for an Euclidean metric and $N = 1$ in unbounded space [199]. As a first step we apply a *general* infinitesimal coordinate transformation $x'^\mu = x^\mu + a^\mu(\mathbf{r})$ (see Eq.(3.26), here $\mathbf{a}(\mathbf{r})$ is *not* conformal) to the Gaussian Hamiltonian $\mathcal{H} = \frac{1}{2}\int d^d r \, (\nabla\Phi)^2$ (see Sec.1.4). From Eq.(3.35) we find the *canonical* stress tensor for the critical Gaussian theory [199, 201]

$$T_{\mu\nu}^{G,c} = \frac{\partial\Phi}{\partial x^\mu}\frac{\partial\Phi}{\partial x^\nu} - \frac{1}{2}\delta_{\mu\nu}\,(\nabla\Phi)^2, \tag{3.40}$$

where Φ is the one–component local magnetization and the Kronecker $\delta_{\mu\nu}$ is the (Euclidean) metric tensor. The symmetry requirement in Eq.(3.36) is already satisfied by Eq.(3.40) as an operator identity but $T^{G,c}_{\mu\nu}$ is neither traceless nor conserved inside cumulants (see Eq.(3.38)). To meet these latter requirements the so called *improvement term* [201]

$$t_{\mu\nu} = \left(\frac{\partial^2}{\partial x^\mu \partial x^\nu} - \delta_{\mu\nu}\Delta \right) \Phi^2 \tag{3.41}$$

which is symmetric and conserved but not traceless is used as an additional contribution to $T^{G,c}_{\mu\nu}$ [199, 201, 202]. One defines

$$T^G_{\mu\nu} = T^{G,c}_{\mu\nu} + \lambda t_{\mu\nu} - \frac{1}{d}\delta_{\mu\nu} \left(T^{\mu G,c}_\mu + \lambda t^\mu_\mu \right), \tag{3.42}$$

which is symmetric and traceless for any parameter λ as an *operator* and therefore inside any cumulant. We can thus choose λ such that Eq.(3.38) holds inside cumulants without violating Eq.(3.36). In a Gaussian theory Eq.(3.38) holds inside any cumulant if it holds inside $\langle \Phi(\mathbf{r_1})\Phi(\mathbf{r_2}) \rangle$, $\mathbf{r_1} \neq \mathbf{r_2}$ [199]:

$$\left\langle \frac{\partial T^G_{\mu\nu}(\mathbf{r})}{\partial x_\nu} \Phi(\mathbf{r_1})\Phi(\mathbf{r_2}) \right\rangle = 0 \quad , \quad \mathbf{r} \neq \mathbf{r_1}, \mathbf{r_2}. \tag{3.43}$$

Equation (3.43) implies $\lambda = -\frac{1}{4}\frac{d-2}{d-1}$, which together with Eq.(3.42) gives the stress tensor for a critical Gaussian theory [199, 201, 202].

For a critical Φ^4–theory $T_{\mu\nu}$ can only be constructed perturbatively. To first order in the Φ^4–coupling constant g one finds for $N = 1$ in unbounded space [201, 202]

$$T_{\mu\nu} = T^G_{\mu\nu} - \delta_{\mu\nu}\frac{g}{4!}\Phi^4. \tag{3.44}$$

Higher orders in g change the value of the parameter λ in Eq.(3.42) [201, 202]. Note, that in contrast to $T^G_{\mu\nu}$ in Eq.(3.42) $T_{\mu\nu}$ is no longer traceless as an operator, but inside cumulants $T^\mu_\mu = 0$ up to and including $\mathcal{O}(g)$.

The scaling dimension of the stress tensor $T_{\mu\nu}$ equals its naive dimension d, i.e., Eq.(3.44) already gives the *renormalized* expression for the stress tensor $T_{\mu\nu}$ to first order in the coupling constant [199, 201, 202]. If there are planar surfaces present in the geometry, any cumulant of the stess tensor $T_{\mu\nu}(\mathbf{r})$ in Eq.(3.44) with scaling operators varies smoothly as the position $\mathbf{r}$ of the stress tensor approaches a surface from the interior of the system [199]. In a Gaussian theory this can be shown explicitly for $T^G_{\mu\nu}$. In general a perturbative approach is needed to handle the cumulants with $T_{\mu\nu}$ in Eq.(3.44) [199].

3.4 The Casimir Effect and the Stress Tensor

The scaling dimension of the stress tensor is given by the spatial dimension d [39, 199]. From dimensional analysis we have for the thermodynamic average of the stress tensor $\langle T_{\mu\nu}\rangle \propto L^{-d}$, where L is a macroscopic length scale. A critical bulk system is macroscopically homogeneous and isotropic, so that $\langle T_{\mu\nu}\rangle = \text{const} \propto L^{-d}$ *independent* of μ and ν. The tracelessness condition in Eq.(3.36) and the absence of a macroscopic length scale L then implies $\langle T_{\mu\nu}\rangle = 0$ in a critical bulk system.

A critical semiinfinite system with a uniform boundary condition is still homogeneous and isotropic parallel to the (plane) surface at $z = 0$ so that for $\mathbf{r} = (\mathbf{r}_{\parallel}, z)$ $\langle T_{\mu\nu}(\mathbf{r})\rangle = \langle T_{\mu\nu}(0, z)\rangle$. Here a distance $z > 0$ of the stress tensor from the surface provides a macroscopic length scale. The $d - 1$ directions parallel to the surface are all equivalent and therefore $\langle T_{\mu\nu}\rangle$ has at most four independent components: $\langle T_{\parallel\parallel}\rangle$, $\langle T_{\parallel z}\rangle$, $\langle T_{z\parallel}\rangle$, and $\langle T_{zz}\rangle$, where $\parallel$ denotes any direction parallel to the surface. From Eq.(3.36) we conclude $\langle T_{\parallel z}\rangle = \langle T_{z\parallel}\rangle$ (symmetry) and $(d-1)\langle T_{\parallel\parallel}\rangle + \langle T_{zz}\rangle = 0$ (tracelessness) so that only two components are actually independent, e.g., $\langle T_{zz}\rangle$ and $\langle T_{\parallel z}\rangle$. Now Eq.(3.38), which holds inside cumulants, can be applied to $\langle T_{\mu\nu}(0, z)\rangle$ giving $\frac{\partial}{\partial z}\langle T_{zz}(0, z)\rangle = \frac{\partial}{\partial z}\langle T_{\parallel z}(0, z)\rangle = 0$, and therefore $\langle T_{\mu\nu}(\mathbf{r})\rangle$ is in fact not position dependent. For $z \to \infty$ $\langle T_{\mu\nu}(\mathbf{r})\rangle$ must approach its bulk value and hence $\langle T_{\mu\nu}(\mathbf{r})\rangle = 0$ in a critical semiinfinite system [199].

In a critical *film* the same arguments as for a critical semiinfinite system hold concerning $\langle T_{\mu\nu}(\mathbf{r})\rangle$ *except* the last one. It is impossible to approach the bulk value of $\langle T_{\mu\nu}(\mathbf{r})\rangle$ because z cannot be arbitrarily far from any of the two surfaces. In a critical film of thickness L, where L sets the macroscopic length scale, therefore $\langle T_{\mu\nu}(\mathbf{r})\rangle = \text{const} \propto L^{-d}$. Moreover, $\langle T_{\parallel z}(\mathbf{r})\rangle = 0$ for symmetry reasons in a film geometry [199]. A change in L can be described by a *nonconformal* mapping of one film of thickness L onto another film of thickness L'. Such a mapping causes a deviation $\delta\mathcal{H}$ between the fixed–point Hamiltonians $\mathcal{H}_L^*$ and $\mathcal{H}_{L'}^*$ of the two films of thickness L and L', respectively. The change in the free energy associated with $\delta\mathcal{H}$ is given by $\langle\delta\mathcal{H}\rangle$ which in turn is related to $\langle T_{\mu\nu}\rangle$ via Eq.(3.35). On the other hand the change in the free energy with the film thickness L is associated with the Casimir effect (see Sec.3.2). Hence $\langle T_{\mu\nu}\rangle$ in a film geometry must be closely related to the Casimir amplitude $\Delta_{a,b}$.

The relation between the stress tensor and the Casimir amplitude can be revealed if a general infinitesimal coordinate transformation $\mathbf{r}' = \mathbf{r} + \mathbf{a}(\mathbf{r})$ is interpreted as a *real–space* renormalization–group transformation applied on the fixed point Ginzburg–Landau Hamiltonian $\mathcal{H}_L^*$ in a film geometry (see Eq.(2.48). Specifically, we use the following linear infinitesimal mapping [120, 199]

$$\mathbf{r}_\parallel' = \mathbf{r}_\parallel(1 + a_\parallel) \quad , \quad |a_\parallel| \ll 1$$
$$z' = z(1 + a_z) \quad , \quad |a_z| \ll 1, a_z \neq a_\parallel \tag{3.45}$$

in order to map a film of thickness L onto a film of thickness $L' = L(1 + a_z)$. Note, that a lateral area A transforms as $A' = A\left(1 + (d-1)a_\parallel\right)$. The fixed–point Hamiltonian $\mathcal{H}_L^*$ of the film shows the behavior

$$\mathcal{H}_L^* = \mathcal{H}_{L'}^* + \delta\mathcal{H}, \tag{3.46}$$

because Eq.(3.45) defines a *nonconformal* mapping (see Eq.(3.35)). The free energy $\mathcal{F}\{\mathcal{H}_L^*\}$ can be written as

$$\mathcal{F}\{\mathcal{H}_L^*\} = \mathcal{F}\{\mathcal{H}_{L'}^*\} + \langle\delta\mathcal{H}\rangle + \mathcal{F}^0, \tag{3.47}$$

where the additive contribution $\mathcal{F}^0$ is a *nonsingular* background term *independent* of the degrees of freedom in the Hamiltonian. $\mathcal{F}^0$ is a byproduct of the real–space renormalization–group transformation Eq.(3.45) [18, 199] due to the overall change in volume $V = AL$ and surface area A of the film. The *leading* finite size behavior of $\mathcal{F}\{\mathcal{H}_L^*\}$ is given by $\langle\delta\mathcal{H}\rangle$, where

$$\langle\delta\mathcal{H}\rangle = \int d^d r \frac{\partial a^\mu}{\partial x_\nu}\langle T_{\mu\nu}\rangle = AL\left(a_z - a_\parallel\right)\langle T_{zz}\rangle \tag{3.48}$$

due to Eqs.(3.35), (3.36), and (3.45). Note, that $\langle T_{\mu\nu}\rangle$ is a *constant* in a film with homogeneous boundary conditions. We define $f(0, L) = \lim_{A\to\infty}\left(\mathcal{F}\{\mathcal{H}_L^*\}/(k_B T_{c,b}A)\right)$, where $f(t = 0, L)$ contains both singular and nonsingular terms. Expanding $\mathcal{F}\{\mathcal{H}_{L'}^*\}$ in Eq.(3.47) around $L = L'$ to first order in a_z leads to the following differential equation for $f(0, L)$ (see Ref.[199])

$$\left((d-1)a_\parallel + a_z L\frac{\partial}{\partial L}\right)f(0, L) \; + \; L\left(a_z - a_\parallel\right)\langle T_{zz}\rangle + L\left((d-1)a_\parallel + a_z\right)f_{bulk}^0$$
$$+ \; (d-1)a_\parallel(f_{s,a}^0 + f_{s,b}^0) = 0, \tag{3.49}$$

where $\mathcal{F}^0/A$ in the limit $A \to \infty$ has been written as a sum of a bulk contribution $L((d-1)a_\parallel + a_z)f_{bulk}^0$ and two surface contributions $(d-1)a_\parallel f_{s,i}^0$, $i = a, b$, respectively. The amplitudes f_{bulk}^0, $f_{s,a}^0$, and $f_{s,b}^0$ are nonuniversal constants. Any physically relevant solution of Eq.(3.49) must be *independent* of both $a_{\parallel}$ and a_z. Furthermore, one must bear in mind that $\langle T_{zz} \rangle \propto L^{-d}$ by dimensional analysis and therefore

$$f(0, L) = -L f_{bulk}^0 - f_{s,a}^0 - f_{s,b}^0 + \Delta_{a,b} L^{-(d-1)} \tag{3.50}$$

with

$$\Delta_{a,b} = \frac{L^d}{d-1} \langle T_{zz} \rangle. \tag{3.51}$$

According to Eq.(3.9) the amplitude $\Delta_{a,b}$ in Eq.(3.50) is the Casimir amplitude. Hence Eq.(3.51) provides the desired relation between the Casimir amplitude $\Delta_{a,b}$ and the thermodynamic average of the stress tensor in a film geometry. Note, that for $a_{\parallel} = a_z$ the coordinate transformation in Eq.(3.45) becomes a uniform scale transformation, which leads to $\langle \delta \mathcal{H} \rangle = 0$ in Eq.(3.47) (see Eq.(3.48)). The solution of Eq.(3.49) is then still given by Eq.(3.50), except that Δ is left *undetermined*. This is in accordance with scale invariance which fixes the power law in terms of L but *not* the *amplitude*. The film average $\langle T_{zz} \rangle$ according to Eq.(3.51) gives the *Casimir force* per unit area between the plates. Using the tracelessness condition $\langle T_{zz} \rangle = -(d-1)\langle T_{\parallel\parallel} \rangle$ (see Eq.(3.36)) and $\langle T_{\parallel z} \rangle = 0$ together with $\Delta_{O,O} = -N\pi^2/1440$ (see Eq.(3.12)) in $d = 4$ we find for a critical film with (O, O) boundary conditions

$$\langle T_{\mu\nu} \rangle = N \frac{\pi^2}{1440} \begin{pmatrix} 1 & & \\ & 1 & \mathbf{0} \\ & \mathbf{0} & 1 \\ & & & -3 \end{pmatrix} \tag{3.52}$$

in striking analogy with Eq.(3.5). The relation given by Eq.(3.51) is of great importance for the so–called *short distance expansion* of scaling operators (see Sec.4.3) close to one of the surfaces of a film [192, 203]. Moreover, Eq.(3.51) provides an alternative way for the field–theoretical calculation of the Casimir amplitudes (see below).

3.4.1 *The Casimir Amplitudes in Critical Strips*

In two dimensions the Casimir amplitudes in a critical strip can be evaluated via Eq.(3.51) without knowing the stress tensor component T_{zz} explicitly [120, 121]. Instead, the transformation law of the stress tensor under *finite* conformal mappings generated by holomorphic and antiholomorphic functions is exploited. To achieve this in a comfortable way the two–dimensional position vector $\mathbf{r} = (r_{\parallel}, z)$ is replaced with the complex coordinates $\zeta = r_{\parallel} + iz$ and $\overline{\zeta} = r_{\parallel} - iz$. The stress tensor transforms accordingly and exhibits the components $T_{\zeta\zeta}$, $T_{\zeta\overline{\zeta}}$, $T_{\overline{\zeta}\zeta}$, and $T_{\overline{\zeta}\overline{\zeta}}$. Symmetry and trace-lessness (see Eq.(3.36)) imply $T_{\zeta\overline{\zeta}} = T_{\overline{\zeta}\zeta} = 0$ and conservation (see Eq.(3.38)) implies that $T_{\zeta\zeta}$ is independent of $\overline{\zeta}$ and that $T_{\overline{\zeta}\overline{\zeta}}$ is independent of ζ inside cumulants. Therefore the notation $T_{\zeta\zeta} = T(\zeta)$ and $T_{\overline{\zeta}\overline{\zeta}} = \overline{T}(\overline{\zeta})$ is often used [39, 115, 116, 117, 199]. Specifically, we find

$$\langle T_{zz} \rangle = \Delta L^{-2} = -\langle T(\zeta) \rangle - \langle \overline{T}(\overline{\zeta}) \rangle, \tag{3.53}$$

where Eq.(3.51) has been used for $d = 2$. Note, that the definition of the stess tensor in Eq.(3.35) for $d = 2$ differs from the definition in Refs.[39, 115, 116, 117] by a factor $-(2\pi)^{-1}$. For an analytic function $\eta = \eta(\zeta)$, which represents a conformal and *finite* coordinate transformation, the thermodynamic averages of the stress tensor components $T(\zeta)$ and $\overline{T}(\overline{\zeta})$ transform according to [115, 199]

$$\langle T(\zeta) \rangle = (\eta'(\zeta))^2 \langle T(\eta) \rangle - \frac{c}{24\pi} \{\eta, \zeta\}$$

$$\langle \overline{T}(\overline{\zeta}) \rangle = \left(\overline{\eta'(\zeta)}\right)^2 \langle \overline{T}(\overline{\eta}) \rangle - \frac{c}{24\pi} \{\overline{\eta}, \overline{\zeta}\}, \tag{3.54}$$

where

$$\{\eta, \zeta\} = \left(\frac{\eta''(\zeta)}{\eta'(\zeta)}\right)' - \frac{1}{2}\left(\frac{\eta''(\zeta)}{\eta'(\zeta)}\right)^2 \tag{3.55}$$

denotes the Schwarz derivative of η with respect to ζ. The number c in Eq.(3.54) is known as the *central charge* [39, 115, 116]. The central charge c depends on the universality class of the two–dimensional system and it cannot be determined by general conformal invariance considerations. However, additional conditions like, e.g., the unitarity of the transfer matrix, pose restrictions on the possible values of c [119] (see Eqs.(2.60) and (2.61)). In view of Eq.(3.25) the transformation law for the averaged stress tensor in Eq.(3.54) states that T is *not* a primary scaling operator. Moreover, the scaling factors $(\eta'(\zeta))^2$ and $(\overline{\eta'(\zeta)})^2$, respectively, contain phase factors $e^{\pm 2i\varphi}$ indicating that the stress tensor is a spin–2 operator.

The Casimir amplitude in a critical strip can now be determined from Eqs.(3.53) and (3.54) by mapping the strip conformally onto a halfplane or the complex plane itself. Let $\zeta = r_{\parallel} + iz$ denote the complex coordinate in the strip such that $-\infty < r_{\parallel} < \infty$ and $0 \leq z \leq L$. For *periodic* boundary conditions the exponential function $\eta(\zeta) = \exp\left(\frac{2\pi}{L}\zeta\right)$ maps the strip onto the complex plane which is parameterized by

η. From $\langle T(\eta) \rangle = \langle \overline{T}(\overline{\eta}) \rangle = 0$ in the unbounded plane we conclude

$$\langle T(\zeta) \rangle = \langle \overline{T}(\overline{\zeta}) \rangle = \frac{\pi c}{12} L^{-2} \qquad (3.56)$$

and Eq.(3.53) implies [118, 120, 121]

$$\Delta_{per} = -\frac{\pi}{6} c \qquad (3.57)$$

for the Casimir amplitude Δ_{per}. Likewise, the exponential function $\eta(\zeta) = \exp\left(\frac{\pi}{L}\zeta\right)$ maps the strip with (O, O) or $(+, +)$ boundary conditions (see Sec.3.2) onto the upper half plane with a *uniform* O or $+$ boundary condition on the real axis. Again $\langle T(\eta) \rangle = \langle \overline{T}(\overline{\eta}) \rangle = 0$ in a half plane with a uniform boundary and from Eq.(3.54) we find

$$\langle T(\zeta) \rangle = \langle \overline{T}(\overline{\zeta}) \rangle = \frac{\pi c}{48} L^{-2}. \qquad (3.58)$$

Equation (3.58) implies via Eq.(3.53) that [118, 120]

$$\Delta_{O,O} = \Delta_{+,+} = -\frac{\pi}{24} c, \qquad (3.59)$$

which explicitly proves the statement that in $d = 2$ the Casimir amplitudes for equal boundary conditions must be equal. The remaining amplitudes Δ_{aper}, $\Delta_{O,+}$, and $\Delta_{+,-}$ (see Table 3.3) require a more sophisticated analysis [118]. Note, that Eqs.(3.56) and (3.58) imply that $\langle T_{\|z} \rangle = i(\langle T(\zeta) \rangle - \langle \overline{T}(\overline{\zeta}) \rangle) = 0$ in the strip.

The universality class of the system under consideration enters Eqs.(3.57) and (3.59) through the value of the central charge c. A Gaussian model ($N = 1$) in $d = 2$ corresponds to $c = 1$, which reduces the results in Eqs.(3.57) and (3.59) to the values displayed in Eq.(3.12) for $d = 2$ and $N = 1$. The critical Ising model is characterized by $c = \frac{1}{2}$ so that Eqs.(3.57) and (3.59) reproduce the corresponding entries in Table 3.3. The critical 3–state Potts model is characterized by $c = \frac{4}{5}$ giving [120, 121]

$$\Delta_{per} = -\frac{2\pi}{15} \quad , \quad \Delta_{O,O} = \Delta_{1,1} = -\frac{\pi}{30}, \qquad (3.60)$$

where $(1, 1)$ stands for equal surface Potts fields, which fix the Potts spins on the boundary to the same value. Note, that $\Delta_{p,p} = \Delta_{1,1}$ for any Potts state p. A complete list of the Casimir amplitudes for the critical 3–state Potts model in strips is given in Ref.[118]. For the tricritical Ising model ($c = \frac{7}{10}$) and the tricritical 3–state Potts model ($c = \frac{6}{7}$) Eqs.(3.57) and (3.59) finally yield the values given by Eq.(3.18).

Returning to the transformation law for the stress tensor given by Eq.(3.54) it is worth noting that the Schwarz derivative $\{\eta, \zeta\}$ (see Eq.(3.55)) vanishes, i.e., the transformation law becomes *homogeneous* (as for primary scaling operators), if and only if $\eta(\zeta)$ is a Möbius transformation (see Eq.(3.22)). Hence $\langle T(\zeta) \rangle = \langle \overline{T}(\overline{\zeta}) \rangle = 0$ for any geometry which is related to either the complex plane or a half plane with a homogeneous boundary via Eq.(3.22).

3.4.2   The Casimir Amplitudes in Critical Films

In general dimensions $d > 2$ the conformal group reduces to the small conformal group given by Eq.(3.19), which is the analogue of the Möbius transformation in $d = 2$ (see Eq.(3.22)). It is therefore impossible to map a film geometry conformally onto the unbounded space or a semiinfinite geometry, so that a d–dimensional generalization of Eq.(3.54) would not facilitate to evaluate Eq.(3.51). The only remaining option therefore is to evaluate $\langle T_{zz} \rangle$ directly using perturbation theory for, e.g., a Φ^4–Hamiltonian which in turn requires some knowledge on the *explicit* form of T_{zz}. For clearity we quote T_{zz} in unbounded space using Eqs.(3.42) and (3.44) for general N

$$T_{zz} = \frac{1}{2}\left(\frac{\partial \Phi}{\partial z}\right)^2 - \frac{1}{2}\left(\nabla_{\parallel}\Phi\right)^2 + \frac{1}{4}\frac{d-2}{d-1}\Delta_{\parallel}\Phi^2 - \frac{d-2}{2d}\Phi\Delta\Phi - \frac{g}{4!}\left(\Phi^2\right)^2 , \quad (3.61)$$

where $\nabla_{\parallel}$ and $\Delta_{\parallel}$ denote the gradient and the Laplacian, respectively, taken with respect to $\mathbf{r}_{\parallel}$. Note, that Eq.(3.61) is only correct in the sense of a first order perturbation theory in the bare coupling g. In order to evaluate Eq.(3.51) $\langle T_{zz} \rangle$ in a *film* geometry is needed. The operator representation of T_{zz} for *interior* points in films is in fact given by Eq.(3.61). However, at a *surface* the representation of T_{zz} in terms of *surface operators* differs slightly from Eq.(3.61), although $\langle T_{zz} \rangle = $ const throughout the film *including* the surfaces (see Eq.(4.38) in Sec.4.3). For the calculation of $\langle T_{zz} \rangle$ the aforementioned modification of Eq.(3.61) is irrelevant so that we conclude

$$\langle T_{zz} \rangle = \frac{1}{2}\left\langle \left(\frac{\partial \Phi}{\partial z}\right)^2 \right\rangle - \frac{1}{2}\left\langle \left(\nabla_{\parallel}\Phi\right)^2 \right\rangle - \frac{g}{4!}\left\langle \left(\Phi^2\right)^2 \right\rangle \quad (3.62)$$

in a film geometry, where T_{zz} may be located at any position in the film including the surfaces. If the $z = 0$ – surface of the film belongs to the ordinary (O) surface universality class, Eq.(3.62) can be simplified to

$$\langle T_{zz}\rangle_{O,a} = \frac{1}{2}\left\langle \left(\frac{\partial \Phi}{\partial z}\right)^2\bigg|_{z=0}\right\rangle_{O,a} = (d-1)\Delta_{O,a}L^{-d}, \qquad (3.63)$$

where Eq.(3.51) has been used and the subscripts O (ordinary) and $a = O, SB$ (surface–bulk or special) denote the boundary conditions at $z = 0$ and $z = L$, respectively (see Sec.3.2). Likewise, we have for a $z = 0$ – surface in the SB surface universality class

$$\langle T_{zz}\rangle_{SB,SB} = -\frac{1}{2}\left\langle \left(\nabla_{\parallel}\Phi\right)^2\bigg|_{z=0}\right\rangle_{SB,SB} - \frac{g}{4!}\left\langle \left(\Phi^2\right)^2\bigg|_{z=0}\right\rangle_{SB,SB} = (d-1)\Delta_{SB,SB}L^{-d}. \qquad (3.64)$$

For periodic and antiperiodic boundary conditions Eq.(3.62) does not simplify, because there are no real surfaces confining the film at $z = 0$ or $z = L$.

The relations given by Eqs.(3.62), (3.63), and (3.64) can be used to calculate the Casimir amplitudes to first order in ε (see Eq.(3.13) and Ref.[199]). This is an alternative to the procedure followed in Refs.[188, 189], where a generalization to *finite* reduced temperatures is used in order to obtain the Casimir amplitudes directly from the singular part of the free energy.

4. Wall Effects in Critical Films

The scaling operators $\Psi(\mathbf{r})$ which govern the field–theoretical description of critical phenomena in the Ginzburg–Landau Hamiltonian (see Sec.1.4) are *local* quantities. The most common examples for such scaling operators in the framework of the ϕ^4–theory are given by the *local magnetization* $\Phi(\mathbf{r})$ and the *energy density* $-\frac{1}{2}\Phi^2(\mathbf{r})$. In a spatially homogeneous and isotropic system the thermal average $\langle\Psi(\mathbf{r})\rangle$ of a single scaling operator $\Psi(\mathbf{r})$ is a constant ψ_b. Surfaces break this spatial symmetry so that the thermal average of $\Psi(\mathbf{r})$ exhibits a spatial dependence. For $\Psi(\mathbf{r}) = \Phi(\mathbf{r})$ one obtains the magnetization density profile $m(\mathbf{r}) = \langle\Phi(\mathbf{r})\rangle$, and setting $\Psi(\mathbf{r}) = -\frac{1}{2}\Phi^2(\mathbf{r})$ yields the energy density profile $e(\mathbf{r}) = -\frac{1}{2}\langle\Phi^2(\mathbf{r})\rangle$. In a semiinfinite geometry, which is bounded by a plane wall such that the z–component of the position vector $\mathbf{r} = (\mathbf{r}_\|, z)$ measures the distance from the wall, these profiles are particularly simple. The system is still homogeneous and isotropic with respect to $\mathbf{r}_\|$ and therefore any scaling density profile is only a function of z.

At a critical point of the bulk system the z–dependence of the scaling density profile $\langle\Psi(\mathbf{r})\rangle_{\infty/2}$ in a corresponding semiinfinite system is already fixed by the requirement that $\langle\Psi(\mathbf{r}_\|, z \to \infty)\rangle_{\infty/2} = \psi_b$ and by the *scaling dimension* x_Ψ of $\Psi(\mathbf{r})$. For *large* distances z from the wall we obtain [20]

$$\langle\Psi(\mathbf{r}_\|, z)\rangle_{\infty/2} - \psi_b = A_\Psi z^{-x_\Psi}, \tag{4.1}$$

where A_Ψ is a nonuniversal amplitude. The scaling dimension x_Ψ in Eq.(4.1) is given by $x_\Psi = \beta/\nu$ for the magnetization density and by $x_\Psi = (1 - \alpha)/\nu$ for the energy density. Note, that Eq.(4.1) predicts a divergence of the scaling density profile for $z \to 0$ which is due to the absence of any *microscopic* length scale in our continuum theory. In fact, the behavior of, say, the magnetization profile near the boundary of a semiinfinite Ising lattice is governed by the microscopic lattice constant, which ensures the existence of a *finite* value of the magnetization density on the boundary of the system. Likewise, in a binary fluid the molecular diameter provides the microscopic scale on which the composition profile crosses over to a *finite* value at a wall immersed into the fluid. The presence of a microscopic length scale therefore serves as a cutoff for the spurious singularity of scaling density profiles as given by Eq.(4.1) at microscopic distances from a wall.

Finally, we note that in the *vicinity* of the critical point the power law in Eq.(4.1) is replaced by the scaling law (see, e.g., Ref.[24])

$$\langle \Psi(\mathbf{r}_\|, z) \rangle_{\infty/2} - \psi_b = A_\Psi z^{-x_\Psi} f_\pm(z/\xi_\pm) \tag{4.2}$$

which again only holds for macroscopic distances z from the wall. Here $f_+(y_+)$ denotes a *universal scaling function* for $T > T_{c,b}$ and $f_-(y_-)$ denotes a corresponding scaling function for $T < T_{c,b}$ with the property $f_\pm(0) = 1$. For short–ranged interactions the macroscopic distance z from the wall in the scaling argument $y_\pm = z/\xi_\pm$ is measured in units of the *bulk correlation length* $\xi_\pm = \xi_0^\pm |t|^{-\nu}$ above or below $T_{c,b}$, respectively (for details see Eqs.(5.10) and (5.11)). Note, that the property $f_\pm(0) = 1$ of the scaling function ensures that Eq.(4.1) is the limit of Eq.(4.2) for $T \to T_{c,b}$. However, the existence of a *finite* bulk correlation length does not cure the spurious singularity of the scaling density profile as given by Eq.(4.2) for $z \to 0$.

There are some examples known in the literature for scaling density profiles in semiinfinite geometries in $d = 4 - \varepsilon$ which have been obtained by field–theoretical renormalization group methods. The energy density profile $e(z)$ has been discussed in Ref.[24] for $T \geq T_{c,b}$ and the ordinary surface universality class (see also Ref.[20]). The order parameter profile $m(z)$ in a semiinfinite geometry is of special interest in the context of critical adsorption [32, 204]. The full profile $m(z)$ above and below bulk criticality (see Eq.(4.2)) has recently been calculated for the extraordinary transition, i.e., in presence of a surface field h_1 within the Ising universality class [22]. For the following considerations we restrict ourselves to the case $T = T_{c,b}$, where in zero external bulk field the constant ψ_b in Eq.(4.1) can be taken to be zero for the local magnetization Φ and the energy density $-\frac{1}{2}\Phi^2$.

4.1   *Distant Wall Corrections*

In a film geometry a second wall is present at $z = L$, where L is the thickness of the film. Looking at scaling density profiles near the first wall, i.e., at $z \ll L$ the presence of the the second wall at $z = L$ should be visible as a small perturbation of the power law given by Eq.(4.1). In the following we will refer to this small perturbation as the *distant wall correction*. Sometimes it is also called the *critical wall perturbation* in the literature [205, 206].

Let, for example, Φ represent the composition of a binary liquid mixture close to its critical demixing point $(T = T_c, \Phi = \Phi_c)$ [205]. A plate which is immersed into the mixture will generally show a preferential affinity for one of the components. *At* the demixing point the local average composition $m_{\infty/2}(z) = \langle \Phi(\mathbf{r}) \rangle - \Phi_c$ at a distance z from the wall will then deviate from zero according to Eq.(4.1) with the exponent $x_\Phi = \beta/\nu$. In the magnetic language of the Ising universality class the plate is characterized by a *surface field* h_1 which induces either $m(0) > 0$ or $m(0) < 0$, i.e., which prefers either one component or the other depending on the sign of h_1 [205]. The *distant wall correction* imposed on the composition profile $m_{\infty/2}(z)$ by the presence of a *second* plate immersed into the critical binary liquid mixture at a distance L from the first plate can be expressed by the corresponding film profile $m_{film}(z)$ at small distances z [203]

$$m_{film}(z \ll L) = m_{\infty/2}(z) \left(1 + B \left(\frac{z}{L} \right)^{d^*} + \dots \right), \qquad (4.3)$$

where $m_{\infty/2}(z)$ is given by Eq.(4.1) and $d^* = 3$ in three dimensions. The value of the constant B depends on the boundary conditions at both the near and the far wall, where the latter can be characterized by a second surface field [205]. Scaling theory does not reveal the form of the correction displayed in Eq.(4.3). The derivation of Eq.(4.3) presented in Ref.[203] (see also Ref.[205]) was based on the postulate that the *local free energy functional*

$$A(L, T = T_c, \Phi = \Phi_c; m_0, m_0') = A_0 \int_0^L |m|^{\delta+1} \left(1 + \left| \frac{\xi(m)}{m} \frac{dm}{dz} \right|^p \right) dz \qquad (4.4)$$

expresses the *total free energy* of the film *at* criticality in terms of the boundary conditions $m_0 = m(0)$, $m_0' = m(L)$, the exponent δ of the critical isotherm, and the local correlation length $\xi(m) = \xi_0 |m_\infty/m(z)|^{\nu/\beta}$ [205]. The exponent p which is left undetermined in Ref.[203] must obey certain restrictions in order to avoid unphysical singularities of $m(z)$ [205]. However, upon minimalization of the free energy functional in Eq.(4.4) the distant wall correction given by Eq.(4.3) with $d^* = (2-\alpha)/\nu$ is obtained *irrespective* of the value of p [203, 205]. If hyperscaling holds then the exponent d^* of the distant wall correction coincides with the *spatial dimension* d.

To this end Eq.(4.3) together with $d^* = d$ for $2 \leq d \leq 4$ is still only a conjecture about the structure of the distant wall correction, because Eq.(4.4) has been *postulated* in Ref.[203]. One should therefore look for confirmations of Eq.(4.3) by independent methods. An explicit calculation of the magnetization profile in an Ising model strip in two dimensions in fact confirms Eq.(4.3) for $d^* = d = 2$ [64]. In a first set of boundary conditions discussed in Ref.[64] a surface field h_1 acts on the spins in the near edge and the spins in the far edge are free and in a second set of boundary conditions *two* surface fields h_1 and h_2 act on the spins in the two edges, respectively. The exponent d^* of the distant wall correction has been examined within the framework of the field–theoretical renormalization group to first order in $\varepsilon = 4 - d$, which confirms $d^* = d$ for distant wall corrections to the order parameter profile $\mathbf{m}(z)$ *along* the direction of $\mathbf{m}(z)$ in the $O(N)$-symmetric case [206]. Corrections transverse to $\mathbf{m}(z)$ are governed by an exponent $d_\perp^* = d - 1$ [206].

4.1.1   *Distant Wall Corrections and Conformal Invariance*

In two dimensions a semiinfinite geometry with a homogeneous or uniform boundary condition can be mapped onto a strip with equal boundary conditions by means of the *conformal* transformation $\zeta = \frac{L}{\pi} \ln \eta$, where $\mathrm{Im}\,\eta \geq 0$ (upper half plane) and $0 \leq \mathrm{Im}\,\zeta \leq L$. The principle of *conformal invariance* (see Sec.3.3) establishes a generalized scaling relation among cumulants of primary scaling operators in these two geometries (see Eq.(3.25)). For a scalar primary scaling operator $\Psi(\zeta,\bar\zeta)$, $\zeta = r_\parallel + iz$ in $d = 2$ Eq.(3.25) can be written as [39, 115, 116, 127]

$$\langle \Psi(\zeta,\bar\zeta)\rangle_{strip} = |\eta'(\zeta)|^{x_\Psi} \langle \Psi(\eta,\bar\eta)\rangle_{\infty/2}, \tag{4.5}$$

where the scaling exponent $x_\Psi = \frac{1}{8}$ for the order parameter and $x_\Psi = 1$ for the energy density in the Ising universality class. The scaling density profiles in the upper half plane with a homogeneous boundary condition along the real axis can be written as $\langle \Psi(\eta,\bar\eta)\rangle_{\infty/2} = A_\Psi[(\eta - \bar\eta)/2i]^{-x_\Psi}$ which follows from pure scaling arguments. The amplitude A_Ψ is nonuniversal and depends on the scaling operator Ψ. From Eq.(4.5) the corresponding scaling density profile in a strip with *equal* boundary conditions can be inferred as [127]

$$\langle \Psi(\zeta,\bar\zeta)\rangle_{strip} = A_\Psi \left[\frac{L}{\pi} \sin \frac{\pi z}{L} \right]^{-x_\Psi}, \tag{4.6}$$

where $z = (\zeta - \bar\zeta)/2i$. For $z \ll L$ the distant wall correction can be read off from an expansion of Eq.(4.6) in powers of z/L and we obtain

$$\langle \Psi(\zeta,\bar\zeta)\rangle_{strip} = \langle \Psi(\zeta,\bar\zeta)\rangle_{\infty/2} \left(1 + \frac{\pi^2}{6} x_\Psi \left(\frac{z}{L}\right)^2 + \dots \right). \tag{4.7}$$

Obviously Eq.(4.7) confirms Eq.(4.3) for *any* scaling operator Ψ in $d = 2$ for strips with *equal* boundary conditions. Furthermore, Eq.(4.7) allows the remarkable conclusion, that the distant wall correction for a given scaling operator Ψ in strips with *equal* boundary conditions is *independent* of the corresponding surface universality class. Note, that an analogous statement has been made for the *Casimir amplitudes* in $d = 2$ (see Eq.(3.59)).

Strips with *different* boundary conditions a and b along the two edges, respectively, can be represented as conformal images of the upper complex half plane with a boundary condition of type a along the positive real axis and a boundary condition of type b along the negative real axis. The scaling density profile of an operator Ψ in such a half plane reads [123, 128, 129]

$$\langle \Psi(\eta, \overline{\eta}) \rangle_{a,b,\infty/2} = \left[\frac{\eta - \overline{\eta}}{2i} \right]^{-x_\Psi} F_{ab}^\Psi \left(\frac{\eta + \overline{\eta}}{2\sqrt{\eta\overline{\eta}}} \right), \qquad (4.8)$$

where the function F_{ab}^Ψ follows from a generalized conformal Ward identity which explicitly depends on the combination (a, b) of the boundary conditions [128, 129]. In terms of the ordinary (O) and the extraordinary $(+$ or $-)$ surface universality classes for an Ising system (see Sec.1.2) the boundary conditions a and $b \neq a$ can be combined in two independent ways, namely $(a, b) = (+, -)$ or $(+, O)$. From Eqs.(4.5) and (4.8) one finds for the order parameter profile $m_{a,b}(z)$ in an Ising strip [123, 128, 129]

$$m_{+,-}(z) = A_m \left(\frac{L}{\pi} \sin \frac{\pi z}{L} \right)^{-1/8} \cos \frac{\pi z}{L},$$

$$m_{+,O}(z) = A_m \left(\frac{L}{\pi} \sin \frac{\pi z}{L} \right)^{-1/8} \left(\cos \frac{\pi z}{2L} \right)^{1/2} \qquad (4.9)$$

which immediately confirms the form of the distant wall correction given by Eq.(4.3) in $d = 2$. A corresponding analysis can be performed for the energy density profiles $e_{a,b}(z)$ giving [128, 129]

$$e_{+,-}(z) = A_e \left(\frac{L}{\pi} \sin \frac{\pi z}{L} \right)^{-1} \left(1 - 4 \sin^2 \frac{\pi z}{L} \right),$$

$$e_{+,O}(z) = A_e \left(\frac{L}{\pi} \sin \frac{\pi z}{L} \right)^{-1} \cos \frac{\pi z}{L} \qquad (4.10)$$

which again confirms Eq.(4.3) in $d = 2$, but for a different scaling operator. A third set of profiles is provided by the Potts magnetization in a two–dimensional strip for various boundary conditions which confirms Eq.(4.3) as well [128, 129]. The striking number of examples, which show the conjectured distant wall correction, of course asks for a general foundation of Eq.(4.3). In $d = 2$ the framework of conformal field theory provides the key to the general structure of the distant wall correction. Writing the scaling density profile $\langle \Psi(\zeta, \bar{\zeta}) \rangle_{a,b}$ for $z = (\zeta - \bar{\zeta})/2i \ll L$ and $d = 2$ in the form

$$\langle \Psi(\zeta, \bar{\zeta}) \rangle_{a,b} = A_\Psi z^{-x_\Psi} \left(1 + B_{a,b}^\Psi \left(\frac{z}{L} \right)^d + \ldots \right) \tag{4.11}$$

the coefficient $B_{a,b}^\Psi$ turns out to be a simple combination of the *scaling exponent* x_Ψ, the *central charge* c, and the *Casimir amplitude* $\Delta_{a,b}$ which reads [128, 129]

$$B_{a,b}^\Psi = -4\pi \frac{x_\Psi}{c} \Delta_{a,b}. \tag{4.12}$$

For the Ising universality class one has $c = \frac{1}{2}$ and the Casimir amplitudes are given by Table 3.3. According to Eq.(4.12) the Ψ–dependence of $B_{a,b}^\Psi$ is absorbed into x_Ψ, and the dependence on the boundary conditions is absorbed into $\Delta_{a,b}$. The remaining factor in $B_{a,b}^\Psi$ only depends on the universality class of the bulk system, i.e., it is *hyperuniversal*. Note, that the structure of the Ψ–dependence of $B_{a,b}^\Psi$ is already indicated by Eq.(4.7) which only holds for equal boundary conditions. Moreover, a general argument for Eq.(4.11) can be found if Eq.(4.11) is interpreted as a film average over the *short distance expansion* for the scaling operator Ψ of the form [192]

$$\Psi(\mathbf{r}_\parallel, z) = A_\Psi z^{-x_\Psi} \left(1 + b^\Psi T_{zz}(\mathbf{r}_\parallel, 0) z^d + \ldots \right) \tag{4.13}$$

for $d = 2$, where the stress tensor component T_{zz} (see Eqs.(3.42) and (3.44) in Sec.3.3) is evaluated *at* the near wall. The scaling dimension of the stress tensor is given by the spatial dimension d so that the coefficient b^Ψ is dimensionless. Note, that in principle $\Psi(\mathbf{r}_\parallel, 0)$ itself is a candidate for a surface operator to occur in the short distance expansion in Eq.(4.13) *before* T_{zz}, because the scaling dimension of the local magnetization or the energy density is less than d. It may also happen that other operators with scaling dimensions close to d like $\frac{g}{4!} (\Phi^2)^2$ in $d = 4 - \varepsilon$ show up in the short distance expansion of the order parameter or the energy density. However, for the *extraordinary* surface universality class there is evidence that these additional surface operators do not contribute in $d = 4 - \varepsilon$ [20] and $d = 2$ [118]. Short distance expansions like Eq.(4.13) give access to Eq.(4.11) and, moreover, to Eq.(4.3) in general dimension d including the coefficient $B_{a,b}^\Psi$. In fact, a statement on $B_{a,b}^\Psi$ in *general* dimension d has already been made in Ref.[192], namely

$$B_{a,b}^{\Psi} = -\frac{2^{d-1}d\pi^{d/2}}{\Gamma(d/2)}\frac{x_{\Psi}}{c}\Delta_{a,b},\qquad(4.14)$$

where c is given by the amplitude of the stress–tensor stress–tensor correlation function in *unbounded* d–dimensional space and therefore provides a possible generalization of the central charge to arbitrary dimension [39, 192]. Again Eq.(4.14) states
that the dependence of $B_{a,b}^{\Psi}$ on Ψ and the boundary conditions factorizes into x_{Ψ} and
$\Delta_{a,b}$, where the remaining factor is *hyperuniversal*, i.e., independent of the surface
universality classes a or b (see Eq.(4.12)). Unfortunately, there are no scaling density
profiles currently available for $d > 2$ in order to check Eq.(4.14) for the extraordinary
surface universality class. For the ordinary (O) and surface–bulk (SB) surface universality classes one must resort to *energy density* profiles because the order parameter
profile vanishes identically at bulk criticality in these cases [207]. Therefore Eqs.(4.13)
and (4.14) must be reconsidered for the energy density scaling operator in a film geometry with symmetry conserving boundary conditions of the type $(a, b) = (O, O)$,
(O, SB), (SB, O), and (SB, SB). As a first step the corresponding energy density
profiles in $d = 4 - \varepsilon$ will be discussed [207]. Moreover, the identification of $B_{a,b}^{\Psi}$ and
c in Eq.(4.14) yields valuable informations for the construction of local free energy
functionals from which the form of the distant wall corrections have first been inferred
[66, 67, 203, 205].

Finally, we would like to remark that distant wall corrections do not only occur in
the vicinity of bulk critical points. The distant wall correction to the magnetization
profile of an Ising system in a film geometry with *opposing* surface fields $h_1 = -h_2$
at the temperature T_w of a *critical wetting transition* can be much more pronounced
than at the bulk critical temperature $T_{c,b}$ [208] (see Eq.(2.56) in Sec.2.3).

4.2   The Energy Density Profiles

In the vicinity of a bulk critical point the *internal energy* $\mathcal{U}(T, L)$ of a film measured
in units of $k_B T_{c,b}$ and per area A can be decomposed into a *regular* part $u^{reg}(t, L)$
and a *renormalized* part $u^R(t, L)$ according to

$$\lim_{A\to\infty}\frac{\mathcal{U}(T, L)}{k_B T_{c,b} A} = u^R(t, L) + u^{reg}(t, L)\qquad(4.15)$$

(see Eq.(3.10)), where $u^R(t, L)$ contains the critical singularities of $\mathcal{U}(T, L)$ and fulfills a certain normalization condition (see below). Using the thermodynamical relation $\mathcal{U} = -T^2 \frac{\partial}{\partial T}(\mathcal{F}/T)$ between the internal energy $\mathcal{U}$ and the free energy $\mathcal{F}$, the renormalized part u^R of the internal energy can be related to the *renormalized free energy* f^R via $u^R = -\frac{\partial}{\partial t} f^R \equiv -f_t^R$ (see Eqs.(5.1) and (5.8)) so that u^R contains the *leading singular behavior* of the internal energy near bulk criticality. The field–theoretical renormalization procedure which gives access to f^R will be described in detail in Sec.5.1 and we therefore only give a brief outline here.

The relation $u^R = -f_t^R$ identifies $u^R(t, L)$ with the *renormalized energy density* of the Ginzburg–Landau Hamiltonian for a critical film (see Eq.(2.48)). At a certain reference reduced temperature $t_1 \neq 0$, which can be chosen as $t_1 = \mathrm{sgn}t$, $u^R(t, L)$ obeys the *normalization* conditions $u^R(t_1, L) = 0$ and $u_t^R(t_1, L) = 0$ (see Refs.[20, 24]). Therefore $u^R(t, L)$ has the representation (see Eq.(5.8))

$$u^R(t, L) = u(t, L) - u(\mathrm{sgn}t, L) - u_t(\mathrm{sgn}t, L)(t - \mathrm{sgn}t), \qquad (4.16)$$

where $u(t, L)$ is the *unnormalized singular energy density* of the film, which in turn is related to the *unnormalized singular free energy* f (see Eq.(5.7)) by $u = -f_t$. The function $u(t, L)$ still displays the same singular behavior as $u^R(t, L)$, but it does *not* fulfill the normalization conditions. Moreover, $u^R(t, L)$ has a finite limit as the spatial dimension d approaches the upper critical dimension, whereas $u(t, L)$ may exhibit singularities with respect to d in that limit (see Eq.(5.16)).

We now define an *energy density profile* $e(t, z, L)$ in terms of the renormalized energy density scaling operator $-\frac{1}{2}\Phi^{2R}(\mathbf{r}_{\parallel}, z)$ of the Ginzburg–Landau Hamiltonian in a film by

$$e(t, z, L) \equiv -\frac{1}{2}\langle\Phi^{2R}(\mathbf{r}_{\parallel}, z)\rangle. \qquad (4.17)$$

Note, that $\langle\Phi^{2R}(\mathbf{r}_{\parallel}, z)\rangle$ is only a function of z, because the film is homogeneous and isotropic with respect to $\mathbf{r}_{\parallel}$. The *renormalized energy density* $u^R(t, L)$ then has a representation of the form

$$u^R(t, L) = \int_0^L e^R(t, z, L)dz, \qquad (4.18)$$

where the *renormalized energy density profile* $e^R(t, z, L)$ can be related to the energy density profile $e(t, z, L)$ by imposing the normalization conditions for $u^R(t, L)$ (see Eq.(4.16)) directly on the profile $e^R(t, z, L)$ according to

$$e^R(t, z, L) = e(t, z, L) - e(\mathrm{sgn}t, z, L) - e_t(\mathrm{sgn}t, z, L)(t - \mathrm{sgn}t) \qquad (4.19)$$

(see Refs.[24] and [207]). In fact, $e^R(t, z, L)$ has a *finite* limit as $z \to 0$ or $z \to L$ at given L and t. This is a consequence of the field–theoretical analysis performed in Ref.[24], where the renormalized energy density profile $e^R_{\infty/2}(t, z)$ of a semiinfinite system has been calculated. The limit $z \to 0$ in Eq.(4.19) first means that z becomes very small compared to L but remains large compared to any *microscopic* length. In this case the renormalized energy density profile $e^R(t, z, L)$ in the film approaches its counterpart $e^R_{\infty/2}(t, z)$ in a semiinfinite geometry. As z is further reduced to zero $e^R_{\infty/2}(t, z)$ remains *finite* [24]. By the same reasoning $e^R(t, z, L)$ approaches a finite value as $z \to L$, because in this case $z' = L - z$ becomes small compared to L. From the short discussion after Eq.(4.1) one may come to the conclusion that $e^R(t, z, L)$ (see Eq.(4.19)) contains a microscopic length scale which manages the crossover to a finite value as $z \to 0$ or $z \to L$. This is in fact the case, because instead of the reduced temperature t the bulk correlation length $\xi_\pm(t) = \xi_0^\pm |t|^{-\nu}$ can serve as the third argument in $e(t, z, L) \equiv \tilde{e}(\xi_\pm, z, L)$. The corresponding renormalized energy density profile $\tilde{e}^R(\xi_\pm, z, L)$ contains the same subtractions as in Eq.(4.19) which depend on the *microscopic* correlation length *amplitude* $\xi_0^\pm = \xi_\pm(t_1 = \mathrm{sgn}t)$ according to

$$\tilde{e}^R(\xi_\pm, z, L) = \tilde{e}(\xi_\pm, z, L) - \tilde{e}(\xi_0^\pm, z, L) + \nu \xi_0^\pm \tilde{e}_{\xi_\pm}(\xi_0^\pm, z, L)\left((\xi_\pm/\xi_0^\pm)^{-1/\nu} - 1\right). \quad (4.20)$$

The desired microscopic scale is therefore set by $\xi_0^\pm$. The profile $e(t, z, L) = \tilde{e}(\xi_\pm, z, L)$, however, depends only on the macroscopic lengths indicated in the argument list. In the limit $z \to 0$, which again means $z \ll L$, $e(t, z, L)$ approaches its counterpart in a semiinfinite geometry (see Ref.[24]) which is of the form indicated by Eq.(4.2). Therefore $e(t, z, L)$ *diverges* like $z^{-(1-\alpha)/\nu}$ as $z \to 0$ and like $(L - z)^{-(1-\alpha)/\nu}$ as $z \to L$. Comparing Eqs.(4.16), (4.18), and (4.19) it is tempting to interpret $e(t, z, L)$ as the energy density profile for the unnormalized singular energy density $u(t, L)$. However, the exponent $(1 - \alpha)/\nu$ is greater than one and therefore the integral $\int_0^L e(t, z, L)dz$ (see Eq.(4.18)) does not exist due to the singularities of $e(t, z, L)$ at $z = 0$ and $z = L$ which require a microscopic cutoff.

In order to study distant wall corrections to the energy density profile *at* bulk criticality ($t = 0$) near the wall at $z = 0$ the distance z from this wall must be small compared to the film thickness L which is a *macroscopic* length. Now $z \ll L$ according to the above discussion within continuum theory means that z is still much larger than any *microscopic* length scale. Although $z \ll L$, we are then actually far away from the limit $z \to 0$ in the energy density profiles $e(t = 0, z, L)$ and $e^R(t = 0, z, L)$, so that their different behavior for $z \to 0$ is irrelevant for the investigation of the distant wall correction. It is therefore sufficient to calculate $e(t = 0, z, L)$ according to Eq.(4.17) within the framework of the field–theoretical renormalization group in order to obtain a test of Eq.(4.3) for the energy density profiles in $d = 4 - \varepsilon$. Finally, we note that the profiles $e(t, z, L)$ and $e^R(t, z, L)$ additionally depend on the combination (a, b) of the boundary conditions at the surfaces $(z = 0, z = L)$ of the film. We will keep track of this dependence by assigning the subscript a, b to the energy density profiles in the following.

4.2.1   The Scaling Functions

The energy density profiles $e_{a,b}(t, z, L)$ can equivalently be written as $\tilde{e}_{a,b}(\xi_\pm, z, L)$, where the bulk correlation length $\xi_\pm = \xi_0^\pm |t|^{-\nu}$ replaces the reduced temperature t (see, e.g., Eq.(4.20)). Scaling arguments then lead to the representation

$$e_{a,b}(t, z, L) = L^{-d} \left(\frac{L}{\xi_0^+} \right)^{1/\nu} f_{\pm a,b} \left(\frac{z}{\xi_\pm}, \frac{z}{L} \right), \tag{4.21}$$

where $f_{\pm a,b}(z_\pm, x)$ is a *universal* scaling function of the two scaling variables $z_\pm = z/\xi_\pm$ and $x = z/L$, i.e., the correlation length amplitude ξ_0^+ absorbs all nonuniversal aspects [209]. Other ratios of the three lengths z, L, and $\xi_\pm$ can equally well be chosen as scaling arguments for $f_{\pm a,b}$. We will be exclusively concerned with the case $T = T_{c,b}$ which eliminates the correlation length $\xi_\pm$ from the argument list. The energy density profile $e_{a,b}(0, z, L)$ has the scaling form [207]

$$e_{a,b}(0, z, L) = L^{-d} \left(\frac{L}{\xi_0^+} \right)^{1/\nu} A_{a,b} \, g_{a,b} \left(\frac{z}{L} \right), \tag{4.22}$$

where $A_{a,b} \, g_{a,b}(x) = f_{\pm a,b}(0, x)$ is again a universal scaling function. The amplitudes $A_{a,b}$ have only been introduced for convenience (see below). Note, that the scaling dimension of the energy density $(1 - \alpha)/\nu = d - 1/\nu$ (see Eq.(1.6)) appears as the exponent of L in Eq.(4.22). For $z/L \to 0$ the film profile $e_{a,b}(t, z, L)$ approaches the energy density profile

$$e_{\infty/2,a}(t = 0, z) = A_a \, z^{-d} \left(\frac{z}{\xi_0^+} \right)^{1/\nu} \tag{4.23}$$

in a semiinfinite geometry with a uniform boundary condition of type a (see Eq.(4.1)). The scaling function $g_{a,b}(x)$ therefore displays the power law behavior

$$g_{a,b}(x) = \frac{A_a}{A_{a,b}} \, x^{-(1-\alpha)/\nu} \tag{4.24}$$

for $x \to 0$. As already indicated in Sec.4.1 $e_{a,b}(0,z,L)$ is only available for symmetry conserving boundary conditions which can be represented by the combinations $(a,b) = (O,O)$, (O,SB), and (SB,SB) of the ordinary and the surface–bulk surface universality classes [207]. The combination $(a,b) = (SB,O)$ is captured by the simple relation $e_{b,a}(t,z,L) = e_{a,b}(t,L-z,L)$. In order to complete the set of symmetry conserving boundary conditions we quote the results for periodic (per) and antiperiodic (aper) boundary conditions as well. In particular, if the amplitudes $A_{a,b}$ for an $O(N)$–symmetric Ginzburg–Landau Hamiltonian are chosen according to [207, 209]

$$A_{O,O} = \frac{N}{2} \frac{\Gamma\left(\frac{d}{2}-1\right)}{2^d \pi^{d/2}} \left(1 - \varepsilon \frac{N+2}{N+8}\gamma\right),$$

$$A_{O,SB} = A_{SB,O} = A_{O,O}\left(1 + \frac{\varepsilon}{2}\frac{N+2}{N+8}\right),$$

$$A_{SB,SB} = -A_{O,O}\left(1 + \varepsilon\frac{N+2}{N+8}\right),$$

$$A_{per} = -N\frac{\Gamma\left(\frac{d}{2}-1\right)}{4\pi^{d/2}}\zeta(2)\left\{1 - \varepsilon\left[\frac{\zeta'(2)}{\zeta(2)} - \frac{N+2}{N+8}\left(\frac{1}{2}+\ln(4\pi)-\gamma\right)\right]\right\},$$

$$A_{aper} = N\frac{\Gamma\left(\frac{d}{2}-1\right)}{4\pi^{d/2}}\eta(2)\left\{1 - \varepsilon\left[\frac{\eta'(2)}{\eta(2)} - \frac{N+2}{N+8}\left(\frac{1}{2}+\ln\pi-\gamma\right)\right]\right\} \tag{4.25}$$

then the scaling functions $g_{a,b}(x)$ are given by [209]

$$g_{O,O}(x) = \left(\frac{\pi^2}{\sin^2\pi x} - \frac{\pi^2}{3}\right)\left(1 + \varepsilon\frac{N+2}{N+8}\ln\frac{\pi}{\sin\pi x}\right)$$

$$- \varepsilon\left(\zeta^{(1)}(2,x) + \zeta^{(1)}(2,1-x) - 2\zeta'(2)\right) - \varepsilon\frac{N+2}{N+8}\frac{\pi^2}{6},$$

$$g_{O,SB}(x) = \frac{\pi^2\cos\pi x}{\sin^2\pi x}\left(1 + \varepsilon\frac{N+2}{N+8}\ln\frac{\pi}{\sin\pi x}\right) + \frac{\pi^2}{6}\left(1 + \varepsilon\frac{N+2}{N+8}\ln\left[\frac{\pi}{2}\cot\frac{\pi}{2}x\right]\right)$$

$$- \ \varepsilon \left(\eta^{(1)}(2, x) - \eta^{(1)}(2, 1 - x) + 2\eta'(2) \right) - \frac{\varepsilon}{2} \frac{N + 2}{N + 8} \frac{\pi^2}{\sin^2 \pi x},$$

$$g_{SB,O}(x) \ = \ g_{O,SB}(1 - x),$$

$$g_{SB,SB}(x) \ = \ g_{O,O}(x) + 4\zeta(2) \left\{ 1 - \varepsilon \left[\frac{\zeta'(2)}{\zeta(2)} + \frac{N + 2}{N + 8} \left(3 - \ln(2\pi) \right) \right] \right\},$$

$$g_{per}(x) \ = \ g_{aper}(x) = 1 \tag{4.26}$$

to two–loop order. In Eq.(4.26) $\zeta^{(1)}(\lambda, x)$ and $\eta^{(1)}(\lambda, x)$ denote the first derivative of the Hurwitz function $\zeta(\lambda, x)$ and the bivariate eta function $\eta(\lambda, x)$, respectively, with respect to λ [210]. The Riemann zeta function $\zeta(\lambda)$ and the eta function $\eta(\lambda)$ can be written as $\zeta(\lambda) = \zeta(\lambda, 1)$ and $\eta(\lambda) = \eta(\lambda, 1)$, where $\zeta'(\lambda) = \zeta^{(1)}(\lambda, 1)$ and $\eta'(\lambda) = \eta^{(1)}(\lambda, 1)$. The scaling functions $g_{a,b}(x)$ in Eq.(4.26) in fact behave as required by Eq.(4.24) for $x \to 0$. We find [209]

$$g_{O,O}(x \to 0) = x^{-(1-\alpha)/\nu}, \ g_{O,SB}(x \to 0) = \left(1 - \frac{\varepsilon}{2} \frac{N + 2}{N + 8} \right) x^{-(1-\alpha)/\nu},$$

$$g_{SB,SB}(x \to 0) = x^{-(1-\alpha)/\nu}, \ g_{SB,O}(x \to 0) = - \left(1 + \frac{\varepsilon}{2} \frac{N + 2}{N + 8} \right) x^{-(1-\alpha)/\nu} \tag{4.27}$$

in partly exponentiated form and therefore the amplitudes $A_{a,b}$ in Eq.(4.25) are related to the amplitudes A_a and A_b of the energy density profiles in critical semiinfinite systems (see Eq.(4.23)) via

$$A_O = A_{O,O} \ , \ A_O = A_{O,SB} \left(1 - \frac{\varepsilon}{2} \frac{N + 2}{N + 8} \right),$$

$$A_{SB} = A_{SB,SB} \ , \ A_{SB} = -A_{O,SB} \left(1 + \frac{\varepsilon}{2} \frac{N + 2}{N + 8} \right), \tag{4.28}$$

where the relation between A_O and $A_{O,SB}$ is equivalent to the relation between $A_{O,O}$ and $A_{O,SB}$ in Eq.(4.25) to order ε.

Due to the logarithmic contributions to the ε–expansion of the scaling functions $g_{a,b}(x)$ the first–order ε–expansion results for $g_{a,b}(x)$ given in Eq.(4.26) grossly misrepresent the shape of the true scaling functions for $x \to 0$ and $x \to 1$. In order to obtain a reasonable representation of the energy density profiles in three dimensions the scaling functions $g_{a,b}(x)$ are written in an exponentiated form which reproduces

Eq.(4.26) when reexpanded to first order in ε and gives the correct limiting behavior for $x \to 0$ (see Eq.(4.27)) and $x \to 1$. We obtain [207, 209]

$$g_{0,0}(x) = [\zeta(d-2,x) + \zeta(d-2,1-x) - 2\zeta(d-2)]$$

$$\times \left[\left(\frac{\pi}{\sin \pi x}\right)^{2-1/\nu} - \frac{(2-1/\nu)\,\zeta(2)}{\zeta(1/\nu,x) + \zeta(1/\nu,1-x) - 2\zeta(1/\nu)}\right]$$

$$g_{0,SB}(x) = [\eta(d-2,x) - \eta(d-2,1-x)]\left(\frac{\pi}{\sin \pi x}\right)^{2-1/\nu}$$

$$+\; 2\eta(d-2)\left(\frac{\pi}{2}\cot\frac{\pi}{2}x\right)^{2-1/\nu} - \frac{2\nu-1}{d\nu-1}\left(\frac{\pi}{\sin \pi x}\right)^{d-1/\nu}$$

$$g_{SB,SB}(x) = g_{0,0}(x) + 4\zeta(2)\left\{1 - \varepsilon\left[\frac{\zeta'(2)}{\zeta(2)} + \frac{N+2}{N+8}(3-\ln(2\pi))\right]\right\}. \quad (4.29)$$

The additive constant in $g_{SB,SB}(x)$ has not been exponentiated. Note, that the Hurwitz function $\zeta(d-2,x)$ does not exist for $d = 3$. However, the expression $\zeta(d-2,x)+\zeta(d-2,1-x)-2\zeta(d-2)$ in Eq.(4.29) has the finite limit $-\psi(x)-\psi(1-x)-2\gamma$ as $d \to 3$, where $\psi(x) = \Gamma'(x)/\Gamma(x)$ is the Digamma function and $\gamma = -\psi(1)$ is Euler's constant [210]. However, the bivariate eta function $\eta(1,x)$ does exist. There are several possibilities to exponentiate an ε-expansion consistently to a certain order in ε. To illustrate this we give a second version of an exponentiated expression for the scaling function $g_{0,0}(x)$ [209]:

$$\tilde{g}_{0,0}(x) = [\zeta(d-2,x) + \zeta(d-2,1-x) - 2\zeta(d-2)]$$

$$\times \left[\zeta\left(2-\tfrac{1}{\nu},x\right) + \zeta\left(2-\tfrac{1}{\nu},1-x\right) - 2\zeta\left(2-\tfrac{1}{\nu}\right) - \frac{\left(2-\tfrac{1}{\nu}\right)\left(3-\tfrac{1}{\nu}\right)\zeta\left(4-\tfrac{1}{\nu}\right)}{\zeta(2,x) + \zeta(2,1-x) - 2\zeta(2)}\right].$$

$$(4.30)$$

The scaling functions $g_{0,0}$ according to Eq.(4.29) and $\tilde{g}_{0,0}$ according to Eq.(4.30) differ most at $x \simeq \frac{1}{2}$, where the mutual deviation is 1%. Thus from both the analytical and numerical point of view $g_{0,0}(x)$ and $\tilde{g}_{0,0}(x)$ are equally well suited for our purposes.

The expressions in Eq.(4.29) give access to the scaling functions for the energy density profiles $e_{a,b}(0,z,L)$ in a critical film in $d = 3$ and $d = 4$. A comparison between $g_{a,b}(x)$ in $d = 3,4$ and a corresponding scaling function in $d = 2$ would now be desirable. The only scaling function, which exists in $d = 2$ for $O(N)$-symmetry with $N \geq 1$

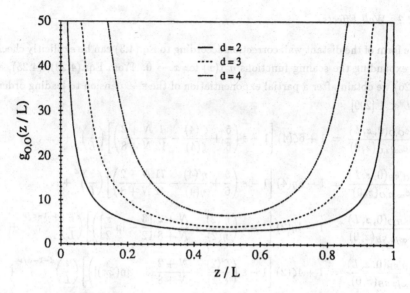

Fig. 4.1:   Scaling functions $g_{O,O}(z/L) = e_{O,O}(0, z, L)/e_{\infty/2,O}(0, L)$ for $d = 2$ (solid line), $d = 3$ (dashed line), and $d = 4$ (dash–dotted line). Numerical evaluations of Eq.(4.31) ($d = 2$) and Eq.(4.29) for $d = 3$ and $d = 4$ are displayed, respectively.

and without surface fields is $g_{O,O}(x)$ for $N = 1$ (Ising universality class). The SB surface universality class has no analogue in this case. From Eqs.(4.22), (4.23), and the identity $A_O = A_{O,O}$ in Eq.(4.28) we have $g_{O,O}(z/L) = e_{O,O}(0, z, L)/e_{\infty/2,O}(0, L)$ and therefore (see Eq.(4.6) and Ref.[127])

$$g_{O,O}(x) = \frac{\pi}{\sin \pi x} \qquad (4.31)$$

is the corresponding scaling function for the energy density profile in two dimensions. The numerical evaluation of $g_{O,O}(x)$ in $d = 2, 3, 4$ is displayed in Fig.4.1 which shows that the curve for $d = 3$ nicely fits in between the other two curves. The divergences at $z = 0$ and $z = L$ are compensated if the subtractions needed for the *renormalized* energy density according to Eq.(4.19) are taken into account (see Sec.4.2).

4.2.2   Wall Effects

The form of the distant wall correction according to Eq.(4.3) can be explicitly checked
by expanding the scaling functions $g_{a,b}(x)$ for $x \to 0$. From Eqs.(4.22), (4.25), and
(4.26) we obtain after a partial exponentiation of the ε–expansion to leading order in
$z/L \ll 1$ [209]

$$\frac{e_{0,0}(0, z, L)}{e_{\infty/2,0}(z, 0)} = 1 + 6\zeta(4) \left[1 - \varepsilon \left(\frac{5}{6} + \frac{\zeta'(4)}{\zeta(4)} - \frac{1}{12} \frac{N+2}{N+8} \right) \right] \left(\frac{z}{L} \right)^d + \dots ,$$

$$\frac{e_{0,SB}(0, z, L)}{e_{\infty/2,0}(z, 0)} = 1 - 6\eta(4) \left[1 - \varepsilon \left(\frac{5}{6} + \frac{\eta'(4)}{\eta(4)} - \frac{71}{42} \frac{N+2}{N+8} \right) \right] \left(\frac{z}{L} \right)^d + \dots ,$$

$$\frac{e_{SB,0}(0, z, L)}{e_{\infty/2,SB}(z, 0)} = 1 - 4\eta(2) \left[1 - \varepsilon \left(\frac{\eta'(2)}{\eta(2)} + \frac{N+2}{N+8} \left[\frac{3}{2} - \ln \frac{\pi}{2} \right] \right) \right] \left(\frac{z}{L} \right)^{d-1-\phi/\nu} + \dots ,$$

$$\frac{e_{SB,SB}(0, z, L)}{e_{\infty/2,SB}(z, 0)} = 1 + 4\zeta(2) \left[1 - \varepsilon \left(\frac{\zeta'(2)}{\zeta(2)} + \frac{N+2}{N+8} [3 - \ln(2\pi)] \right) \right] \left(\frac{z}{L} \right)^{d-1-\phi/\nu} + \dots ,$$

$$(4.32)$$

where $\phi = \frac{1}{2} - \frac{\varepsilon}{4} \frac{N+2}{N+8} + \mathcal{O}(\varepsilon^2)$ is the *crossover exponent* (see Eq.(1.12) in Sec.1.2) [20]
and $e_{\infty/2,a}(0, z)$ is given by Eq.(4.23). First, we should note that the exponentiation
leading to Eq.(4.32) relies on the assumption that distant wall corrections of higher
order than displayed in Eq.(4.32) are governed by exponents, which exceed d or $d-1-$
ϕ/ν, respectively, by an amount of order ε^0 [209]. This assumption remains unproved
within the framework of the ε–expansion of the scaling functions $g_{a,b}(x)$ according to
Eq.(4.26), because for the presently available results it is impossible to disentangle
the contributions of two powers of z/L to the ε–expansion if their exponents only
differ by terms of order ε. Second, the exponent $x_{\phi^2}^{(s)} = d-1-\phi/\nu = 2 - \frac{6}{N+8}\varepsilon + \mathcal{O}(\varepsilon^2)$
is the scaling dimension of the *surface* energy density $e_s = -\frac{1}{2}\langle \Phi^{2R}(\mathbf{r}_\parallel, 0) \rangle$ [20, 211].
Although the answers offered by Eq.(4.26) concerning the structure of the distant
wall corrections remain incomplete, there is one rigorous conclusion to be drawn.
The distant wall correction to the energy density profile near a wall of the *SB-*
type has certainly not the form anticipated in Eq.(4.3), because the leading distant
wall correction is governed by the exponent $x_{\phi^2}^{(s)} < d$. It is therefore also clear that
Eq.(4.13) will not hold in this case. However, if the near wall belongs to the ordinary
surface universality class, Eq.(4.32) seems to support Eqs.(4.3) and (4.13) for the
energy density profile. If the aforementioned assumption for the exponentiation is

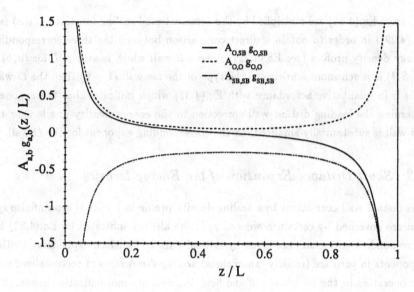

Fig. 4.2:   Scaling functions $A_{O,O}\, g_{O,O}(z/L)$ (dashed line), $A_{O,SB}\, g_{O,SB}(z/L)$ (solid line; the O wall is on the left side), and $A_{SB,SB}\, g_{SB,SB}(z/L)$ (dash–dotted line) in $d = 3$ ($\varepsilon = 1$). The amplitudes $A_{a,b}$ and the scaling functions $g_{a,b}(z/L)$ are taken from Eq.(4.25) and (4.29), respectively.

correct, Eq.(4.32) gives direct access to the coefficients $B_{O,O}^{\phi^2}$ and $B_{O,SB}^{\phi^2}$ (see Eqs.(4.11) and (4.14)) to order ε. The general relation in Eq.(4.14) states that $B_{a,b}^{\Psi}/\Delta_{a,b}$ is independent of the boundary condition b. Using Eq.(4.32) and taking $\Delta_{O,O}$ and $\Delta_{O,SB}$ from Eq.(3.13) we indeed obtain

$$-\frac{2^{d-1}d\pi^{d/2}}{\Gamma(d/2)}\frac{\Delta_{O,O}}{B_{O,O}^{\phi^2}}x_{\phi^2} = -\frac{2^{d-1}d\pi^{d/2}}{\Gamma(d/2)}\frac{\Delta_{O,SB}}{B_{O,SB}^{\phi^2}}x_{\phi^2} = \frac{2}{3}N\left[1+\varepsilon\left(\frac{1}{12}-\frac{5N+2}{6\,N+8}\right)\right]$$

(4.33)

for $x_{\phi^2} = 2-\varepsilon+\varepsilon\frac{N+2}{N+8}$ and $d = 4-\varepsilon$ to first order in ε. Taking Eq.(4.14) literally the right hand side of Eq.(4.33) offers a generalization of the central charge c to $d = 4-\varepsilon$. However, a corresponding test with $B_{SB,O}^{\phi^2}$ and $B_{SB,SB}^{\phi^2}$ is impossible at the moment, so that the right hand side of Eq.(4.33) may still be specific for the ordinary surface universality class which characterizes the near wall in this case.

A graphical view of the distant wall corrections is given in Fig.4.2, where $A_{a,b}\, g_{a,b}$ is shown in $d = 3$ ($\varepsilon = 1$) for $(a,b) = (O,O)$, (O,SB), and (SB,SB). The coefficients

$A_{a,b}$ from Eq.(4.25) are multiplied by the exponentiated scaling functions $g_{a,b}(x)$ (see
Eq.(4.29)) in order to obtain a direct comparison between the three corresponding
energy density profiles (see Eq.(4.22)). The SB wall which is at $z = L$ for $(a, b) =$
(O, SB) is much more sensitive to the type of the far wall ($z = 0$) than the O wall.
This is in qualitative accordance with Eq.(4.32) which indicates that the exponent
governing the leading distant wall correction to the energy density profile near the
SB wall is substantially smaller than the corresponding exponent for the O wall.

4.3   Short Distance Expansion of the Energy Density

The distant wall corrections to a scaling density profile in a critical semiinfinite system are governed by certain powers of z/L. As already indicated by Eq.(4.32) the
corresponding exponents are given by critical bulk and surface exponents. Critical
exponents in turn are (usually anomalous) *scaling dimensions* of renormalized scaling operators in the framework of the field–theoretical renormalization group. The
expansion of the energy density *profile* according to Eq.(4.32) can then be regarded
as the film average over a *short distance expansion* of the energy density *scaling operator* in the vicinity of one of the surfaces of the film. The powers of the distance
z from this surface reflect the scaling dimensions of the operators in the short distance expansion. Following Eqs.(4.3), (4.13), and (4.32) we write the short distance
expansion of a renormalized scaling operator $\Psi(\mathbf{r}_{\|}, z)$ near a surface of type a in the
normalized form [207, 212]

$$\Psi^R(\mathbf{r}_{\|}, z) = \langle \Psi^R(\mathbf{r}_{\|}, z) \rangle \left(1 + \sum_i \mathcal{C}_a^i \, \mathcal{O}_i^R(\mathbf{r}_{\|}) \, z^{x_i^{(s)}} \right), \qquad (4.34)$$

where $x_i^{(s)}$ with $x_1^{(s)} \leq x_2^{(s)} \leq \ldots$ denotes the scaling dimension of the *renormalized
surface operator* $\mathcal{O}_i^R(\mathbf{r}_{\|})$. Note, that the expansion coefficients $\mathcal{C}_a^i$ depend on the type
a of the boundary condition at the near surface. However, the second boundary
condition in a film geometry does not enter Eq.(4.34), because the short distance expansions are *local* relations [20], and therefore their operator form is fixed *irrespective*
of the geometry (semiinfinite or film) once the type a of the near wall is specified.
From the technical point of view the short distance expansion in Eq.(4.34) is the same
as the well known *operator product expansion* (see, e.g., Ref.[36]), where the positions
of two operators in space come close to each other.

For the ordinary and surface–bulk surface universality class those surface opera-tors which contribute to the short distance expansion of the energy density operator $-\frac{1}{2}\Phi^{2R}(\mathbf{r}_{\|}, z)$ can be identified from $O(N)$–symmetry requirements and the type O or SB of the surface. The corresponding scaling dimensions $x_i^{(s)}$ can be determined by a field–theoretical renormalization group analysis in a semiinfinte geometry with the desired surface [20, 24, 211, 213]. In a following step Eq.(4.34) with known sur-face operators $\mathcal{O}_i^r(\mathbf{r}_{\|})$ and scaling exponents $x_i^{(s)}$ is inserted into thermal averages (see Ref.[36]) which are then determined perturbatively in order to obtain the short distance coefficients C_a^i. This analysis can be performed in a semiinfinite geometry with a surface of type a, where $a = O, SB$ in our case. The resulting short distance expansions will give a definite proof of Eq.(4.32) and furthermore yield expressions for $B_{SB,O}^{\phi^2}$ and $B_{SB,SB}^{\phi^2}$ which cannot be determined from the ε–expansion of the scaling functions $g_{SB,O}(x)$ and $g_{SB,SB}(x)$ (see Eq.(4.26)) alone.

4.3.1 Surface Operators

The spectrum of those surface operators which enter the short distance expansion for the energy density scaling operator $\Psi^R(\mathbf{r}_{\|}, z) = -\frac{1}{2}\Phi^{2R}(\mathbf{r}_{\|}, z)$ (see Eq.(4.34)) depends on the symmetry required by the energy density and the surface universality class, where we will only be concerned with the *ordinary* (O) and *surface–bulk* (SB) *fixed–point* properties of the surface [207, 212]. Furthermore, we will truncate the expansion in Eq.(4.34) after those surface operators, which have the surface scaling dimension $x^{(s)} = 4$ in $d = 4$. The energy density scaling operator is $O(N)$–symmetric and thus only $O(N)$–symmetric surface operators contribute to the short distance expansion.

In an O–surface the truncated spectrum of $O(N)$–symmetric surface operators is particularly simple. The boundary condition which represents to the ordinary surface universality class is of the Dirichlet type so that any cumulant which contains the local magnetization $\Phi(\mathbf{r}_{\|}, z)$ at a point in the surface $z = 0$ vanishes identically. By the same argument cumulants containing *derivatives* of $\Phi(\mathbf{r}_{\|}, z)$ *parallel* to the surface vanish as $z \to 0$. The only $O(N)$–symmetric surface operator with $x^{(s)} \le 4$, which enters the short distance expansion of the energy density near an O–wall, can be written as $\frac{1}{2}\left(\frac{\partial \Phi}{\partial z}(\mathbf{r}_{\|}, 0)\right)^2$ [207, 212]. This surface operator does not require renormalization and its scaling dimension coincides with the spatial dimension d [213]. In an O–surface the above operator actually is the zz–component of the *surface stress tensor* (see Refs.[199, 207, 212] and Eq.(3.61))

$$T_{zz}(\mathbf{r}_{\|}, 0) = \frac{1}{2} \left(\frac{\partial \Phi}{\partial z}(\mathbf{r}_{\|}, 0) \right)^2 . \tag{4.35}$$

The short distance expansion of the renormalized energy density scaling operator in the vicinity of an O–surface then takes the simple form [199, 207, 212]

$$-\frac{1}{2}\Phi^{2R}(\mathbf{r}_{\|}, z) = e_{\infty/2, O}(z, 0) \left(1 + C_O^T\, T_{zz}(\mathbf{r}_{\|}, 0)\, z^d + \ldots \right) \tag{4.36}$$

which coincides with Eq.(4.13). Furthermore, Eq.(4.36) proves that the form of the distant wall correction for the energy density profiles $e_{O,O}(0, z, L)$ and $e_{O,SB}(0, z, L)$ as anticipated by Eq.(4.32) is correct. The coefficients $B_{O,O}^{\phi^2}$ and $B_{O,SB}^{\phi^2}$ (see Eqs.(4.11), (4.12) and (4.14)) of the distant wall correction can therefore indeed be read off from Eq.(4.32). Note, that Eq.(4.33) then gives the value of c according to Eq.(4.14) in $d = 4 - \varepsilon$ to order ε.

In an SB–surface there are much more $O(N)$–symmetric surface operators than in an O–surface. One can show order by order in perturbation theory that the Neumann boundary condition permits to disregard those operators which contain derivatives *normal* to the surface. Altogether the SB–surface allows four $O(N)$–symmetric surface operators with scaling dimensions $x^{(s)} \le 4$ in $d = 4$ which can be subdivided into two groups according to [207, 212]

$$x^{(s)} = 2 \ : \ -\frac{1}{2}\Phi^2(\mathbf{r}_{\|}, 0)$$

$$x^{(s)} = 4 \ : \ \frac{1}{2}\left(\nabla_{\|}\Phi(\mathbf{r}_{\|}, 0)\right)^2 \ , \quad -\frac{1}{2}\Delta_{\|}\Phi^2(\mathbf{r}_{\|}, 0) \ , \quad \frac{g}{4!}\left(\Phi^2(\mathbf{r}_{\|}, 0)\right)^2, \quad (4.37)$$

where g is the bare coupling constant in the Ginzburg–Landau Hamiltonian. In contrast to an O–surface the *surface energy density* operator $-\frac{1}{2}\Phi^2(\mathbf{r}_{\|}, 0)$ contributes to the short distance expansion in Eq.(4.34) for the energy density. This proves that the form of the direct expansion of the energy density profiles $e_{SB,O}(0, z, L)$ and $e_{SB,SB}(0, z, L)$ according to Eq.(4.32) is indeed correct. The surface operators listed in Eq.(4.37) are not renormalized. In order to construct a *renormalized* version of Eq.(4.37) it is very desirable to identify one of these renormalized operators as a component of the *surface stress tensor*. In fact, the zz–component of the stress tensor in an SB–surface, which is the only relevant component of $T_{\mu\nu}$ in this case, can be written as a linear combination of the three operators in the second row of Eq.(4.37). One obtains [207, 212]

$$T_{zz}(\mathbf{r}_{\|},0) = -\frac{1}{2}\left(\nabla_{\|}\Phi(\mathbf{r}_{\|},0)\right)^2 + \frac{d-2}{2(d-1)}\left[1+\frac{N+2}{6}u\right]\frac{1}{2}\Delta_{\|}\Phi^2(\mathbf{r}_{\|},0)$$

$$-2^d\pi^{\frac{d}{2}}\frac{u}{4!}\mu^\varepsilon\left(\Phi^2(\mathbf{r}_{\|},0)\right)^2 + \mathcal{O}(u^2)\,, \tag{4.38}$$

where u is the renormalized coupling constant given by $g = 2^d\pi^{d/2}\mu^\varepsilon u + \mathcal{O}(u^2)$ and μ is a momentum scale in units of inverse lengths (see Eq.(5.2)). Note, that Eq.(4.38) is not simply the evaluation of Eq.(3.61) for $z = 0$ with $\frac{1}{2}\left(\frac{\partial\Phi}{\partial z}\right)^2$ omitted. The operator $T_{zz}(\mathbf{r}_{\|},0)$ given by Eq.(4.38) is that specific combination of surface operators which describes the *limit* of $T_{zz}(\mathbf{r}_{\|},z)$ for $z \to 0$ *inside cumulants* for a SB–surface [207, 212]. In the same sense we have Eq.(4.35) in an O–surface. As before $T_{zz}(\mathbf{r}_{\|},0)$ is already renormalized and Eq.(4.38) shows this renormalized form to order u. The bare surface energy density operator $-\frac{1}{2}\Phi^2(\mathbf{r}_{\|},0)$ and its second parallel derivative $-\frac{1}{2}\Delta_{\|}\Phi^2(\mathbf{r}_{\|},0)$ are renormalized multiplicatively [20, 207, 211, 212] and $g\left(\Phi^2(\mathbf{r}_{\|},0)\right)^2$ can be regarded as renormalized to first order in u if g is replaced with u via $g = 2^d\pi^{d/2}\mu^\varepsilon u$. To order u we therefore obtain the following spectrum of renormalized surface operators in a SB–surface for $d = 4 - \varepsilon$ [207, 212]

$$x_0^{(s)} = d - 1 - \frac{\phi}{\nu} \qquad : \mathcal{O}_0^R = -\frac{1}{2}\Phi^{2R}(\mathbf{r}_{\|},0)$$

$$x_1^{(s)} = d \qquad : \mathcal{O}_1^R = T_{zz}(\mathbf{r}_{\|},0)$$

$$x_2^{(s)} = d + 1 - \frac{\phi}{\nu} \qquad : \mathcal{O}_2^R = -\frac{1}{2}\Delta_{\|}\Phi^{2R}(\mathbf{r}_{\|},0)$$

$$x_3^{(s)} = 4\left(1 + \frac{3\varepsilon}{N+8} + \mathcal{O}(\varepsilon^2)\right) : \mathcal{O}_3^R = 2^d\pi^{\frac{d}{2}}\mu^\varepsilon\frac{u}{4!}\left(\Phi^2(\mathbf{r}_{\|},0)\right)^2 + \mathcal{O}(u^2). \tag{4.39}$$

Note, that the representation of the *unrenormalized* surface operator $g\left(\Phi^2(\mathbf{r}_{\|},0)\right)^2$ by *renormalized* surface operators has several components. The *principal* component is an operator with the scaling dimension $x_3^{(s)}$ (see Eq.(4.39)). A second component is given by $T_{zz}(\mathbf{r}_{\|},0)$ which contributes to higher orders in u than displayed in Eq.(4.39) [207, 212]. The short distance expansion of the renormalized energy density scaling operator in the vicinity of an SB–surface can then be cast into the form [207, 212]

$$-\frac{1}{2}\Phi^{2R}(\mathbf{r}_{\|}, z) = e_{\infty/2,SB}(z,0)\left\{1 - C_{SB}^{\phi^2}\frac{1}{2}\Phi^{2R}(\mathbf{r}_{\|},0)\, z^{d-2}(\mu z)^{1-\phi/\nu}\right.$$

$$+ C_{SB}^{T}\, T_{zz}(\mathbf{r}_{\|},0)\, z^d - C_{SB}^{\Delta\phi^2}\frac{1}{2}\Delta_{\|}\Phi^{2R}(\mathbf{r}_{\|},0)\, z^d(\mu z)^{1-\phi/\nu}$$

$$\left.+ C_{SB}^{\phi^4}\, 2^d\pi^{\frac{d}{2}}\mu^\varepsilon\frac{u}{4!}\left(\Phi^2(\mathbf{r}_{\|},0)\right)^2 z^d(\mu z)^{\frac{N+20}{N+8}\varepsilon} + \ldots\right\}\;. \quad (4.40)$$

Note, that the last contribution to the right hand side of Eq.(4.40) is only displayed to order u. The coefficients C_a^i in the short distance expansions can now be obtained by inserting Eqs.(4.36) and (4.40) into thermal averages.

4.3.2   The Expansion Coefficients

The short distance expansions given by Eq.(4.36) and (4.40) are local operator identities. The short distance coefficients can therefore be determined from any thermal average with the energy density operator in a semiinfinite geometry. In view of Eqs.(4.25) and (4.26) which summarize our present knowledge on energy density profiles in critical films in $d = 4 - \varepsilon$ we can alternatively think of averaging the short distance expansion itself in a *film geometry*. A comparison with the direct expansion of the energy density profiles which is shown in Eq.(4.32) for the *leading* distant wall corrections then provides information on the short distance coefficients *at the renormalization group fixed points*. In particular we obtain from Eqs.(4.36) and (4.40) at the fixed point $u = u^* = \frac{3\varepsilon}{N+8} + \mathcal{O}(\varepsilon^2)$ [207, 212]

$$\frac{e_{O,a}(z,L,0)}{e_{\infty/2,O}(z,0)} = 1 + C_O^T\,(d-1)\,\Delta_{O,a}\left(\frac{z}{L}\right)^d + \ldots,$$

$$\frac{e_{SB,a}(z,L,0)}{e_{\infty/2,SB}(z,0)} = 1 + C_{SB}^{\phi^2}\left\langle -\frac{1}{2}\Phi^{2R}(\mathbf{r}_{\|},0)\right\rangle_{SB,a} z^{d-2}(\mu z)^{1-\phi/\nu}$$

$$+ C_{SB}^T\,(d-1)\,\Delta_{SB,a}\left(\frac{z}{L}\right)^d$$

$$+ C_{SB}^{\phi^4}\left[2^d\pi^{\frac{d}{2}}\frac{u^*}{4!}\mu^\varepsilon\left\langle\left(\Phi^2(\mathbf{r}_{\|},0)\right)^2\right\rangle_{SB,a} z^d + \mathcal{O}(\varepsilon^2)\right] + \ldots, \quad (4.41)$$

where Eq.(3.51) has been used to relate the film average $\langle T_{zz}\rangle_{a,b}$ to the *Casimir amplitudes* $\Delta_{a,b}$ (see Eqs.(3.13) and (3.16) in Sec.3.2), and $a = O, SB$ indicates the surface universality class of the second wall. Note, that averages in a film geometry are invariant under translations with respect to $r_\parallel$ and therefore $\langle \Delta_\parallel \Phi^2(r_\parallel, 0) \rangle_{SB,a} = 0$ so that the short distance expansion of the energy density profile in a film no longer depends on $C_{SB}^{\Delta \phi^2}$. The scaling dimensions $x_1^{(s)} = d$ of $T_{zz}(r_\parallel, 0)$ and $x_3^{(s)} = d + \frac{N+20}{N+8}\varepsilon + \mathcal{O}(\varepsilon^2)$ of $\mu^\varepsilon \left(\Phi^2(r_\parallel, 0) \right)^2$ in Eq.(4.41) can only be distinguished if the ε–expansion is carried to the order ε^2 which requires a three–loop calculation of the energy density profiles and the surface averages involved in Eq.(4.41). However, the present knowledge of both is limited to two–loop order and on this level of approximation $x_3^{(s)}$ can be regarded to be the same as $x_1^{(s)} = d$. Note, that $x_3^{(s)}$ can be determined *independent* of the above considerations in a *semiinfinite* geometry.

A two–loop calculation yields the following results for the thermal averages appearing in Eq.(4.41) [209, 212]

$$\left\langle -\frac{1}{2}\Phi^{2R}(r_\parallel, 0) \right\rangle_{SB,SB} = -2N \frac{\Gamma\left(\frac{d}{2} - 1\right)}{2^d \pi^{\frac{d}{2}}} \zeta(d-2)$$

$$\times \left[1 - \frac{N+2}{N+8}\varepsilon \left(\frac{3}{2} + \frac{\gamma}{2} - \ln(2\pi) \right) \right] L^{-(d-2)}(\mu L)^{-(1-\phi/\nu)},$$

$$\left\langle -\frac{1}{2}\Phi^{2R}(r_\parallel, 0) \right\rangle_{SB,O} = 2N \frac{\Gamma\left(\frac{d}{2} - 1\right)}{2^d \pi^{\frac{d}{2}}} \eta(d-2)$$

$$\times \left[1 - \frac{N+2}{N+8}\varepsilon \left(\frac{\gamma}{2} - \ln\frac{\pi}{2} \right) \right] L^{-(d-2)}(\mu L)^{-(1-\phi/\nu)}, \quad (4.42)$$

$$2^d \pi^{\frac{d}{2}} \frac{u^*}{4!} \mu^\varepsilon \left\langle \left(\Phi^2(r_\parallel, 0) \right)^4 \right\rangle_{SB,SB} = (d-1)N \frac{\Gamma\left(\frac{d}{2}\right)\zeta(d)}{2^d \pi^{\frac{d}{2}}} \frac{N+2}{N+8} \frac{5}{3}\varepsilon L^{-d},$$

$$2^d \pi^{\frac{d}{2}} \frac{u^*}{4!} \mu^\varepsilon \left\langle \left(\Phi^2(r_\parallel, 0) \right)^4 \right\rangle_{SB,O} = (d-1)N \frac{\Gamma\left(\frac{d}{2}\right)\eta(d)}{2^d \pi^{\frac{d}{2}}} \frac{N+2}{N+8} \frac{10}{21}\varepsilon L^{-d}, \quad (4.43)$$

where $\zeta(z)$ and $\eta(z)$ denote the Riemann zeta–function and the eta–function, respectively, and γ is Euler's constant. The short distance coefficients C_O^T and $C_{SB}^{\phi^2}$ can now be directly inferred from a comparison between Eq.(4.41) and Eq.(4.32) for $a = O$ or $a = SB$ using Eqs.(4.42) and (3.16). In order to determine the coefficients C_{SB}^T and $C_{SB}^{\phi^4}$ the *next to leading* distant wall correction for $e_{SB,O}(0, z, L)$ and $e_{SB,SB}(0, z, L)$ must first be calculated by expanding the corresponding scaling functions $g_{SB,a}(x)$

in Eq.(4.26) for small x. The comparison with Eq.(4.41) for $a = O$ and $a = SB$ using Eq.(4.43) and (3.16) gives a system of two linear equations for C_{SB}^T and $C_{SB}^{\phi^4}$. In summary, the results read [207, 212]

$$C_O^T = -\frac{1}{N}2^{d-1}(d-2)S_d\left(1+\frac{4}{3}\varepsilon\,\frac{N+2}{N+8}+\mathcal{O}(\varepsilon^2)\right),$$

$$C_{SB}^{\phi^2} = -\frac{1}{N}2^{d-1}(d-2)S_d\left(1+\frac{\gamma-3}{2}\varepsilon\,\frac{N+2}{N+8}+\mathcal{O}(\varepsilon^2)\right),$$

$$C_{SB}^T = -\frac{1}{N}2^{d-1}(d-2)S_d\left(1-\frac{1}{3}\varepsilon\,\frac{N+2}{N+8}+\mathcal{O}(\varepsilon^2)\right),$$

$$C_{SB}^{\phi^4} = \frac{1}{N}2^{d-1}(d-2)S_d\left(1+\mathcal{O}(\varepsilon)\right), \tag{4.44}$$

where $S_d = 2\pi^{d/2}/\Gamma(d/2)$ denotes the area of the d–dimensional unit sphere. We note again that we restrict our considerations to the renormalization group fixed point. Thus the meaning of Eqs.(4.36) and (4.40) together with the coefficients given by Eq.(4.44) is, that the insertion of the short distance expansion into thermal averages yields results which are correct to order ε at the fixed point [207].

Using the short distance coefficient C_{SB}^T given by Eq.(4.44) the second test of Eq.(4.14) (see Ref.[192]) for $(a,b) = (SB,O)$ and (SB,SB) is possible. First, we note that in general [207]

$$B_{a,b}^{\phi^2} = C_a^T\,(d-1)\,\Delta_{a,b} \tag{4.45}$$

(see Eq.(4.41)) so that the dependence of $B_{a,b}^{\phi^2}$ on b is in fact absorbed by the Casimir amplitude $\Delta_{a,b}$. Second, we have to order ε (see Eq.(4.33))

$$-\frac{2^{d-1}d\pi^{d/2}}{\Gamma(d/2)}\frac{x_{\phi^2}}{B_{SB,a}^{\phi^2}}\Delta_{SB,a} = -\frac{2^{d-2}S_d x_{\phi^2}}{(d-1)C_{SB}^T} = \frac{2}{3}N\left[1+\varepsilon\left(\frac{1}{12}+\frac{5}{6}\frac{N+2}{N+8}\right)\right], \tag{4.46}$$

which is obviously at variance with Eq.(4.33). According to Eq.(4.14) both Eq.(4.33) and (4.46) should give a number c which is *independent* of the surface universality class of the near wall [192]. The right hand sides of Eqs.(4.33) and (4.46) show that the generalization of the central charge c to d dimensions suggested by Eq.(4.14) is in fact *surface specific* contrary to the behavior of c in $d = 2$ [128, 192, 207] (for details see Sec.4.4).

The film average of Eq.(4.40) like any other laterally invariant thermal average with Eq.(4.40) is independent of the short distance coefficient $C_{SB}^{\Delta\phi^2}$. Looking at the film averages has just been most convenient here in order to use the information contained in the scaling functions of the energy density profiles $e_{SB,0}(z, L, 0)$ and $e_{SB,SB}(z, L, 0)$ (see Eqs.(4.25) and (4.26)) for the calculation of the short distance coefficients. In general, however, one must resort to correlation functions which are not laterally invariant. As an example we discuss the cumulant of the *bulk* energy density scaling operator $-\frac{1}{2}\Phi^{2R}(\mathbf{r}_{\|}, z)$ with the *surface* energy density scaling operator $-\frac{1}{2}\Phi^{2R}(0, 0)$ in a semiinfinite geometry bounded by a SB–wall. From conformal invariance (see Refs.[39] and [199] and Sec.3.3) we have the *othogonality relations*

$$\left\langle -\frac{1}{2}\Phi^{2R}(\mathbf{r}_{\|}, 0) \, 2^d \pi^{\frac{d}{2}} \frac{u}{4!} \mu^\varepsilon \left(\Phi^2(\mathbf{r}'_{\|}, 0)\right)^2 \right\rangle_{\infty/2, SB} = 0,$$

$$\left\langle -\frac{1}{2}\Phi^{2R}(\mathbf{r}_{\|}, 0) \, T_{zz}(\mathbf{r}'_{\|}, 0) \right\rangle_{\infty/2, SB} = 0,$$

$$\left\langle 2^d \pi^{\frac{d}{2}} \frac{u}{4!} \mu^\varepsilon \left(\Phi^2(\mathbf{r}'_{\|}, 0)\right)^2 T_{zz}(\mathbf{r}'_{\|}, 0) \right\rangle_{\infty/2, SB} = 0 \qquad (4.47)$$

among the surface operators in Eq.(4.40), where $\langle \dots \rangle$ denotes the *cumulant*. The additional orthogonality relations

$$\left\langle -\frac{1}{2}\Delta_{\|}\Phi^{2R}(\mathbf{r}_{\|}, 0) \, 2^d \pi^{\frac{d}{2}} \frac{u}{4!} \mu^\varepsilon \left(\Phi^2(\mathbf{r}'_{\|}, 0)\right)^2 \right\rangle_{\infty/2, SB} = 0,$$

$$\left\langle -\frac{1}{2}\Delta_{\|}\Phi^{2R}(\mathbf{r}_{\|}, 0) \, T_{zz}(\mathbf{r}'_{\|}, 0) \right\rangle_{\infty/2, SB} = 0 \qquad (4.48)$$

follow directly from Eq.(4.47) by taking $\Delta_{\|}$ out of the cumulant. From Eqs.(4.40) and (4.47) we find the short distance expansion

$$\frac{\left\langle \Phi^{2R}(\mathbf{r}_{\|}, z) \, \Phi^{2R}(0, 0) \right\rangle_{\infty/2, SB}}{e_{\infty/2, SB}(z, 0)} = C_{SB}^{\phi^2} \left\langle \Phi^{2R}(\mathbf{r}_{\|}, 0) \, \Phi^{2R}(0, 0) \right\rangle_{\infty/2, SB} z^{d-2}(\mu z)^{1-\phi/\nu}$$

$$+ C_{SB}^{\Delta\phi^2} \left\langle \Delta_{\|}\Phi^{2R}(\mathbf{r}_{\|}, 0) \, \Phi^{2R}(0, 0) \right\rangle_{\infty/2, SB} z^d (\mu z)^{1-\phi/\nu}$$

$$+ \dots \qquad (4.49)$$

for the cumulant $\left\langle \Phi^{2R}(\mathbf{r}_{\|}, z) \, \Phi^{2R}(0, 0) \right\rangle_{\infty/2, SB}$. The functional form of the cumulants in Eq.(4.49) is again fixed by the principle of conformal invariance (see especially

Eq.(3.32) in Sec.3.3 and Refs.[39] and [199]), and we conclude from Eq.(3.32) with $x_1 = d - 1/\nu$ and $x_2^{(s)} = d - 1 - \phi/\nu$

$$\left\langle \Phi^{2R}(\mathbf{r}_{\parallel}, z)\, \Phi^{2R}(0,0) \right\rangle_{\infty/2,SB} = A_{SB}^{(bs)}\, r^{-2(d-2)}(\mu r)^{-(3-(1+\phi)/\nu)}\left(\frac{z}{r}\right)^{(1-\nu-\phi)/\nu},$$

$$\left\langle \Phi^{2R}(\mathbf{r}_{\parallel}, 0)\, \Phi^{2R}(0,0) \right\rangle_{\infty/2,SB} = A_{SB}^{(ss)}\, r_{\parallel}^{-2(d-2)}(\mu r_{\parallel})^{-2(1-\phi/\nu)}, \qquad (4.50)$$

where $r = \sqrt{r_{\parallel}^2 + z^2}$, $A_{SB}^{(bs)}$ and $A_{SB}^{(ss)}$ are constant amplitudes, and μ is the arbitrary inverse length scale in Eq.(4.40). Writing the energy density profile $e_{\infty/2,a}(0,z)$ (see Eq.(4.23)) in the form

$$e_{\infty/2,a}(0,z) = A_a^{(b)} z^{-(d-2)}(\mu z)^{-(2-1/\nu)} \qquad (4.51)$$

for $a = O, SB$ we find a relation between the short distance coefficients $C_{SB}^{\phi^2}$ and $C_{SB}^{\Delta\phi^2}$ and the amplitudes $A_{SB}^{(b)}$, $A_{SB}^{(bs)}$, and $A_{SB}^{(ss)}$ if the left hand side of Eq.(4.49) is expanded in powers of $z/r_{\parallel}$ using Eq.(4.50). As one might have guessed from the structure of Eq.(4.49) the short distance coefficients $C_{SB}^{\phi^2}$ and $C_{SB}^{\Delta\phi^2}$ are *not* independent. We find the *exact* relation [212]

$$C_{SB}^{\Delta\phi^2} = \frac{-C_{SB}^{\phi^2}}{2(d+1-2\,\phi/\nu)} \qquad (4.52)$$

which determines the coefficient $C_{SB}^{\Delta\phi^2}$ missing in Eq.(4.44) in terms of $C_{SB}^{\phi^2}$, the crossover exponent ϕ (see Eq.(4.32)), and the exponent ν of the bulk correlation length.

The short distance expansions given by Eq.(4.36) and (4.40) together with the expansion coefficients in Eqs.(4.44) and (4.52) determine the *general* structure of the distant wall corrections to the energy density profiles. On the other hand the real meaning of Eq.(4.14) (see Ref.[192]) which involves a generalization of the *central charge c* in view of Eqs.(4.33) and (4.46) is not clear. However, Eqs.(4.45) and (4.46) indicate that the short distance coefficients C_O^T and C_{SB}^T of the *stress tensor* play the key part in the answer to this question.

4.4   The Short Distance Expansion and the Central Charge

The distant wall corrections to a scaling density profile in a semiinfinite geometry are governed by those surface operators which are compatible with the required symmetry and the boundary condition at the surface. One of these surface operators is the *surface stress tensor* which has the special properties to exhibit no anomalous scaling dimension, and that *both* the bulk and the surface scaling dimension are given by the spatial dimension d. Interestingly, only the stress tensor is common to the two different short distance expansions for the energy density scaling operator given by Eqs.(4.36) and (4.40) for an $O-$ and a SB–surface, respectively. As Eq.(4.41) shows, the contribution $B_{a,b}^{\phi^2}(z/L)^d$ to the distant wall corrections is therefore common to all energy density profiles considered here. The coefficient $B_{a,b}^{\phi^2}$ obeys Eq.(4.45) which confirms Eq.(4.14) in so far as the Casimir amplitude $\Delta_{a,b}$ in fact absorbs the dependence of $B_{a,b}^{\phi^2}$ on b. However, after combining Eq.(4.14) with Eq.(4.45) we are left with a statement on the *short distance coefficient* C_a^T which, in view of the explicit results for C_O^T and C_{SB}^T given by Eq.(4.44), must be reconsidered.

The orthogonality relations displayed by Eqs.(4.47) and (4.48) allow one to express the short distance coefficient C_a^T in terms of cumulants with the stress tensor *in* the surface of a semiinfinite geometry. From Eqs.(4.36) and (4.40) one obtains [192]

$$\frac{\left\langle -\frac{1}{2}\Phi^{2R}(\mathbf{r}_{\|}, z)\, T_{zz}(\mathbf{r}_{\|}', 0)\right\rangle_{\infty/2,a}}{e_{\infty/2,a}(z,0)} = C_a^T \left\langle T_{zz}(\mathbf{r}_{\|}, 0) T_{zz}(\mathbf{r}_{\|}', 0)\right\rangle_{\infty/2,a} z^d + \dots, \qquad (4.53)$$

where $a = O, SB$. The functional form of the stress–tensor stress–tensor correlation function on the right hand side of Eq.(4.53) is already fixed by scale invariance. We define an amplitude factor c_a by writing [192, 207]

$$\left\langle T_{zz}(\mathbf{r}_{\|}, 0)\, T_{zz}(\mathbf{r}_{\|}', 0)\right\rangle_{\infty/2,a} = 2c_a\, S_d^{-2}\left(2 - \frac{2}{d}\right) \left|\mathbf{r}_{\|} - \mathbf{r}_{\|}'\right|^{-2d}, \qquad (4.54)$$

where the extra factor two in front of c_a accounts for the contribution of the image term to the correlation function in the surface [192]. Note, that our definition of the stress tensor follows Ref.[202] which differs from the definition used on Ref.[192] by a factor $-S_d^{-1}$ (see also Eq.(3.35)). The bulk analogue of Eq.(4.54) reads [192]

$$\left\langle T_{zz}(\mathbf{r})\, T_{zz}(\mathbf{r}')\right\rangle_{bulk} = c_{bulk}\, S_d^{-2}\left(2 - \frac{2}{d}\right) \left|\mathbf{r} - \mathbf{r}'\right|^{-2d} \qquad (4.55)$$

by which an amplitude c_{bulk} is defined. In *two* dimensions the amplitudes c_a and c_{bulk} are the same and equal the *central charge* c of the theory [39, 192]. Therefore Eqs.(4.54) and (4.55) both offer a generalization of the central charge to general dimension d. For a Gaussian model with $N = 1$ in d dimensions one obtains [214]

$$c_{bulk} = c_a \equiv c^{(G)} = \frac{d}{2(d-1)}, \qquad (4.56)$$

which is in accordance with the behavior of the central charge c in two dimensions. However, a two–loop calculation for a Ginzburg–Landau model in a semiinfinite geometry and in unbounded space gives [207, 212]

$$(c_O, c_{SB}, c_{bulk}) = Nc^{(G)} \left[1 + \frac{5}{6} \varepsilon \frac{N+2}{N+8} (-1, 1, 0) + \mathcal{O}(\varepsilon^2) \right]. \qquad (4.57)$$

The d–dimensional analogue of the central charge c according to Eq.(4.57) takes a different value in the bulk than in a surface and, moreover, it is sensitive to the surface universality class. The amplitudes c_O, c_{SB}, and c_{bulk} of the stress–tensor stress–tensor correlation function therefore do not completely correspond to the central charge c in $d = 2$. The independence of c of the surface universality classes ("hyperuniversality") in two dimensions, which has implicitly been generalized to $d > 2$ in Eq.(4.14), is in fact violated between $d = 2$ and $d = 4$. Nonetheless the amplitudes c_O, c_{SB}, and c_{bulk} are *universal* which is still a special property among the amplitudes of other correlation functions. The results for c_O and c_{SB} in Eq.(4.57) have been confirmed by explicit calculations of the full stress–tensor stress–tensor correlation function in a semiinfinite geometry [215]. The resulting scaling functions provide a crossover from the amplitudes c_O or c_{SB} in the surface to the amplitude c_{bulk} away from the surface.

From the comparison between Eqs.(4.33) and (4.46) on one hand and the results for c_O and c_{SB} in Eq.(4.57) on the other hand one might suspect that Eq.(4.14) does hold, but with c replaced by c_O or c_{SB}, respectively. Following the line of argument of Ref.[192] one can show that this is in fact the case. First, we note that conformal invariance considerations again fix the functional form of the left hand side of Eq.(4.53). Analogous to Eq.(3.32) one finds [192, 212]

$$\left\langle -\frac{1}{2} \Phi^{2R}(\mathbf{r}_{\|}, z) \, T_{zz}(\mathbf{r}'_{\|}, 0) \right\rangle_{\infty/2, a} = A_a^{\phi^2, T} z^{-(d-2)} (\mu z)^{-(2-1/\nu)} \left(\frac{z}{(\mathbf{r}_{\|} - \mathbf{r}'_{\|})^2 + z^2} \right)^d, \qquad (4.58)$$

where $A_a^{\phi^2, T}$ for $a = O, SB$ is a constant amplitude and μ is the inverse length scale introduced in Eq.(4.40). The energy density profile $e_{\infty/2, a}(z, 0)$ can be written in the form of Eq.(4.51). Second, we observe that the cumulant on the left hand side of Eq.(4.58) appears as the integrand in the conformal Ward identity for the energy density profile $e_{\infty/2, a}(z, 0)$ (see Eq.(3.39)). Especially for the nonconformal mapping

$$\mathbf{r}' \left(\mathbf{r}_{\|}, z \right) = \left(\mathbf{r}'_{\|}, z' \right) = \left(\mathbf{r}_{\|}, z + z_0 \left(1 + \tanh \left(\tfrac{z}{\alpha} \right) \right) \right), \qquad (4.59)$$

where α and z_0 are positive constants this conformal Ward identity takes the limiting form [207, 212]

$$\frac{\partial}{\partial z}e_{\infty/2,a}(z,0) = \int d^{d-1}r'_{\|} \left\langle -\frac{1}{2}\Phi^{2R}(\mathbf{r}_{\|},z)\,T_{zz}(\mathbf{r}'_{\|},0)\right\rangle_{\infty/2,a} \tag{4.60}$$

for $\alpha \to 0$ and $z > 0$. We evaluate Eq.(4.60) directly by using Eqs.(4.51) and (4.58), and we find

$$A_a^{\phi^2,T} = -2^d\,S_d^{-1}\,x_{\phi^2}\,A_a^{(b)} \tag{4.61}$$

for the amplitudes $A_a^{\phi^2,T}$ and $A_a^{(b)}$, where $x_{\phi^2} = d - 1/\nu$ is the bulk scaling dimension of the energy density. The short distance coefficient $\mathcal{C}_a^T$ can now be inferred from Eq.(4.53) as [192, 207]

$$\mathcal{C}_a^T = -\frac{2^{d-1}d\pi^{d/2}}{(d-1)\Gamma(d/2)}\frac{x_{\phi^2}}{c_a} \tag{4.62}$$

which immediately leads to [192, 207]

$$B_{a,b}^{\phi^2} = -\frac{2^{d-1}d\pi^{d/2}}{\Gamma(d/2)}\frac{x_{\phi^2}}{c_a}\Delta_{a,b}, \tag{4.63}$$

where Eq.(4.45) has been used. The *form* of Eq.(4.14) is thus proved to be correct, and indeed c has to be replaced with c_a for $a = O, SB$. For $d = 2$, however, we have $c_a = c$ and therefore Eq.(4.63) gives a second proof of Eq.(4.12) which has been derived in Ref.[128] following a different line of argument.

The conformal invariance considerations presented here are by no means limited to the short distance expansion of the energy density scaling operator. The same arguments can be given for any scaling operator $\Psi(\mathbf{r}_{\|}, z)$ with a bulk scaling dimension x_Ψ and therefore Eq.(4.63) holds in general.

5. The Finite Size Scaling Functions

5.1 Universality and Scaling Functions

In the thermodynamic limit critical points are characterized by *singularities* in the thermodynamic functions such as the free energy and by an infinite correlation length ξ. In a fully *finite* system the correlation length will clearly not become infinite and therefore the singularities in the free energy are replaced by *rounded* extrema located at somewhat *shifted* positions as compared to the positions of the aforementioned singularities. The former finite size effect has become known as *fractional rounding*, the latter as *fractional shift* [48, 51, 52] (see Eqs.(2.5) and (2.7) in Sec.2.1). In practice any system is finite, but nonetheless often still large enough so that the concept of critical *singularities* governed by universal critical exponents remains valid in the sense that in the *vicinity* of a critical point the thermodynamic functions of the system display the expected singular behavior to a very high accuracy. In such systems both the fractional rounding and the fractional shift cannot be resolved experimentally. In the following we will therefore still refer to "regular" and "singular" parts of the free energy even for such finite samples.

A second characteristic feature of critical points is *scaling*. If a thermodynamic function is considered in terms of several variables then the *singular* part of this function only depends on a certain combinations of the variables, where, e.g., in the case of two variables one is scaled by a certain power of the other. In experiments scaling manifests itself as *data collapse*, if the properly scaled thermodynamic function is plotted versus a properly scaled variable (see Eqs.(1.3) and (1.4) in Sec.1.1). The resulting curves then display the shape of the *scaling functions*. The finite size dependence of the singular part of thermodynamic functions like the free energy follows this scheme, i.e., the size dependence has the scaling form [51, 52] (see Eqs.(2.3) and (2.4)). If L is a typical macroscopic length of the system then L only enters the list of variables of the singular part of the free energy in the combination L/ξ. Moreover, the *finite size scaling functions* which govern the singular finite size behavior of the free energy are *universal*. The only nonuniversality enters through *metric factors* (nonuniversal amplitudes) multiplying the scaling arguments [43] (see also Refs.[56] and [38]). The finite size scaling behavior of the singular free energy or, equivalently, the singular specific heat has been studied explicitly for an Ising model on a finite rectangular lattice [49] and on a strip [50] in $d = 2$, for a film of ideal Bosons in $d = 3$ [53], and for a spherical model for ferromagnetism in a slab in $2 < d < 4$ [91]. Other investigations have been performed on the basis of field theory for $O(N)$–symmetric systems in, e.g., a spherical geometry with Dirichlet boundary conditions in $d = 4 - \varepsilon$ [63], a hypercube with periodic boundary conditions in $d = 4$ and $d = 4 - \varepsilon$ [107], a hypercube with Dirichlet boundary conditions as a model for ^{4}He close to T_λ in pores [108], for mixed Dirichlet and periodic boundary conditions in the same geometry [109], and for a film geometry for all possible symmetry–conserving boundary conditions in $d = 4 - \varepsilon$ [188, 189]. A recent study of the finite–size behavior of an $O(N)$–symmetric model in a $L \times L_\perp^{d-1}$-geometry with helical boundary conditions along L and otherwise periodic boundary conditions is described in Ref.[216] for $T < T_c$ (for more details see Sec.5.5). A review on the concept of finite size scaling can be found in Refs.[42] and [56]. For the following considerations concerning universality we will restrict ourselves to the film geometry without external fields.

In an experiment a system forming a film is characterized by the thickness L which is much smaller than the lateral extensions of the system. The experimental setup must be designed such, that the correlation length of the system can be increased up to the order of magnitude of the film thickness L, because otherwise only the usual three–dimensional critical behavior will be visible without any resolvable finite–size effects. The field–theoretical treatment of such a system is based on the Ginzburg–Landau Hamiltonian for a film given by Eq.(2.48). In order to derive the universal properties of the finite size scaling function of the singular part of the free energy in the sense of Ref.[43] all information about the free energy contained in the *renormalized* field–theoretical model must be collected in the *renormalized free energy*. Besides the multiplicative renormalization of the bare reduced temperature τ and the bare coupling constant g (see Eq.(1.28)) by the *bulk* renormalization factors Z_t and Z_u, respectively, the free energy requires an *additive* renormalization. In units of $k_B T_{c,b}$ and per cross–section area A (see Eq.(3.10)) the renormalized free energy f^R in terms of the *bare* free energy $\bar{f}$ and its first and second derivative $\bar{f}_\tau$ and $\bar{f}_{\tau\tau}$ reads [188]

$$f^R(t, L, u, \mu) = \bar{f}(\tau, L, g) - \bar{f}(\tau_1, L, g) - \bar{f}_\tau(\tau_1, L, g)(\tau - \tau_1) - \tfrac{1}{2}\bar{f}_{\tau\tau}(\tau_1, L, g)(\tau - \tau_1)^2,$$
(5.1)

where t and u are the renormalized dimensionless reduced temperature and coupling constant, respectively, given by [188]

$$\tau = \mu^2 Z_t t \quad , \quad \tau_1 = \mu^2 Z_t \text{sgn} t,$$
$$g = 2^d \pi^{d/2} \mu^\epsilon Z_u u \quad , \quad \epsilon = 4 - d. \tag{5.2}$$

The parameter μ is an arbitrary momentum (wave vector) scale which absorbs the naive dimensions of τ and g. The bare form of the free energy $\bar{f}$ depends on the calculation scheme, e.g., cutoff or dimensional regularization which is used to calculate Feynman diagrams for $\bar{f}$. For more details consult Ref.[188]. Note, that in comparison with Eq.(3.10) the momentum scale μ and the renormalized coupling u appear in the list of variables of f^R given by Eq.(5.1) only for reasons of bookkeeping. Note further, that $t = 0$ corresponds to *bulk* criticality $T = T_{c,b}$ (and $H = H_c = 0$). If there is still a critical point present in the phase diagram of the *film*, it is in general shifted to a new critical temperature $T_c(L) \neq T_{c,b}$ and a new critical field $H_c \neq 0$ [40, 41, 47, 71]. The renormalization prescription in Eq.(5.1) normalizes the singular part of the free energy of the film such, that at a certain reference temperature $T_1 \neq T_{c,b}$ the free energy $f^R(t, L, u, \mu)$, the energy density $u^R(t, L, u, \mu) = -f_t^R(t, L, u, \mu)$, and the specific heat $c^R(t, L, u, \mu) = -f_{tt}^R(t, L, u, \mu)$ vanish simultaneously. This is achieved by subtracting a second order Taylor polynomial in τ (i.e., an analytic function of τ) from the bare free energy $\bar{f}(L, \tau, g)$. The implication of this procedure on the finite size behavior of f^R will be discussed below.

From Eq.(5.1) it follows immediately that the third derivative of $\bar{f}$ with respect to τ is renormalized multiplicatively according to [188]

$$f_{ttt}^R(t, L, u, \mu) = \mu^6 Z_t^3 \bar{f}_{\tau\tau\tau}(\tau, L, g) \tag{5.3}$$

which corresponds to the multiplicative renormalization of the first derivative $\bar{c}_\tau$ of the bare specific heat $\bar{c}(\tau, L, g) = -\bar{f}_{\tau\tau}(\tau, L, g)$ with respect to τ. The variation of f^R and f_{ttt}^R under rescaling, i.e., a change in μ at *fixed* τ, g, and L is then governed by the following set of *renormalization group equations* for f^R and f_{ttt}^R [188]:

$$\left[\mu \frac{\partial}{\partial \mu} + \beta(u) \frac{\partial}{\partial u} - \frac{t}{\nu(u)} \frac{\partial}{\partial t} \right] f^R(t, L, u, \mu) = -\frac{(t - \text{sgn} t)^2}{2\nu(u)} f_{ttt}^R(\text{sgn} t, L, u, \mu)$$

$$\left[\mu \frac{\partial}{\partial \mu} + \beta(u) \frac{\partial}{\partial u} - \frac{t}{\nu(u)} \frac{\partial}{\partial t} - \frac{3}{\nu(u)} \right] f_{ttt}^R(t, L, u, \mu) = 0, \tag{5.4}$$

where $\beta(u)$ and $\nu(u)$ are the usual Wilson functions [34, 35, 36, 37]. The structure of Eq.(5.4) is similar to the structure of flow equations, where $(u(\mu), t(\mu))$ represent a trajectory along which f^R and f^R_{ttt} are carried towards their fixed point behavior as $\mu \to 0$, i.e., in the limit of *macroscopic* length scales. Both f^R and f^R_{ttt} are measured in units of inverse $(d-1)$–dimensional areas and therefore the fixed point solution of Eq.(5.4) can be written as [188]

$$f^R(t, L, u, \mu) = L^{-(d-1)} \left\{ H_\pm(z) - H_\pm(z_1) - H'_\pm(z_1)(z - z_1) - \tfrac{1}{2} H''_\pm(z_1)(z - z_1)^2 \right\},$$

$$f^R_{ttt}(t, L, u, \mu) = L^{-(d-1)} \frac{\partial^3}{\partial t^3} H_\pm(z), \qquad (5.5)$$

where $H_+(z)$ denotes the scaling function for $t > 0$, and $H_-(z)$ denotes the scaling function for $t < 0$. The scaling argument z and the reference value z_1 are given by [188]

$$z = E^*_\nu(u)(\mu L)^{1/\nu}|t|, \quad z_1 = E^*_\nu(u)(\mu L)^{1/\nu}, \quad E^*_\nu(u) = \exp \int_u^{u^*} \left(\frac{1}{\nu} - \frac{1}{\nu(u')} \right) \frac{du'}{\beta(u')},$$
$$(5.6)$$

where u^* denotes the infrared stable fixed point value of the renormalized coupling constant u (see Refs.[34, 35, 36, 37]), and $\nu \equiv \nu(u^*)$ is the critical exponent of the bulk correlation length. Note, that $E^*_\nu(u)$ only exists if $d = 4 - \epsilon < 4$. In $d = 4$ logarithmic corrections in terms of $\ln t$ modify Eq.(5.6) (see Refs.[20] and [34]).

The renormalized coupling constant u, which here has the meaning of the initial value of the flowing coupling constant $u(\mu)$ [188], depends on the microscopic parameters of the system and is therefore *nonuniversal*. Thus the scaling variable z and its reference value z_1 may take different values for different systems even if t and L are the same, and so they are nonuniversal as well. On the other hand Eq.(5.4) determines the whole u–dependence of f^R and f^R_{ttt} and thus H_+ and H_- as *functions* do *not* depend on u. In other words, H_+ and H_- are the *universal scaling functions* of the singular energy of the film. If we define an *unnormalized* singular free energy f by

$$f(t, L, u, \mu) = L^{-(d-1)} H_\pm \left(E^*_\nu(u)(\mu L)^{1/\nu}|t| \right), \qquad (5.7)$$

the renormalization prescription in Eq.(5.1) can be written in the form

$$f^R(t, L, u, \mu) = f(t, L, u, \mu) - f(\mathrm{sgn} t, L, u, \mu)$$
$$- f_t(\mathrm{sgn} t, L, u, \mu)(t - \mathrm{sgn} t) - \tfrac{1}{2} f_{tt}(\mathrm{sgn} t, L, u, \mu)(t - \mathrm{sgn} t)^2. \quad (5.8)$$

If we finally identify the nonuniversal metric factor C_1 and the singular part of the free energy f_s defined in Ref.[43] according to $C_1 = E^*_\nu(u)\mu^{1/\nu}$ and $f_s = f/L$, the statement concerning universality as given in Ref.[43] is explicitly proved by Eqs.(5.5) and (5.6) for a vanishing external field. Note, that *no* renormalization trajectory originating from $t \neq 0$ passes through $t = 0$, and one therefore has to distinguish between H_+ and H_- as *independent* scaling functions.

5.1.1   Finite Size Scaling and the Bulk Correlation Length

The scaling argument z given by Eq.(5.6) contains the critical exponent ν of the bulk correlation length ξ for short-ranged interactions. From Eq.(5.6) one may now guess that $z^\nu \propto L/\xi$ and that therefore L/ξ is an equivalent and more intuitive choice for the scaling variable. However, for $O(N)$-systems with $N \geq 2$ (XY or Heisenberg) the intuition is misleading, because the bulk correlation length is infinite for *all* $t < 0$ (see Sec.5.4 and Refs.[44] and [217]). In order to achieve the full equivalence between z and L/ξ we resrict the following considerations to the Ising universality class ($N = 1$) for short–ranged interactions.

The bulk two–point correlation function G, which is defined as the two–point cumulant of the local magnetization Φ at different positions $\mathbf{r}$ and $\mathbf{r}'$ [34, 35, 36, 37], displays scaling behavior in terms of $r = |\mathbf{r} - \mathbf{r}'|$ and the reduced renormalized temperature t governed by a universal scaling function K_+ for $t > 0$ and K_- for $t < 0$. From the solution of the renormalization group equation for G we find [188]

$$G_\pm(\mu, r, |t|, u) = (E_\eta^*(u))^{-2} r^{-(d-2)} (\mu r)^{-\eta} K_\pm \left(E_\nu^*(u)(\mu r)^{1/\nu} |t| \right), \qquad (5.9)$$

where η is the critical exponent of the bulk correlation function at criticality. The nonuniversal amplitude factor $E_\eta^*(u)$ is defined in terms of the Wilson functions $\beta(u)$ and $\eta(u)$ according to [34, 35, 36, 37]

$$E_\eta^*(u) = \exp \frac{1}{2} \int_u^{u^*} \frac{\eta - \eta(u')}{\beta(u')} du',$$

where $\eta \equiv \eta(u^*)$. A common definition of the bulk correlation length involves the second moment of $G_\pm$ according to (see also Ref.21 in Ref.[22])

$$\xi_\pm^2 = \frac{1}{2d} \frac{\int d^d r \, r^2 G_\pm(\mu, r, |t|, u)}{\int d^d r \, G_\pm(\mu, r, |t|, u)} \qquad (5.10)$$

which shows that a distinction between ξ_+ for $t > 0$ and ξ_- for $t < 0$ is neccessary. For the Ising universality class G_+ and G_- decay exponentially for $t > 0$ and $t < 0$, respectively, as $r \to \infty$, so that Eq.(5.10) is well defined. For the XY– and Heisenberg universality classes the definition in Eq.(5.10) becomes meaningless for $t < 0$ [44, 217]. Using Eq.(5.9) the bulk correlation length $\xi_\pm$ can be represented as [188]

$$\xi_\pm = \left(\tfrac{k_\pm}{2d}\right)^{1/2} \mu^{-1} \left[E_\nu^*(u)|t| \right]^{-\nu} \equiv \xi_0^\pm |t|^{-\nu}, \tag{5.11}$$

where

$$k_\pm = \frac{\int_0^\infty x^{(4-\eta)\nu-1} K_\pm(x)dx}{\int_0^\infty x^{(2-\eta)\nu-1} K_\pm(x)dx}$$

defines a *universal* ratio which depends on the sign of t. Note, that Eq.(5.11) only holds for $d < 4$. In $d = 4$ the correlation length $\xi_\pm$ is modified by $|\ln|t||$–corrections not captured in Eq.(5.11) (see Eq.(2.13) and Refs.[20] and [34]). Introducing a new scaling variable $y_\pm = L/\xi_\pm$ we find the desired equivalence between z and $L/\xi_\pm$ from Eq.(5.6) as

$$z = \left(\tfrac{k_\pm}{2d}\right)^{1/2\nu} y_\pm^{1/\nu} \tag{5.12}$$

and equivalent universal scaling functions $\tilde{H}_\pm(y_\pm)$ given by

$$\tilde{H}_\pm(y_\pm) \equiv H_\pm \left(\left(\tfrac{k_\pm}{2d}\right)^{1/2\nu} y_\pm^{1/\nu} \right). \tag{5.13}$$

For the Ising universality class the film thickness L can therefore always be scaled with the bulk correlation length $\xi_\pm$. The nonuniversality in the scaling variable z (see Eq.(5.6)) is now captured by the bulk correlation length amplitude $\xi_0^\pm$ (see Eq.(5.11)). Note, that $\xi_0^\pm$ depends on the *definition* of the bulk correlation length $\xi_\pm$, and therefore the *shape* of the scaling functions $\tilde{H}_\pm$ is universal, but it depends on the definition of $\xi_\pm$. In contrast the scaling functions $H_\pm$ do not show such a dependence, but on the other hand the amplitude factor $E_\nu^*(u)$ in the scaling argument z lacks the illustrative interpretation offered by Eq.(5.11). The scaling functions $\tilde{H}_\pm$ allow for an alternative representation of the unnormalized singular free energy f (see Eq.(5.7)) by

$$f(t, L, u, \mu) = L^{-(d-1)} \tilde{H}_\pm(L/\xi_\pm) \tag{5.14}$$

from which the renormalized free energy f^R can be deduced in terms of $\tilde{H}_\pm$ using Eq.(5.8) [188].

5.1.2   Bulk, Surface, and Finite Size Contributions

The renormalized free energy f^R and the unnormalized singular free energy f express the singular behavior of the free energy of the *film* in units of $k_B T_{c,b}$ and per cross–section area A. According to Eq.(3.8) both f and f^R and therefore the scaling function $\tilde{H}_\pm(y_\pm)$ can be decomposed into a bulk contribution, two surface contributions, and a finite size contribution. Specifically, we find [188]

$$f(t, L, u, \mu) = L^{-(d-1)} \tilde{H}_\pm(y_\pm)$$
$$= L f_{bulk}(t, u, \mu) + f_{s,a}(t, u, \mu) + f_{s,b}(t, u, \mu) + L^{-(d-1)} \Theta_\pm(y_\pm), \qquad (5.15)$$

where $\Theta_\pm$ denotes the *finite size scaling function* of the *finite–size part* of the unnormalized singular free energy f of the film for $t > 0$ or $t < 0$, respectively. The singular bulk free energy density $f_{bulk}(t, u, \mu)$ and the singular surface free energies $f_{s,i}(t, u, \mu)$, $i = a, b$ are given by simple power laws for the renormalized reduced temperature t characterized by the critical exponents α of the bulk specific heat and $\alpha_s = \alpha + \nu$ of the excess specific heat, respectively [20]. One finds [34, 35]

$$f_{bulk}(t, u, \mu) = -\frac{A_{bulk}^\pm}{\alpha(1 - \alpha)(2 - \alpha)} |t|^{2-\alpha} \qquad (5.16)$$

and [24, 211]

$$f_{s,i}(t, u, \mu) = -\frac{A_{s,i}^\pm}{\alpha_s(1 - \alpha_s)(2 - \alpha_s)} |t|^{2-\alpha_s}, \quad i = a, b, \qquad (5.17)$$

where the *nonuniversal* amplitudes $A_b^\pm$ and $A_{s,i}^\pm$ depend on the momentum scale μ and the renormalized coupling constant u. Note, that $f_{bulk}(t, u, \mu)$ in Eq.(5.16) only exists for $\alpha \neq 0, 1, 2$. Especially at the upper critical dimension $d = 4$ for usual critical behavior one has $\alpha = 0$, so that Eq.(5.16) becomes meaningless. However, the *renormalized* bulk free energy density f_b^R which can be constructed from Eq.(5.16) using the renormalization prescription given by Eq.(5.8) displays a $t^2 \ln |t|$–behavior in $d = 4$ [188] [7]. The same holds for the surface free energies $f_{s,i}(t, u, \mu)$ in Eq.(5.17) at the upper critical dimension $d = 3$ for tricritical phenomena, where $\alpha_s = 1$. In this case the renormalized surface free energies $f_{s,i}^R(t, u, \mu)$ exhibit $t \ln |t|$–behavior.

The hyperscaling relation $2 - \alpha = d\nu$ enables us to write the decomposition of the scaling function $\tilde{H}_\pm(y_\pm)$ given by Eq.(5.15) exclusively in terms of the scaling variable $y_\pm$. The result reads [188]

$$\tilde{H}_\pm(y_\pm) = -\frac{a_{bulk}^\pm}{\alpha(1 - \alpha)(2 - \alpha)} y_\pm^d - \frac{a_{s,a}^\pm + a_{s,b}^\pm}{\alpha_s(1 - \alpha_s)(2 - \alpha_s)} y_\pm^{d-1} + \Theta_\pm(y_\pm), \qquad (5.18)$$

---

[7]The logarithmic correction in the bulk correlation length $\xi_\pm$ in $d = 4$ (see Eq.(2.13)) has been disregarded here

where the amplitudes $a_b^\pm = A_a^\pm (\xi_0^\pm)^d$ and $a_{s,i}^\pm = A_{s,i}^\pm (\xi_0^\pm)^{d-1}$, $i = a, b$ are *universal* [34, 38, 188, 218] which is a direct consequence of the universality of $\tilde{H}_\pm$ as a *function*. Likewise, the scaling function $\Theta_\pm$ is universal as a *function*, but it depends on the definition of the bulk correlation length $\xi_\pm$ in the same way as $\tilde{H}_\pm$. Note again, that the decomposition of $\tilde{H}_\pm$ according to Eq.(5.18) only holds for $N = 1$; for $N = 2, 3$ the reduced temperature t is restricted to positive values. Following Eq.(5.7) we define the unnormalized singular finite size part δf of f by

$$\delta f(t, L, u, \mu) = L^{-(d-1)} \Theta_\pm(y_\pm). \tag{5.19}$$

The finite size contribution δf^R to the renormalized free energy f^R is then given by the renormalization prescription Eq.(5.8), where f is replaced by δf taken from Eq.(5.19). We obtain (see Ref.[188])

$$\delta f^R(t, L, u, \mu) = L^{-(d-1)} \left\{ \Theta_\pm(y_\pm) - \Theta_\pm(y_{1\pm}) - \nu \Theta_\pm'(y_{1\pm}) y_{1\pm}^{1-1/\nu} (y_\pm^{1/\nu} - y_{1\pm}^{1/\nu}) \right.$$

$$\left. - \frac{\nu}{2} \left[y_{1\pm}^{1-2/\nu} \left(\nu y_{1\pm} \Theta_\pm''(y_{1\pm}) + (\nu - 1) \Theta_\pm'(y_{1\pm}) \right) \right] (y_\pm^{1/\nu} - y_{1\pm}^{1/\nu})^2 \right\}, \tag{5.20}$$

where $y_{1\pm}$ is the reference value of $y_\pm$ corresponding to z_1 (see Eq.(5.6)) given by $y_{1\pm} = L/\xi_0^\pm$. At bulk criticality, i.e., $y_+ = y_- = 0$ the finite size scaling functions $\Theta_\pm$ reduce to the Casimir amplitude $\Delta = \Theta_+(0) = \Theta_-(0)$ (see Sec.3.2), and $\Theta_+(y_{1+})$ is exponentially small (see Sec.5.3). For macroscopically thick films we have $y_{1+} \gg 1$ and therefore $\Theta_+(0) = \Delta$ dominates all the other terms in Eq.(5.20) giving $\delta f^R(0, L, u, \mu) = \Delta L^{-(d-1)}$ in accordance with the semi–quantitative discussion in Sec.3.2 (see Eq.(3.11)).

5.2 Surface Tensions near Critical End Points

The finite size scaling functions $\tilde{H}_\pm(y_\pm)$ of the unnormalized singular part of the film free energy f (see Eq.(5.7)) do not only depend on the scaling argument $y_\pm$, they also explicitly display a dependence on the *boundary conditions* at the two walls. Whereas the bulk contribution to $\tilde{H}_\pm$ in Eq.(5.18), which is characterized by the universal *bulk* amplitude $a_{bulk}^\pm$ [188, 218], is common to all scaling functions $\tilde{H}_\pm$, the universal *surface* amplitudes $a_{s,i}^\pm$, $i = a, b$ which characterize the two surface contributions to $\tilde{H}_\pm$ do depend on the boundary conditions, i.e., the surface universality classes. For periodic and antiperiodic boundary conditions the amplitudes $a_{s,i}^\pm$ vanish, because there are no real surfaces in the system. For other boundary conditions the amplitudes $a_{s,i}^{\pm}$ are finite and multiply the singular part of the *surface* free energy which can be interpreted as the *surface tension*. Surface tensions and their singular behavior near critical points can be probed experimentally, and therefore the universal amplitudes $a_{s,i}^{\pm}$ are of experimental relevance [219]. The singular behavior of liquid–vapor surface tensions near a *bulk* critical point is typically realized near *critical end points* in the bulk phase diagram (see Figs.1.5, 1.6, and 1.7) which terminate a *line* of second order phase transitions within the liquid phase at a liquid–vapor coexistence line or surface. The most prominent examples of critical end points are given by the lower λ–point in ^{4}He which terminates the line of superfluid transitions at the liquid–vapor coexistence line (see Fig.1.6) and the critical end point of the demixing transitions in binary liquid mixtures which terminates the line of second order demixing transitions at the liquid–vapor coexistence surface (see Fig.1.5). The singular behavior of the surface tensions in these two examples coincides with the singular behavior of the surface free energy at the ordinary and the extraordinary surface transition, respectively, which takes place at the bulk critical point in a semi–infinite system. Moreover, these two cases are characterized by symmetry conserving (ordinary) and symmetry breaking (extraordinary) boundary conditions and therefore the critical end point itself has been termed *symmetric* in the first case and *nonsymmetric* in the second case [220, 221]. Note, that the vapor phase is treated as a rigid inert wall (see Sec.6.3).

The universal amplitudes $a_{s,i}^{\pm}$ can be found as by–products of the field–theoretical calculation of the renormalized free energy f^R (see Eq.(5.1)). Surface contributions can be easily identified as the L–independent terms in Eq.(5.15). Specifically, we find for the singular surface free energy above the ordinary surface transition [188] (see Eq.(5.17))

$$f_{s,0}(t, u^*, \mu) = -\frac{A_{s,0}^+}{\alpha_s(1-\alpha_s)(2-\alpha_s)}t^{2-\alpha_s}$$

$$= \frac{\mu^{d-1}N}{2^d\pi^{d/2}}\frac{\pi}{3}\left\{1 + \varepsilon\left(\frac{4}{3} - \ln 2 - \frac{\gamma}{2} + \frac{3}{4}\frac{N+2}{N+8}(\gamma-1)\right)\right\}t^{2-\alpha_s}, \quad (5.21)$$

where

$$\alpha_s = \alpha + \nu = \frac{1}{2} + \varepsilon\left(\frac{1}{2} - \frac{3}{4}\frac{N+2}{N+8}\right) + \mathcal{O}(\varepsilon^2). \quad (5.22)$$

$A_{s,0}^+$ denotes the nonuniversal amplitude of the singular free energy near the ordinary surface transition and $\gamma = 0.57721566\ldots$ is Euler's constant. The *universal* amplitude

$a_{s,O}^+$ is found to be [188]

$$a_{s,O}^+ = A_{s,O}^+ (\xi_0^+)^{d-1} \simeq -\frac{N}{2^d \pi^{d/2}} \frac{\pi}{8} \left\{ 1 + \varepsilon \left(1 - \ln 2 - \frac{\gamma}{2} + \frac{1}{2} \frac{N+2}{N+8} \right) \right\} \qquad (5.23)$$

in a partly reexponentialized form or

$$a_{s,O}^+ = -\frac{N}{256\pi} \left\{ 2 + \varepsilon \left(2 + \ln \pi - \gamma + \frac{N+2}{N+8} \right) + \mathcal{O}(\varepsilon^2) \right\} \qquad (5.24)$$

in strict ε–expansion. For the special surface transition we find [188]

$$f_{s,SB}(t, u^*, \mu) = -f_{s,O}(t, u^*, \mu) + \mathcal{O}(\varepsilon^2) \qquad (5.25)$$

for $t \geq 0$ which implies $A_{s,SB}^+ = -A_{s,O}^+ + \mathcal{O}(\varepsilon^2)$ and thus

$$a_{s,SB}^+ = -a_{s,O}^+ + \mathcal{O}(\varepsilon^2). \qquad (5.26)$$

Besides $a_{s,O}^\pm$ and $a_{s,SB}^\pm$ the *ratios* $A_{s,O}^+/A_{s,O}^-$ and $A_{s,SB}^+/A_{s,SB}^-$ are universal [38, 222]. These ratios are usually denoted by Q [220, 221], and their field–theoretical estimates for the ordinary and the special surface transition, respectively, are given in Ref.[222] as follows

$$Q_O = \frac{A_{s,O}^+}{A_{s,O}^-} = -\frac{\pi}{2^{3/2}} \frac{N}{N+8} \varepsilon + \mathcal{O}(\varepsilon^2), \quad Q_{SB} = \frac{A_{s,SB}^+}{A_{s,SB}^-} = -N 2^{-3/2} + \mathcal{O}(\varepsilon). \qquad (5.27)$$

For $N = 1$ Eq.(5.27) can be exploited to determine estimates for $a_{s,O}^-$ and $a_{s,SB}^-$ from $a_{s,O}^+$ and $a_{s,SB}^+$, respectively, by

$$\frac{a_{s,O}^-}{a_{s,O}^+} = \frac{1}{Q_O} \left(\frac{\xi_0^-}{\xi_0^+} \right)^{d-1} \quad \text{and} \quad \frac{a_{s,SB}^-}{a_{s,SB}^+} = \frac{1}{Q_{SB}} \left(\frac{\xi_0^-}{\xi_0^+} \right)^{d-1} .$$

The universal ratio of the bulk correlation length amplitudes below and above $T_{c,b}$ is well known [218]

$$R \equiv \frac{\xi_0^-}{\xi_0^+} = 2^{-\nu} \left(1 - \tfrac{5}{24} \varepsilon + \mathcal{O}(\varepsilon^2) \right), \qquad (5.28)$$

and we finally have [188]

$$a_{s,O}^- = \frac{9}{128\pi^2} \frac{1}{\varepsilon} + \mathcal{O}(\varepsilon^0) \quad \text{and} \quad a_{s,SB}^- = -\frac{1}{128\pi} + \mathcal{O}(\varepsilon) \qquad (5.29)$$

for the universal amplitudes $a_{s,O}^-$ and $a_{s,SB}^-$ of the singular surface tension for the ordinary and the special surface transition *below* a critical end point in the Ising universality class ($N = 1$). A realization of a *symmetric* critical end point for $N = 1$ to which the ordinary surface transition and the universal amplitude $a_{s,O}^-$ correspond is provided by binary alloys with order – disorder transitions [220].

The expression for $a_{s,O}^+$ has recently been rederived in Ref.[223] in view of a theoretical interpretation of experimental data of the liquid–vapor surface tension of ^{4}He at the lower λ–point. The analysis indicates that the partly reexponentialized form of $a_{s,O}^+$ given by Eq.(5.23) is consistent with the experimental data for $d = 3$ ($\varepsilon = 1$) and $N = 2$. In the notation of Ref.[223] one has

$$R_{\sigma\xi}^+ = -\frac{a_{s,O}^+}{\alpha_s(1 - \alpha_s)(2 - \alpha_s)} \simeq 0.071, \qquad (5.30)$$

where $\alpha_s \simeq 0.66$ has been used. At the extraordinary transition the universal amplitude ratio Q_E has been calculated by field–theoretical methods for $N = 1$ giving [224]

$$Q_E = -\sqrt{2}\left\{1 + \varepsilon\left(\frac{1}{4} - \frac{5\pi}{36} - \frac{\pi}{6\sqrt{3}} + \frac{\ln 2}{4} + \frac{\ln(2 - \sqrt{3})}{\sqrt{3}}\right) + \mathcal{O}(\varepsilon^2)\right\}. \qquad (5.31)$$

An alternative way to obtain the above universal amplitude ratios using *local free-energy functionals* has been suggested in Refs.[221] and [224]. These functionals are designed such that nonclassical critical exponents and certain properties of the order parameter profile are incorporated (see Eq.(4.4)). These results can be shown to be consistent with the ε–expansion (see, e.g., Eq.(5.31)) [224], but the numerical predictions for $d = 3$ (i.e., $\varepsilon = 1$) deviate strongly. For example, one obtains $Q_E = -0.83$ in $d = 3$ [221] compared to $Q_E = 0.1$ from Eq.(5.31) for $\varepsilon = 1$. However, one expects that the local free–energy functional yields an estimate for the value of the universal amplitude ratios discussed above with an accuracy of a few percent in $d = 3$ [221].

5.3 $T > T_{c,b}$: *The Scaling Functions* Θ_+

The scaling functions $\tilde{H}_+(y_+)$ govern the finite size scaling behavior of the unnormalized singular free energy f (see Eq.(5.7)) and the renormalized free energy f^R (see Eq.(5.8)) in terms of the scaling variable $y_+ = L/\xi_+$ *above* bulk criticality. In this case ξ_+ denotes the *bulk correlation length* $\xi_+ = \xi_0^+|t|^{-\nu}$ (see Eq.(5.11)). Likewise, the *finite-size* contribution $\Theta_+(y_+)$ to $\tilde{H}_+(y_+)$ which has been introduced in Eq.(5.15)

(see also Eq.(5.18)) governs the scaling behavior of the finite–size part δf of f (see
Eq.(5.19)) and of the finite–size part δf^R of the renormalized free energy f^R (see
Eq.(5.20)). Like $\tilde{H}_+$, Θ_+ and therefore δf and δf^R depend on the boundary conditions at the two walls in the film geometry. In the following this additional dependence
will be referenced by either two subscripts a and b or a single subscript per or $aper$ assigned to Θ_+. As in Sec.3.2 per and $aper$ denote periodic and antiperiodic boundary
conditions, respectively, whereas a and b indicate the surface universality classes to
which the two walls belong. For the five symmetry–conserving boundary conditions
$(a, b) = (O, O)$, (O, SB), and (SB, SB) and (per), $(aper)$ (see Sec.3.2) the finite–size
scaling functions $\Theta_{+a,b}(y_+)$ have been calculated in $d = 4 - \varepsilon$ for $O(N)$–symmetric
critical systems using the field–theoretical renormalization group [188, 189]. At the
upper critical dimension $d = 3$ of a tricritical point or end point corresponding results for the finite–size scaling functions of tricritical films are available [188]. For
symmetry–breaking boundary conditions like $(a, b) = (+, +)$ or $(+, -)$, however, the
scaling functions $\Theta_{+a,b}(y_+)$ in $d = 3$ are still unknown. In two dimensions the scaling
functions $\Theta_{\pm a,b}$ can be determined for an Ising strip with $(a, b) = (O, O)$, $(+, +)$, and
$(+, -)$ by means of $exact$ transfer matrix calculations [225].

The field–theoretical calculation of Θ_+ starts from the $bare$ free energy $\bar{f}(\tau, L, g)$
of the film. The finite–size contribution to $\bar{f}$ can be easily identified by inspecting
the L–dependence of $\bar{f}$. The terms proportional to L give the bulk contribution, and
the L–independent terms form the two surface contributions, where the latter ones
are absent for periodic and antiperiodic boundary conditions. The remaining terms
in $\bar{f}$ can be renormalized (see Eq.(5.20)) and thus give access to $\Theta_{+a,b}(y_+)$ [188].
The dimensional regularization procedure used in Ref.[188] to calculate $\bar{f}$ is closely
related to the zeta–function regularization, where in this case the so called Epstein
zeta–function [179] as a function of the spatial dimension d is involved. The zeta-
function regularization and the Epstein zeta–functions have been widely used for the
calculation of the Casimir effect and for the solution of related finite–size problems
(see Sec.3.1 and Refs.[180, 181, 182, 226, 227, 228]).

5.3.1   Above Bulk Criticality

For an $O(N)$–symmetric Gaussian Ginzburg–Landau Hamiltonian at $T > T_{c,b}$ in a
d–dimensional film geometry with symmetry conserving boundary conditions a one–

loop calculation yields the following results for the finite–size scaling functions Θ_+ [188]

$$\Theta^{(1)}_{+O,O}(y_+) = -\frac{N}{2^d\pi^{d/2}}\frac{2\sqrt{\pi}y_+^d}{\Gamma\left(\frac{d+1}{2}\right)}\int_1^\infty \frac{(x^2-1)^{(d-1)/2}}{e^{2xy_+}-1}dx = \Theta^{(1)}_{+SB,SB}(y_+),$$

$$\Theta^{(1)}_{+O,SB}(y_+) = \frac{N}{2^d\pi^{d/2}}\frac{2\sqrt{\pi}y_+^d}{\Gamma\left(\frac{d+1}{2}\right)}\int_1^\infty \frac{(x^2-1)^{(d-1)/2}}{e^{2xy_+}+1}dx,$$

$$\Theta^{(1)}_{+per}(y_+) = -\frac{N}{2^d\pi^{d/2}}\frac{2\sqrt{\pi}y_+^d}{\Gamma\left(\frac{d+1}{2}\right)}\int_1^\infty \frac{(x^2-1)^{(d-1)/2}}{e^{xy_+}-1}dx,$$

$$\Theta^{(1)}_{+aper}(y_+) = \frac{N}{2^d\pi^{d/2}}\frac{2\sqrt{\pi}y_+^d}{\Gamma\left(\frac{d+1}{2}\right)}\int_1^\infty \frac{(x^2-1)^{(d-1)/2}}{e^{xy_+}+1}dx. \tag{5.32}$$

For $y_+ \to 0$ the scaling functions $\Theta^{(1)}_{+a,b}$ approach the corresponding Casimir amplitudes $\Delta^{(1)}_{a,b}$ (see Eq.(3.12)). The simple relations among the Casimir amplitudes $\Delta^{(1)}_{a,b}$ given by Eq.(3.12) reflect the following relations between $\Theta^{(1)}_{+a,b}$:

$$\Theta^{(1)}_{+SB,SB}(y_+) = \Theta^{(1)}_{+O,O}(y_+),$$
$$\Theta^{(1)}_{+O,SB}(y_+) = 2^{1-d}\Theta^{(1)}_{+O,O}(2y_+) - \Theta^{(1)}_{+O,O}(y_+),$$
$$\Theta^{(1)}_{+per}(y_+) = 2^d\Theta^{(1)}_{+O,O}\left(\frac{y_+}{2}\right),$$
$$\Theta^{(1)}_{+aper}(y_+) = 2^d\Theta^{(1)}_{+O,SB}\left(\frac{y_+}{2}\right). \tag{5.33}$$

Note, that Eq.(5.33) is only correct within a Gaussian model or a one–loop calculation. Non–Gaussian fluctuations which become relevant in a $(\Phi^2)^2$–theory in $d = 4-\varepsilon$ destroy these relations as will be shown below. *At* the upper critical dimension $d = 4$, however, the scaling functions $\Theta^{(1)}_{+a,b}$ given by Eq.(5.32) are exact apart from logarithmic corrections in the scaling argument (see Eq.(2.13)), and the relations in Eq.(5.33) hold rigorously. The exponential decay of the scaling functions $\Theta_{+a,b}$ for $y_+ \to \infty$ which has been anticipated in Secs.3.2 and 5.1 is now directly visible in Eq.(5.32). As we will see below the exponential is multiplied by *powers* of the scaling variable y_+ so that the decay is not purely exponential. The overall signs of the scaling functions $\Theta^{(1)}_{+a,b}$ are in accordance with the signs of the corresponding Casimir amplitudes $\Delta^{(1)}_{a,b}$ (see Eq.(3.12)).

In order to estimate the scaling functions $\Theta_{+a,b}$ given by Eq.(5.32) in $d = 3$ the perturbation theory for the bare free energy $\bar{f}(\tau, L, g)$ of the film has been carried to two–loop order using dimensional regularization and ε–expansion techniques. To first order in ε we obtain the following results for the finite–size scaling functions $\Theta_{+a,b}$ of a critical film with symmetry–conserving boundary conditions in $d = 4 - \varepsilon$ [188, 189]:

$$\Theta_{+0,0}(y_+) = \frac{N y_+^4}{8\pi^2} \left\{ \left[-2 + \varepsilon \left(\gamma - \frac{8}{3} + \ln \frac{y_+^2}{\pi} \right) \right] g_{3/2,0}(y_+) \right.$$
$$+ \left. \varepsilon \left[g_{3/2,1}(y_+) + \frac{N+2}{N+8} \left(\frac{\pi}{y_+} g_{1/2,0}(y_+) + \left(g_{1/2,0}(y_+) \right)^2 \right) \right] \right\},$$

$$\Theta_{+SB,SB}(y_+) = \frac{N y_+^4}{8\pi^2} \left\{ \left[-2 + \varepsilon \left(\gamma - \frac{8}{3} + \ln \frac{y_+^2}{\pi} \right) \right] g_{3/2,0}(y_+) \right.$$
$$+ \left. \varepsilon \left[g_{3/2,1}(y_+) + \frac{N+2}{N+8} \left(-\frac{\pi}{y_+} g_{1/2,0}(y_+) + \left(g_{1/2,0}(y_+) \right)^2 \right) \right] \right\},$$

$$\Theta_{+0,SB}(y_+) = \frac{N y_+^4}{8\pi^2} \left\{ \left[2 - \varepsilon \left(\gamma - \frac{8}{3} + \ln \frac{y_+^2}{\pi} \right) \right] h_{3/2,0}(y_+) \right.$$
$$- \left. \varepsilon \left[h_{3/2,1}(y_+) - \frac{N+2}{N+8} \left(h_{1/2,0}(y_+) \right)^2 \right] \right\},$$

$$\Theta_{+per}(y_+) = \frac{N y_+^4}{8\pi^2} \left\{ \left[-2 + \varepsilon \left(\gamma - \frac{8}{3} + \ln \frac{y_+^2}{\pi} \right) \right] g_{3/2,0} \left(\frac{y_+}{2} \right) \right.$$
$$+ \left. \varepsilon \left[g_{3/2,1} \left(\frac{y_+}{2} \right) + \frac{N+2}{N+8} \left(g_{1/2,0} \left(\frac{y_+}{2} \right) \right)^2 \right] \right\},$$

$$\Theta_{+aper}(y_+) = \frac{N y_+^4}{8\pi^2} \left\{ \left[2 - \varepsilon \left(\gamma - \frac{8}{3} + \ln \frac{y_+^2}{\pi} \right) \right] h_{3/2,0} \left(\frac{y_+}{2} \right) \right.$$
$$- \left. \varepsilon \left[h_{3/2,1} \left(\frac{y_+}{2} \right) - \frac{N+2}{N+8} \left(h_{1/2,0} \left(\frac{y_+}{2} \right) \right)^2 \right] \right\}, \tag{5.34}$$

where the functions $g_{a,b}(y)$ and $h_{a,b}(y)$ are defined by [188]

$$g_{a,b}(y) = \frac{1}{a} \int_1^\infty \frac{(x^2-1)^a}{e^{2xy}-1} \left(\ln(x^2-1) \right)^b dx, \quad h_{a,b}(y) = \frac{1}{a} \int_1^\infty \frac{(x^2-1)^a}{e^{2xy}+1} \left(\ln(x^2-1) \right)^b dx. \tag{5.35}$$

The scaling functions $\Theta_{+a,b}$ given by Eq.(5.34) form the basis of our following discussion of experiments at $T > T_{c,b}$ (see Secs.6.1, 6.3, and 6.4). At $T = T_{c,b}$, i.e., $y_+ = 0$ these scaling functions reduce to the corresponding ε–expanded Casimir amplitudes $\Delta_{a,b}$ given by Eq.(3.13). Unfortunately, any attempt to construct partly reexponentialized scaling functions $\Theta_{+a,b}$ from Eq.(5.34) which are *consistent* with the partly reexponentialized Casimir amplitudes in Eq.(3.16) has so far been unsuccessful. In order to make predictions for $d = 3$ we must therefore rely on the ε-expansion in Eq.(5.34) evaluated at $\varepsilon = 1$. As it has been pointed out in Sec.3.2 we cannot expect good numerical accuracy from a first order ε-expansion. However, in view of the new information contained in Eq.(5.34) its limited numerical accuracy is only a minor drawback.

First, we observe that the "power–times–exponential" behavior of the scaling functions $\Theta_{+a,b}(y_+)$ for $y_+ \to \infty$ suggested by Eq.(5.32) is confirmed by the two–loop results in Eq.(5.34). An asymptotic expansion of the functions $g_{a,b}(y)$ and $h_{a,b}(y)$ given by Eq.(5.35) for $y_+ \to \infty$ yields the following asymptotic behavior of the scaling functions for $y_+ \to \infty$ [188]

$$\Theta_{+O,O}(y_+) \simeq -\frac{N}{16\pi^{3/2}}\left[1 + \varepsilon\left(\ln(2\sqrt{\pi}) - \frac{N+2}{N+8}\pi\right)\right]y_+^{(d-1)/2}e^{-2y_+},$$

$$\Theta_{+SB,SB}(y_+) \simeq -\frac{N}{16\pi^{3/2}}\left[1 + \varepsilon\left(\ln(2\sqrt{\pi}) + \frac{N+2}{N+8}\pi\right)\right]y_+^{(d-1)/2}e^{-2y_+},$$

$$\Theta_{+O,SB}(y_+) \simeq \frac{N}{16\pi^{3/2}}\left[1 + \varepsilon\ln(2\sqrt{\pi})\right]y_+^{(d-1)/2}e^{-2y_+},$$

$$\Theta_{+per}(y_+) \simeq -\frac{N}{(2\pi)^{3/2}}\left[1 + \frac{\varepsilon}{2}\ln(2\pi)\right]y_+^{(d-1)/2}e^{-y_+},$$

$$\Theta_{+aper}(y_+) \simeq \frac{N}{(2\pi)^{3/2}}\left[1 + \frac{\varepsilon}{2}\ln(2\pi)\right]y_+^{(d-1)/2}e^{-y_+}, \tag{5.36}$$

where only the *leading* behavior is shown in Eq.(5.36). The power $y_+^{(d-1)/2}$ has been obtained from a partial reexponentialzation of the corresponding ε-expansion result. For $y_+ \to \infty$ the scaling functions $\Theta_{+O,O}$, $\Theta_{+SB,SB}$, and $\Theta_{+O,SB}$ obviously decay twice as fast as Θ_{+per} and Θ_{+aper}. Note, that due to the universality of $\Theta_{+a,b}(y_+)$ as a *function* the prefactors in Eq.(5.36) define a new set of *universal amplitudes* which depend on the *definition* of the correlation length (see Eq.(5.10)). Second, the differences between the scaling functions which are already visible in Eq.(5.32) are enhanced by the two–loop contributions so that especially $\Theta_{+O,O}(y_+) \neq \Theta_{+SB,SB}(y_+)$ although $\Theta_{+O,O}(0) = \Theta_{+SB,SB}(0)$ and $\Theta_{+O,O}^{(1)}(y_+) = \Theta_{+SB,SB}^{(1)}(y_+)$. Simple relations like Eq.(3.14) for the Casimir amplitudes or like Eq.(5.33) for the one–loop scaling functions $\Theta_{+a,b}^{(1)}$ can therefore not be expected to hold for the full scaling functions

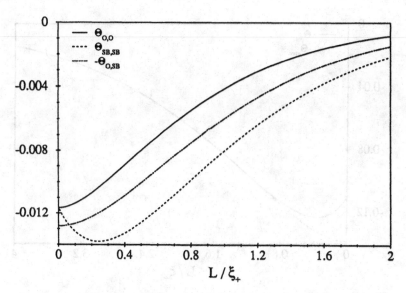

Fig. 5.1: Scaling functions $\Theta_{+O,O}$ (solid line), $\Theta_{+SB,SB}$ (dashed line) and $-\Theta_{+O,SB}$ (dash–dotted line) as functions of $y_+ = L/\xi_+$ for $N = 1$ and $d = 3$. The curves display the numerical evaluation of Eq.(5.34) for $\varepsilon = 1$. ξ_+ is defined in Eq.(5.10). At $y_+ = 0$ the functions reduce to the corresponding Casimir amplitudes (see Eq.(3.13) for $\varepsilon = 1$).

$\Theta_{+a,b}$. Third, Eq.(5.36) suggests that the overall sign of the scaling functions at the two–loop level corresponds to the sign of the one–loop results given by Eq.(5.32). A numerical evaluation of Eq.(5.34) for $\varepsilon = 1$ $(d = 3)$ confirms this conjecture for $\Theta_{+O,O}$, $\Theta_{+SB,SB}$, and $\Theta_{+O,SB}$ as shown in Fig.5.1 and for Θ_{+per} and Θ_{+aper} as shown in Fig.5.2 [188]. Thus for *any* temperature $T > T_{c,b}$ with $T - T_{c,b} \ll T_{c,b}$ critical fluctuations induce an *additional* force between the boundaries of a film which forms a correction to other *noncritical* forces (e.g., van–der–Waals forces, see Sec.3.1). For *like* boundaries the correction is *attractive*, for *unlike* it is *repulsive* (see also Ref.[174]). Although $\Theta_{+O,O}$ and $\Theta_{+SB,SB}$ approach the same Casimir amplitude as $y_+ \to 0$, the *shapes* of $\Theta_{+O,O}$ and $\Theta_{+SB,SB}$ are qualitatively different. $\Theta_{+SB,SB}$ displays a minimum at $T > T_{c,b}$ for any L so that $\Theta_{+SB,SB}$ seems to be *shifted* to the right with respect to $\Theta_{+O,O}$. Apart from the overall sign the shape of $\Theta_{O,SB}$ is similar to the shape of $\Theta_{+O,O}$, but the two corresponding curves roughly maintain their mutual

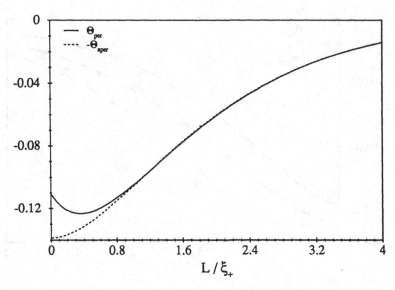

Fig. 5.2:   Scaling functions Θ_{+per} (solid line) and $-\Theta_{+aper}$ (dashed line) as functions of $y_+ = L/\xi_+$ for $N = 1$ and $d = 3$. The curves display the numerical evaluation of Eq.(5.34) for $\varepsilon = 1$. ξ_+ is defined in Eq.(5.10). Note the difference in scale of both axes compared with Fig.5.1. At $y_+ = 0$ the functions reduce to the corresponding Casimir amplitudes (see Eq.(3.13) for $\varepsilon = 1$).

distance over the whole range of y_+ displayed in Fig.5.1. Likewise, the shapes of the scaling functions Θ_{+per} and $\Theta_{+SB,SB}$ shown in Figs.5.2 and 5.1, respectively, coincide, where the magnitude of the former exceeds the magnitude of the latter by about one order of magnitude. The same statement can be made for Θ_{+aper} (see Fig.5.2) in comparison with $\Theta_{+O,SB}$ (see Fig.5.1). The difference in magnitude between Θ_{+per} and $\Theta_{+SB,SB}$ on one hand and Θ_{+aper} and $\Theta_{+O,SB}$ on the other hand is in accordance with the difference between the corresponding Casimir amplitudes given by Eq.(3.13) (see Table 3.1). It is interesting to note that periodic and antiperiodic boundary conditions which are often used in computer simulations cause a finite–size effect in the singular part of the free energy which is an order of magnitude larger than for any other symmetry–conserving boundary condition. Unlike $\Theta_{+SB,SB}$ and $-\Theta_{+O,SB}$ and in accordance with Eq.(5.36) Θ_{+per} and $-\Theta_{+aper}$ become almost equal for $y_+ > 1$.

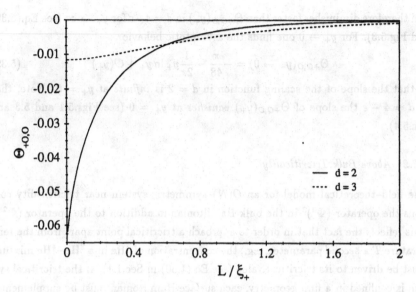

Fig. 5.3: Scaling functions $\Theta_{+0,0}(y_+)$ for $N = 1$ in $d = 2$ (solid line) and in $d = 3$ (dashed line). The scaling function $\Theta_{+0,0}(y_+)$ is exactly known in $d = 2$ and for $N = 1$ [225] (see Eq.(5.37)). The dashed line shows $\Theta_{+0,0}(y_+)$ according to Eq.(5.34) for $\varepsilon = 1$. At $y_+ = 0$ the scaling functions reduce to the corresponding Casimir amplitudes.

Note, that $\Theta_{+0,0}(y_+)$ for $N = 2$ applies to films of ^{4}He for $T > T_\lambda$ concerning the film specific heat (see Sec.6.1), wetting films (see Sec.6.3), and the Casimir force (see Sec.6.4).

In $d = 2$ and for $N = 1$ the *exact* scaling function $\Theta_{+0,0}(y_+)$ can be derived from the transfer matrix calculations presented in Ref.[225]. One obtains

$$\Theta_{+0,0}(y_+) = -\frac{1}{4\pi}\int_{2y_+}^{\infty}\frac{x}{\sqrt{x^2 - 4y_+^2}}\ln\left(1 + \frac{x - 2y_+}{x + 2y_+}e^{-x}\right)dx, \qquad (5.37)$$

where $y_+ = L/\xi_+$, and the bulk correlation length ξ_+ has been chosen according to Ref.[50]. The scaling functions $\Theta_{+0,0}(y_+)$ in $d = 2$ according to Eq.(5.37) and in $d = 3$ according to Eq.(5.34) for $\varepsilon = 1$ are displayed in Fig.5.3. For $y_+ \to \infty$ the scaling function in two dimensions decays as

$$\Theta_{+0,0}(y_+ \to \infty) \simeq \frac{-1}{32\sqrt{\pi}}y_+^{-1/2}e^{-2y_+} \qquad (5.38)$$

and therefore diminishes faster than $\Theta_{+0,0}(y_+)$ in $d = 4-\epsilon$ for $y_+ \rightarrow \infty$ (see Eq.(5.36) and Fig.5.3). For $y_+ \rightarrow 0$ one finds the asymptotic behavior

$$\Theta_{+0,0}(y_+ \rightarrow 0) = -\frac{\pi}{48} - \frac{1}{2\pi}y_+ \ln y_+ + \mathcal{O}(y_+) \tag{5.39}$$

so that the slope of the scaling function in $d = 2$ is *infinite* at $y_+ = 0$. Note, that in $d = 4 - \epsilon$ the slope of $\Theta_{+0,0}(y_+)$ *vanishes* at $y_+ = 0$ (see Figs.5.1 and 5.3 and Sec.5.4).

5.3.2 Above Bulk Tricriticality

The field–theoretical model for an $O(N)$–symmetric system near tricriticality contains the operator $(\Phi^2)^3$ in the bulk Hamiltonian in addition to the operator $(\Phi^2)^2$. This reflects the fact that in order to approach a tricritical point apart from the temperature T a *second* parameter, e.g., the concentration of ^{3}He in a ^{3}He $-^4$He mixture, must be driven to its tricritical value (see Eq.(1.30) in Sec.1.1). If the tricritical system is confined in a film geometry, each surface Hamiltonian must be supplemented by the quartic surface operator $(\Phi_s^2)^2$ which makes the fieldtheoretical analysis even more involved (see Eq.(1.31) in Sec.2.3). However, *at* the upper critical dimension $d = 3$ which is the interesting case for a film geometry the field theory of a tricritical system basically reduces to a Gaussian theory. Therefore the *tricritical* finite–size scaling functions $\Theta_{+a,b}$ are already given by Eq.(5.32) for $d = 3$. The relations in Eq.(5.33) now hold rigorously and we find [188]

$$\Theta_{+0,0}(y_+) = -\frac{N}{16\pi}\left[\mathcal{L}_3\left(e^{-2y_+}\right) + 2y_+\mathcal{L}_2\left(e^{-2y_+}\right)\right],$$

$$\Theta_{+SB,SB}(y_+) = \Theta_{+0,0}(y_+),$$

$$\Theta_{+0,SB}(y_+) = \frac{1}{4}\Theta_{+0,0}(2y_+) - \Theta_{+0,0}(y_+),$$

$$\Theta_{+per}(y_+) = 8\,\Theta_{+0,0}\left(\tfrac{y_+}{2}\right),$$

$$\Theta_{+aper}(y_+) = 8\,\Theta_{+0,SB}\left(\tfrac{y_+}{2}\right), \tag{5.40}$$

where $\mathcal{L}_k(x)$ denotes the polylogarithm given by $\mathcal{L}_1 = -\ln(1-x)$ and the recursion relation $\mathcal{L}_k(x) = \int_0^x t^{-1}\mathcal{L}_{k-1}(t)dt$. The bulk correlation length ξ_+ above the bulk

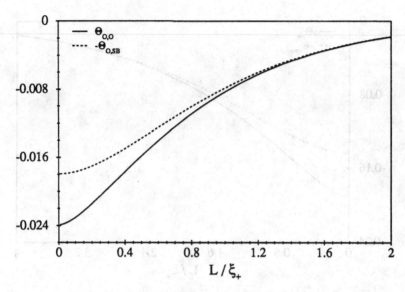

Fig. 5.4:   Scaling functions $\Theta_{+O,O} = \Theta_{+SB,SB}$ (solid line) and $-\Theta_{+O,SB}$ (dashed line) as functions of $y_+ = L/\xi_+$ for $N = 1$. The curves display the numerical evaluation of Eq.(5.40), where ξ_+ is given by Eq.(5.41). At $y_+ = 0$ the functions reduce to the corresponding Casimir amplitudes (see Eq.(3.5)). Note the difference in scale on the Θ–axis compared with Fig.5.1

tricritical temperature $T_{t,b}$ which enters the scaling variable $y_+ = L/\xi_+$ is given by the Gaussian (or mean–field) expression [229, 230] (see Eqs.(5.10) and (5.11))

$$\xi_+ = \mu^{-1}t^{-1/2}, \tag{5.41}$$

where μ is the momentum scale defined in Eq.(5.2), and t denotes the renormalized reduced temperature with respect to the bulk tricritical temperature $T_{t,b}$.

From Eq.(5.32) we immediately conclude that the tricritical scaling functions given by Eq.(5.40) reduce to the corresponding *tricritical* Casimir amplitudes as $y_+ \to 0$ (see Table 3.5). The asymptotic behavior of the scaling functions Θ_+ is in accordance with Eq.(5.36) as far as the "power–times–exponential" decay for $y_+ \to \infty$ is concerned. Specifically, we find from Eq.(5.40)

$$\Theta_{+O,O}(y_+) = \Theta_{+SB,SB}(y_+) \simeq -\frac{N}{8\pi}y_+e^{-2y_+},$$

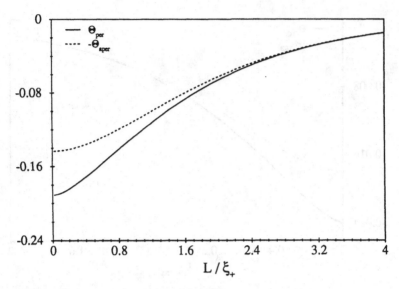

Fig. 5.5:   Scaling functions Θ_{+per} (solid line) and $-\Theta_{+aper}$ (dashed line) as functions of $y_+ = L/\xi_+$ for $N = 1$. The curves display the numerical evaluation of Eq.(5.40), where ξ_+ is given by Eq.(5.41). At $y_+ = 0$ the functions reduce to the corresponding Casimir amplitudes (see Eq.(3.5)). Note the difference in scale on both axes compared with Figs.5.1 and 5.2

$$\Theta_{+O,SB}(y_+) \simeq \frac{N}{8\pi}y_+e^{-2y_+},$$

$$\Theta_{+per}(y_+) \simeq -\frac{N}{2\pi}y_+e^{-y_+},$$

$$\Theta_{+aper}(y_+) \simeq \frac{N}{2\pi}y_+e^{-y_+}, \tag{5.42}$$

where again only the *leading* term is shown. According to Eq.(5.42) the scaling functions $\Theta_{+O,O},\Theta_{+SB,SB}$, and $\Theta_{+O,SB}$ decay twice as fast as Θ_{+per} and Θ_{+aper} (see Eq.(5.36)). The overall sign of the scaling functions agrees with the sign of the tricritical Casimir amplitudes in Table 3.5 so that the Casimir forces are attractive for like boundaries and repulsive for different boundaries. A numerical evaluation of $\Theta_{+O,O}(y_+) = \Theta_{+SB,SB}(y_+)$ and $\Theta_{O,SB}(y_+)$ is shown in Fig.5.4, $\Theta_{+per}(y_+)$ and

$\Theta_{+aper}(y_+)$ are shown in Fig.5.5. As we can infer from Eq.(5.40) the shapes of $\Theta_{+O,O}$ and Θ_{+per} on one hand and $\Theta_{+O,SB}$ and Θ_{+aper} on the other hand coincide if the scaling argument y_+ is properly rescaled. Contrary to Figs.5.1 and 5.2 there is no minimum at $y_+ > 0$ in any of the tricritical scaling functions shown in Figs.5.4 and 5.5. This in turn implies that the minimum of $\Theta_{+SB,SB}$ and Θ_{+per} shown in Fig.5.1 and 5.2, respectively, is a distinguished two–loop effect. Note the difference in scale between Figs.5.4 and 5.5 which reflects the relations among the scaling functions given by Eq.(5.40). Furthermore, the tricritical scaling functions are approximately twice as large as the corresponding critical ones over the whole range of the scaling variable y_+ (compare Figs.5.4 and 5.5 with Figs.5.1 and 5.2). This is in accordance with the difference in magnitude between the tricritical Casimir amplitudes (see Table 3.5) and their critical counterparts (see Table 3.1). Finally, we remark that for $N = 2$ the tricritical scaling function $\Theta_{+O,O}(y_+)$ applies to films of ^{3}He–^{4}He mixtures above tricriticality (see Secs.6.1, 6.3, and 6.4).

5.4 $T < T_{c,b}$: The Scaling Functions Θ_-

The finite–size dependence of the singular part of the free energy of a film has the scaling form (see Eq.(5.7) and Refs.[51, 52, 43]), where the scaling argument z is in general given by $z = C_1|t|L^{1/\nu}$ [43] (see Eq.(5.6)). The scaling variable z only depends on the absolute value $|t|$ of the reduced temperature t so that we can assume z to be nonnegative. On the other hand scaling functions are in general very sensitive to the sign of t. For a full description of the finite–size scaling behavior of the singular part of the free energy (see Eqs.(5.7) and (5.8)) or the scaling behavior of the bulk two–point correlation function (see Eq.(5.9)) the knowledge of two *independent* scaling functions is neccessary, one for $t > 0$ and another one for $t < 0$ (see Sec.5.1).

For $t > 0$ the bulk two–point correlation function $G_+(\mu, r, |t|, u)$ (see Eq.(5.9)) of $O(N)$–symmetric systems with short–ranged interactions decays exponentially as $r \to \infty$. The decay length of G_+ is the bulk correlation length ξ_+ which can be defined via the second moment of G_+ with respect to r (see Eq.(5.10)). The scaling variable $y_+ = L/\xi_+$ which has been used in Sec.5.3 in order to display the L–dependence of the scaling functions clearly is fully equivalent to z given by Eq.(5.6) (see Eq.(5.12) for $t > 0$). As already indicated in the discussion following Eq.(5.10) the situation is different for $t < 0$. In the Ising universality class ($N = 1$) the two–point correlation function $G_-(\mu, r, |t|, u)$ again decays exponentially for $r \to \infty$ (short–ranged interactions) so that the corresponding bulk correlation length ξ_- can be defined by Eq.(5.10) rendering the full equivalence between the scaling variables z and $y_\pm$ (see Eq.(5.12)). In the XY– and Heisenberg universality classes ($N = 2, 3$) the symmetry of the Hamiltonian (see Eq.(2.48)) is continuous which leads to the occurence of *Goldstone modes* for $t < 0$ [34, 35, 36, 37]. For XY– and Heisenberg systems Goldstone modes can be visualized as *spin waves*. Generally speaking these spin waves are collective excitations of the system which lead to long–ranged correlations of the order parameter and therefore to a *power–law* decay of the correlation function $G_-(\mu, r, |t|, u)$ for *any* $t < 0$ as r increases [44, 217]. In this case the definition of ξ_- given by Eq.(5.10) becomes meaningless and the bulk correlation length ξ_- is infinite. Nonetheless $z^\nu = C_1^\nu L |t|^\nu$ (see Eq.(5.6) and Ref.[43]) is a valid replacement of the scaling variable z in the finite size scaling function H_- (see Eq.(5.5)). Only the length scale $C_1^{-\nu} |t|^{-\nu}$ is *no longer* given by a bulk correlation length $\xi_- = \xi_0^- |t|^{-\nu}$ for $N = 2, 3$. For the discussion of the scaling functions Θ_- the universal amplitude ratio $R = \xi_0^- / \xi_0^+$ given by Eq.(5.28) will be used in various places [188]. Because R is not defined for $N = 2, 3$, we restrict the following considerations to the Ising universality class.

In the film geometry studied here at least two distinct types of correlations must be distinguished due to the anisotropy of the film: correlations parallel to the walls characterized by a correlation length $\xi_{pa}(t, L)$ and correlations orthogonal to the walls characterized by a correlation length $\xi_{or}(t, L)$ [188]. The growth of ξ_{or} is clearly limited by the film thickness L, whereas ξ_{pa} can grow without limit. As soon as $\xi_{or}(t, L)$ becomes comparable to L the system crosses over from d–dimensional to $(d-1)$–dimensional critical behavior. If the surface couplings are not too strong, the $(d-1)$–dimensional phase transition will take place at a *shifted* critical temperature $T_c(L) < T_{c,b}$ (see Eq.(2.5)), because in low–dimensional systems fluctuations are more violent. In the presence of surface fields the critical point is additionally shifted to a finite value of a critical field $H_c(L)$ (see Eq.(2.5)) [40, 41, 47, 71]. However, in the case of symmetry–conserving boundary conditions as considered here the critical point of the film is still located at $H = 0$. For $N = 1$ and $d = 3$ the critical phenomena near $T = T_c(L)$ are governed by the two–dimensional critical exponents of the Ising universality class. Note, that systems belonging to the XY– or Heisenberg universality class ($N = 2, 3$) do not exhibit long–range order in three–dimensional films for *any finite* L [45]. The Kosterlitz–Thouless transition [46] observed for $N = 2$ and in $d = 2$ (or finite L) is a phase transition without long–range order, and it has no analogue for $N = 3$.

5.4.1   Below Bulk Criticality

If we assume that the critical temperature $T_c(L)$ of the film is smaller than $T_{c,b}$ then $T = T_{c,b}$ marks a temperature *above* the actual critical point. Therefore the unnormalized singular free energy $f(t, L, u, \mu)$ defined by Eq.(5.7) and the renormalized free energy $f^R(t, L, u, \mu)$ given by Eq.(5.8) are *analytic* functions of the reduced temperature t in a certain interval around $t = 0$. The finite–size scaling functions $\tilde{H}_\pm$ have the same property so that the finite–size scaling functions $\Theta_{\pm a, b}(y_\pm)$ in Eq.(5.18) must *compensate* the singular bulk and surface contributions for $t \to 0$. Therefore the following expansion of $\Theta_{\pm a, b}(y_\pm)$ holds for $y_\pm \to 0$ [188, 189]:

$$\Theta_{\pm a,b}(y_\pm) = \frac{a^\pm_{bulk}}{\alpha(1-\alpha)(2-\alpha)} y^d_\pm + \frac{a^\pm_{s,a} + a^\pm_{s,b}}{\alpha_s(1-\alpha_s)(2-\alpha_s)} y^{d-1}_\pm + \sum_{n=0}^\infty \Delta^\pm_{n(a,b)} y^{n/\nu}_\pm, \quad (5.43)$$

where $\Delta^\pm_{0(a,b)} = \Delta_{a,b}$ is the Casimir amplitude (see Sec.3.2). From the analyticity of the power series in Eq.(5.43) the simple relation [188, 189]

$$\Delta^-_{n(a,b)} = (-1)^n \Delta^+_{n(a,b)} R^{n/\nu} \qquad (5.44)$$

follows immediately, where $R = \xi^-_0 / \xi^+_0$ is given by Eq.(5.28). Note, that $\Delta^+_{n(a,b)}$ and $\Delta^-_{n(a,b)}$ are *universal* amplitudes. If the bulk amplitude a^-_{bulk} [38, 218] and the surface amplitudes $a^-_{s,a}$ and $a^-_{s,b}$ (see Eq.(5.29) and Ref.[222]) are fixed, the shape of the scaling function $\Theta_{-a,b}(y_-)$ is completely determined by the shape of $\Theta_{+a,b}(y_+)$, provided $T_c(L) < T < T_{c,b}$. At the one–loop level the field–theoretical analysis shows that in the case of $(a, b) = (O, O)$, (O, SB), and antiperiodic boundary conditions $T_c(L)$ lies below $T_{c,b}$ separated by a finite amount of the order L^{-2} so that the condition $T_c(L) < T < T_{c,b}$ can be fulfilled. For $(a, b) = (SB, SB)$ and periodic boundary conditions the same analysis suggests $T_c(L) = T_{c,b}$, so that the bulk critical temperature is at least in close vicinity of the $(d-1)$–dimensional phase transition. In fact, $\Theta_{+SB,SB}$ and Θ_{+per} cannot be extrapolated to $T < T_{c,b}$ in an analytical way using our field–theoretical method [188], and we therefore restrict the following analysis of Eq.(5.43) to $\Theta_{\pm O,O}$, $\Theta_{\pm O,SB}$, and $\Theta_{\pm aper}$. In order to decide whether $\Theta_{\pm SB,SB}$ and $\Theta_{\pm per}$ exhibit the structure shown in Eq.(5.43) one must resort to such methods in statistical physics which are able to cope with the dimensional crossover (see Sec.5.5).

The desired information about $\Theta_{-O,O}$, $\Theta_{-O,SB}$, and Θ_{-aper} is provided by calculating the expansion coefficients $\Delta^+_{n(a,b)}$ for $(a,b) = (O,O)$, (O,SB), and antiperiodic boundary conditions to two–loop order and for general N. The corresponding amplitudes $\Delta^-_{n(a,b)}$ for $N = 1$ follow from Eq.(5.44). The leading order terms $\propto y^{1/\nu}_\pm$ in Eq.(5.43) are governed by the amplitudes [188]

$$\Delta^+_{1(O,O)} = \frac{N}{96}\left\{1 + \varepsilon\left[\ln(2\sqrt{\pi}) + \frac{\gamma}{2} - \frac{\zeta'(2)}{\zeta(2)} + \frac{N+2}{N+8}\left(\ln(2\pi) - \frac{5}{2} - \gamma\right)\right]\right\},$$

$$\Delta^-_{1(O,O)} = -\frac{1}{192}\left\{1 + \varepsilon\left[\frac{4}{3}\ln 2 + \frac{5}{6}\ln\pi + \frac{\gamma}{6} - \frac{\zeta'(2)}{\zeta(2)} - \frac{5}{4}\right]\right\},$$

$$\Delta^+_{1(O,SB)} = -\frac{N}{192}\left\{1 + \varepsilon\left[\frac{1}{2}\ln\pi + \frac{\gamma}{2} - \frac{\zeta'(2)}{\zeta(2)} + \frac{N+2}{N+8}\left(\ln\frac{\pi}{2} + \frac{1}{2} - \gamma\right)\right]\right\},$$

$$\Delta^-_{1(O,SB)} = \frac{1}{384}\left\{1 + \varepsilon\left[-\frac{1}{3}\ln 2 + \frac{5}{6}\ln\pi + \frac{\gamma}{6} - \frac{\zeta'(2)}{\zeta(2)} - \frac{1}{4}\right]\right\},$$

$$\Delta^+_{1(aper)} = -\frac{N}{48}\left\{1 + \varepsilon\left[\frac{1}{2}\ln\pi - \ln 2 + \frac{\gamma}{2} - \frac{\zeta'(2)}{\zeta(2)} + \frac{N+2}{N+8}\left(\ln\pi + \frac{1}{2} - \gamma\right)\right]\right\},$$

$$\Delta^-_{1(aper)} = \frac{1}{96}\left\{1 + \varepsilon\left[-\ln 2 + \frac{5}{6}\ln\pi + \frac{\gamma}{6} - \frac{\zeta'(2)}{\zeta(2)} - \frac{1}{4}\right]\right\}, \qquad (5.45)$$

where $\zeta(2) = \frac{\pi^2}{6}$, $\zeta'(2) \simeq -0.937548$. Guided by Eq.(3.14) and equipped with the ε–expansion for the correlation length exponent $\nu = \frac{1}{2} + \frac{\varepsilon}{4}\frac{N+2}{N+8}$ we find that the relations

$$\frac{\Delta^+_{1(aper)}}{\Delta^+_{1(O,SB)}} = 2^{d-1/\nu} \quad \text{and} \quad \frac{\Delta^-_{1(aper)}}{\Delta^-_{1(O,SB)}} = 2^{d-1/\nu} \quad \text{for } N = 1 \qquad (5.46)$$

are in accordance with the ε–expansion results in Eq.(5.45). Unfortunately the counterparts of $\Delta^+_{1(O,O)}$ and $\Delta^-_{1(O,O)}$ are not available. The behavior of $\Theta_{\pm O,O}$, $\Theta_{\pm O,SB}$, and $\Theta_{\pm aper}$ for $y_\pm \to 0$ can now be cast into the form $\Theta_{\pm a,b}(y_\pm) = \Delta_{a,b} + \Delta^\pm_{1(a,b)}y^{1/\nu}_\pm + \mathcal{O}(y^{d-1}_\pm)$ (see Eq.(5.43)) which demonstrates that the leading correction to the Casimir amplitude Δ is *linear* in t for small $|t|$. In contrast the scaling functions $\Theta_{+SB,SB}$ and Θ_{+per} for $y_+ \to 0$ behave according to [188]

$$\Theta_{+SB,SB}(y_+) = \Delta_{SB,SB} - \frac{\pi}{48}N\frac{N+2}{N+8}\varepsilon y_+ + \mathcal{O}(y^2_+ \ln y_+),$$

$$\Theta_{+per}(y_+) = \Delta_{per} - \frac{\pi}{12}N\frac{N+2}{N+8}\varepsilon y_+ + \mathcal{O}(y_+^2 \ln y_+) \tag{5.47}$$

which yields a *singular* leading correction $\propto |t|^\nu$ to the Casimir amplitude as $t \searrow 0$. However, as indicated above Eq.(5.47) may be an artefact of the perturbative nature of the ε-expansion.

The expansion coefficients $\Delta_{2(a,b)}^\pm$ are of special interest, because the correction governed by them is $y_\pm^{2/\nu} = (L/\xi_0^\pm)^{2/\nu}t^2$. Therefore these coefficients appear as *universal factors* in the amplitude of the singular part of the film specific heat *at* bulk criticality $t = 0$ [188, 189] (see Sec.6.1). For the three symmetry–conserving boundary conditions under consideration the coefficients $\Delta_{2(a,b)}^\pm$ can be written as [188]

$$\Delta_{2(O,O)}^+ = -\frac{\nu}{\alpha}\frac{N}{32\pi^2}\left\{1 + \varepsilon\left[\frac{\gamma}{2} - \frac{1}{2}\ln\pi + \frac{N+2}{N+8}\left(1 - 2\gamma + 2\ln(2\pi)\right)\right]\right\},$$

$$\Delta_{2(O,O)}^- = -\frac{\nu}{\alpha}\frac{1}{128\pi^2}\left\{1 + \varepsilon\left[\frac{2}{3}\ln 2 + \frac{1}{6}\ln\pi - \frac{1}{2} - \frac{\gamma}{6}\right]\right\},$$

$$\Delta_{2(O,SB)}^+ = -\frac{\nu}{\alpha}\frac{N}{32\pi^2}\left\{1 + \varepsilon\left[\frac{\gamma}{2} - \frac{1}{2}\ln\pi + 2\ln 2 + \frac{N+2}{N+8}\left(1 - 2\gamma + 2\ln\frac{\pi}{2}\right)\right]\right\},$$

$$\Delta_{2(O,SB)}^- = -\frac{\nu}{\alpha}\frac{1}{128\pi^2}\left\{1 + \varepsilon\left[\frac{4}{3}\ln 2 + \frac{1}{6}\ln\pi - \frac{1}{2} - \frac{\gamma}{6}\right]\right\},$$

$$\Delta_{2(aper)}^+ = -\frac{\nu}{\alpha}\frac{N}{32\pi^2}\left\{1 + \varepsilon\left[\frac{\gamma}{2} - \frac{1}{2}\ln\pi + \ln 2 + \frac{N+2}{N+8}\left(1 - 2\gamma + 2\ln\pi\right)\right]\right\},$$

$$\Delta_{2(aper)}^- = -\frac{\nu}{\alpha}\frac{1}{128\pi^2}\left\{1 + \varepsilon\left[\ln 2 + \frac{1}{6}\ln\pi - \frac{1}{2} - \frac{\gamma}{6}\right]\right\}, \tag{5.48}$$

where ν and α are the critical exponents of the bulk correlation length and the bulk specific heat, respectively. In the limit $d \to 4$ the exponent α vanishes so that the amplitudes $\Delta_{2(a,b)}^\pm$ in Eq.(5.48) diverge. Nonetheless the scaling functions $\Theta_{+a,b}(y_+)$ have a finite limit as $d \to 4$ (see Eq.(5.32)) which can be recovered from Eq.(5.43) by the observation that the terms $\Delta_{2(a,b)}^\pm y_+^{2/\nu}$ and $a_{bulk}^+ y_+^d/(\alpha(1-\alpha)(2-\alpha))$ merge as $d \to 4$ to give contributions $\sim y_+^4$ and $\sim y_+^4 \ln y_+$ [188]. Note, that the logarithmic term can be decomposed according to

$$y_+^4 \ln y_+ = \tfrac{1}{2}(L/\xi_0^+)^4 t^2 \ln t + (L/\xi_0^+)^4 t^2 \ln(L/\xi_0^+),$$

where the first term compensates the bulk singularity $\sim t^2 \ln t$ and the second term is *analytic* in t and displays the growth $\sim \ln(L/\xi_0^+)$ of the singular part of the film specific heat per unit volume in $d = 4$ as $L \to \infty$ (see Sec.6.1). For $\Theta_{-a,b}(y_-)$ the limit $d \to 4$ does not exist [188] (see Eq.(5.29)) which is due to additional logarithmic corrections not captured by Eq.(5.43) in $d = 4$. In analogy with Eq.(5.46) we note that the relations

$$\frac{\Delta_{2(aper)}^+}{\Delta_{2(O,SB)}^+} = 2^{d-2/\nu} \quad \text{and} \quad \frac{\Delta_{2(aper)}^-}{\Delta_{2(O,SB)}^-} = 2^{d-2/\nu} \quad \text{for } N = 1 \qquad (5.49)$$

are in accordance with Eq.(5.48). The amplitude $\Delta_{2(O,O)}^+$ for $N = 2$ applies to the specific heat of ^{4}He–films at $T = T_\lambda$ as a function of the film thickness L (see Sec.6.1). The ε–expansion of the higher coefficients indicates that $\Delta_{k(aper)}^\pm / \Delta_{k(O,SB)}^\pm = 2^{d-k/\nu}$ holds for general k. However, the ε–expansion becomes less reliable as one proceeds to larger k so that we refrain from giving explicit results for these coefficients (see also Ref.[188]). The results up to $k = 2$ are summarized in Fig.5.6 [188, 189] which displays $\Theta_{+O,O}(y_+)$ and the expansion of $\Theta_{-O,O}(y_-)$ according to Eq.(5.43) up to $k = 2$ (dashed–dotted line) for $d = 3$ and $N = 1$. The reliability of the expansion of $\Theta_{+O,O}$ according to Eq.(5.43) can be tested by comparing the dashed and the dotted lines to the ε–expansion of $\Theta_{+O,O}$ given by Eq.(5.34) (solid line). The inset in Fig.5.6 displays a magnification of the region $y_\pm \leq 0.05$ which indicates that the true minimum of $\Theta_{\pm O,O}$ is located at a temperature T_{min} *below* $T_{c,b}$ and that the value of $\Theta_{-O,O}$ at that temperature T_{min} lies slightly *below* $\Delta_{O,O}$.

For completeness we quote the expansion of the corresponding *tricritical* scaling functions $\Theta_{+(O,O)}(y_+)$, $\Theta_{+(O,SB)}(y_+)$, and $\Theta_{+(aper)}(y_+)$ in $d = 3$ according to Eq.(5.43) for $y_+ \to 0$. From Eq.(5.40) we find

$$\Theta_{+(O,O)}(y_+) = -\frac{N}{16\pi}\left\{\zeta(3) + 2y_+^2 \ln y_+ + (2\ln 2 - 1)y_+^2 - \frac{4}{3}y_+^3 + \frac{1}{6}y_+^4 + \mathcal{O}(y_+^6)\right\},$$

$$\Theta_{+(O,SB)}(y_+) = \frac{N}{16\pi}\left\{\frac{3}{4}\zeta(3) - 2\ln 2\, y_+^2 + \frac{4}{3}y_+^3 - \frac{1}{2}y_+^4 + \mathcal{O}(y_+^6)\right\},$$

$$\Theta_{+(aper)}(y_+) = \frac{N}{16\pi}\left\{6\zeta(3) - 4\ln 2\, y_+^2 + \frac{4}{3}y_+^3 - \frac{1}{4}y_+^4 + \mathcal{O}(y_+^6)\right\}. \qquad (5.50)$$

For the tricritical scaling function $\Theta_{+(O,O)}$ the expansion given by Eq.(5.50) displays a logarithmic contribution, which is not visible in Eq.(5.43). The logarithm emerges from the terms $\Delta_1^+ y_+^{1/\nu}$ and $2a_{s,O}^+ y_+^{d-1}/(\alpha_s(1 - \alpha_s)(2 - \alpha_s))$ in Eq.(5.43) in the limit

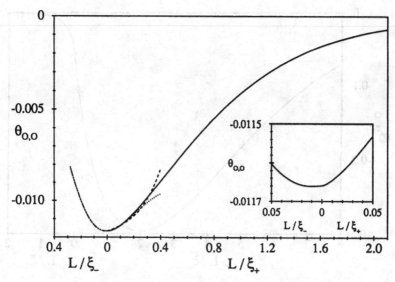

Fig. 5.6: Scaling function $\Theta_{+O,O}$ as a function of L/ξ for $N = 1$ and $d = 3$.
The solid line displays the ε-expansion of Θ_+. The dotted line shows the
ε-expansion of $\Theta_+(y_+ \to 0)$, including the leading two terms $\propto y_+^{1/\nu}$ and
$\propto y_+^{d-1}$. The dashed and dashed-dotted lines show the exponentiated form of
$\Theta_\pm(y_\pm)$ around $y_\pm = 0$, including the terms $\propto y_\pm^{1/\nu}$, $y_\pm^{d-1}$, $y_\pm^d$, and $y_\pm^{2/\nu}$ using
all available analytic results for both the amplitudes and the exponents. The
dotted and solid lines agree up to $L/\xi_+ \approx 0.2$ giving the range within which
the asymptotic behavior $y_+ \to 0$ is captured correctly by the dotted and the
dashed line. By construction $\Theta_-(y_- \to \infty) \to 0$.

$d \to 3$. Note, that in $d = 3$ the tricritical exponents α and ν are both given by their
Landau values $\frac{1}{2}$ and therefore $\alpha_s = \alpha + \nu = 1$. This is analogous to the interference
between the power laws $y_+^{2/\nu}$ and y_+^d as $d \to 4$ near a usual *critical* point, where
$\alpha \to 0$ (see Eq.(5.48)). For (O, SB)-boundary conditions the surface amplitudes $a_{s,O}^+$
and $a_{s,SB}^+$ are equal in opposite and for antiperiodic boundary conditions the surface
amplitudes $a_{s,i}^+$, $i = a, b$ vanish so that a corresponding $y_+^2 \ln y_+$-term does not occur
in the expansions of $\Theta_{+(O,SB)}(y_+)$ and $\Theta_{+(aper)}(y_+)$.

   The expansion coefficients Δ_2^+ again appear as universal amplitude factors in the
L-dependence of the singular film specific heat at bulk tricriticality (see Sec.6.1).

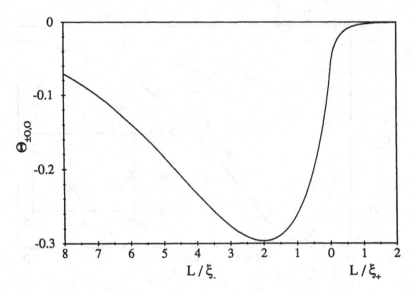

Fig. 5.7: Scaling function $\Theta_{\pm 0,0}(y_{\pm})$ in $d = 2$ according to Eqs.(5.37) and (5.52) (see also Ref.[225]). The function $\Theta_{-0,0}(y_-)$ is shown in the range $8 \geq y_- \geq 0$, and $\Theta_{+0,0}(y_+)$ is shown in the range $0 \leq y_+ \leq 2$. The slope of $\Theta_{\pm 0,0}(y_{\pm})$ at $y_{\pm} = 0$ is infinite.

These amplitudes can be read off from Eq.(5.50) as [188]

$$\Delta_{2(0,0)}^+ = -\frac{N}{96\pi} \simeq -0.00332\,N,$$

$$\Delta_{2(0,SB)}^+ = -\frac{N}{32\pi} \simeq -0.00995\,N,$$

$$\Delta_{2(aper)}^+ = -\frac{N}{64\pi} \simeq -0.00497\,N. \tag{5.51}$$

For $N = 2$ the amplitude $\Delta_{2(0,0)}^+$ in Eq.(5.51) applies to the singular specific heat of a tricritical ^{3}He–^{4}He–film (see Fig.1.9 in Sec.6.1). Like the critical scaling functions $\Theta_{-a,b}$ in $d = 4$ the corresponding *tricritical* scaling functions $\Theta_{-a,b}$ in $d = 3$ contain logarithmic corrections which are not covered by Eqs.(5.43) and (5.50). We will therefore refrain from discussing the amplitudes $\Delta_{n(a,b)}^-$ for the tricritcal scaling functions $\Theta_{-a,b}$.

In $d = 2$ and for $N = 1$ the *exact* scaling function $\Theta_{-0,0}(y_-)$ can be obtained from Ref.[225] in the form

$$\Theta_{-O,O}(y_-) = -\frac{1}{4\pi} \int_{y_-}^{\infty} \frac{x}{\sqrt{x^2 - y_-^2}} \ln\left(1 + \frac{x + y_-}{x - y_-}e^{-x}\right) dx, \qquad (5.52)$$

where the bulk correlation length ξ_- in the scaling argument $y_- = L/\xi_-$ has been chosen according to Ref.[50] (see also Eq.(5.37)). The exact finite–size scaling function $\Theta_{\pm O,O}(y_\pm)$ for a strip in two dimensions with Dirichlet boundary conditions (see Eqs.(5.37) and (5.52)) is displayed in Fig.5.7. Due to the absence of a phase transition in a two–dimensional Ising model confined to a strip $\Theta_{-O,O}(y_-)$ is an *analytic* function of y_- for $y_- > 0$. In the limit $y_- \to 0$ one finds

$$\Theta_{-O,O}(y_- \to 0) = -\frac{\pi}{48} + \frac{1}{4\pi}y_- \ln y_- + \mathcal{O}(y_-) \qquad (5.53)$$

so that the slope of $\Theta_{-O,O}(y_-)$ at $y_- = 0$ is infinite (see also Eq.(5.39)). As shown in Fig.5.7 $\Theta_{\pm O,O}(y_\pm)$ posesses a very pronounced minimum for $T < T_c$ at $y_- \simeq 2.0$. A corresponding although much less pronounced minimum has also been found for $\Theta_{\pm O,O}(y_\pm)$ in $d = 4 - \varepsilon$ (see Fig.5.6).

5.4.2 Limitations of the ε–Expansion

The $(d-1)$–dimensional phase transition at $T = T_c(L)$ for $N = 1$ and in the presence of symmetry–conserving boundary conditions marks a barrier which cannot be overcome by means of the field–theoretical ε–expansion used here. The reason is that the upper critical dimension d_c of the phase transition remains at $d_c = 4$ although the dimensionality of the system is reduced by one. Therefore the character of the phase transition does not change for the crossover from the bulk system ($L = \infty$) to the film system ($L < \infty$). It is clearly impossible to keep both d and $d-1$ close to $d_c = 4$ simultaneously so that the field–theoretical renomalization group can only account for *one* of the fixed points at a time (see Sec.5.5). Therefore the field–theoretical knowledge of the scaling functions $\Theta_{-a,b}$ so far is limited to $T > T_c(L)$ in the Ising universality class. For $\Theta_{\pm SB,SB}$ and $\Theta_{\pm per}$ the expansion according to Eq.(5.43) is rather delicate, because to one–loop order $T_c(L) = T_{c,b}$ (see Ref.[188]). This raises the question to which extent the Casimir amplitudes $\Delta_{SB,SB} = \Theta_{+SB,SB}(0)$ and $\Delta_{per} = \Theta_{+per}(0)$ as obtained from the ε–expansion (see Eq.(3.13) and Table 3.1) are numerically reliable if $T_c(L)$ is close or even equal to $T_{c,b}$. From the point of view of the ε–expansion used here one is on the safe side whenever $T > T_{c,b} \geq T_c(L)$. Any of the scaling functions $\Theta_{+a,b}(y_+)$ given by Eq.(5.34) can be *uniquely* extrapolated to $y_+ = 0$ ($T = T_{c,b}$) and the results of this extraplation are in fact given by Eq.(3.13). The finite size scaling functions $\Theta_{\pm a,b}$ govern the scaling behavior of the finite–size part of the renormalized free energy (see Eq.(5.20)) and therefore $\Theta_{+a,b}(y_+ \to 0)$ must match $\Theta_{-a,b}(y_- \to 0)$ for *any* boundary conditions. Hence the presence of the $(d-1)$–dimensional phase transition does not disturb the validity of Eq.(3.13). However, the construction of $\Theta_{-SB,SB}(y_-)$ and $\Theta_{-per}(y_-)$ from the universal expansion coefficients Δ_n^+ according to Eq.(5.43) is *not* possible [188].

At $T = T_c(L)$ and for $N = 1$ the free energy of the film shows the singularity of the $(d-1)$–dimensional phase transition. According to Eqs.(5.7) and (5.18) only the scaling functions $\Theta_{-a,b}(y_-)$ can account for this singularity which occurs at $y_- = y_c$ (i.e., $T = T_c(L)$) and is characterized by the $(d-1)$–dimensional critical exponents. We therefore have

$$\Theta_-(y_- \to y_c) \propto \left| y_-^{1/\nu_d} - (y_c)^{1/\nu_d} \right|^{2-\alpha_{d-1}}, \qquad (5.54)$$

where ν_d denotes the d–dimensional correlation length exponent and α_{d-1} is the $(d-1)$–dimensional specific heat exponent. The condition $y_- = y_c$ for $(d-1)$–dimensional criticality can then be rewritten as [188]

$$t_c(L) = \left| \frac{T_c(L) - T_{c,b}}{T_{c,b}} \right| = (y_c)^{1/\nu_d} \left(\frac{L}{\xi_0^-} \right)^{-1/\nu_d}. \qquad (5.55)$$

The exponent λ of the *fractional shift* $t_c(L \to \infty) \propto L^{-\lambda}$ [51, 52] is therefore identified by $\lambda = 1/\nu_d$. Recalling that $\Theta_{-a,b}(y_-)$ is *universal* as a *function* then immediately implies that the position y_c of its $(d-1)$–dimensional critical singularity is a *universal* number. The fractional shift given by Eq.(5.55) is thus governed by the *universal amplitude* $(y_c)^{1/\nu_d}$ if the film thickness L is scaled with the correlation length amplitude ξ_0^- which absorbs any nonuniversality in Eq.(5.55).

The parallel film correlation length $\xi_{pa}(t, L)$ can be discussed in analogy with the renormalized free energy of the film. The general scaling theory (see Ref.[43]) can be applied in the form [188]

$$\xi_{pa}(t, L) = \xi_\pm \Xi_\pm(L/\xi_\pm) = \xi_\pm \Xi_\pm(y_\pm), \qquad (5.56)$$

where $\Xi_+(y_+)$ and $\Xi_-(y_-)$ are universal scaling functions. For $L \gg \xi_\pm$ the parallel correlation length $\xi_{pa}(t, L) \simeq \xi_{pa}(t, \infty) = \xi_\pm$ and for $t \to 0$ (i.e., $T \to T_{c,b}$) at finite

L we have $\xi_{pa}(t \rightarrow 0, L) \sim L$. At $T_c(L)$ the parallel correlation length shows the characteristic power–law divergence governed by the critical exponent ν_{d-1}. The scaling functions $\Xi_{\pm}(y_{\pm})$ have the properties [188]

$$\Xi_{\pm}(\infty) = 1,$$
$$\Xi_{\pm}(y_{\pm} \rightarrow 0) \sim y_{\pm},$$
$$\Xi_{-}(y_{-} \rightarrow y_c) \propto \left| y_{-}^{1/\nu_d} - (y_c)^{1/\nu_d} \right|^{-\nu_{d-1}}, \tag{5.57}$$

where y_c is the same as in Eqs.(5.54) and (5.55). The choice $x = L/\xi_{pa}(t, L)$ as a new scaling variable via Eq.(5.56) only amounts to a reparametrisation $x = y_{\pm}/\Xi_{\pm}(y_{\pm})$ of the scaling functions $\Theta_{\pm a, b}$ and therefore does not provide any progress towards a description of the dimensional crossover. However, despite the aforementioned difficulties several attempts have been made to tackle the dimensional crossover within the framework of the field–theoretical renormalization group for various geometries.

5.5 The Dimensional Crossover

Critical behavior in its literal sense concerning critical *singularities* is an effect of the thermodynamic limit. Depending on the range of the interactions and internal symmetries the dimension d of the system must be sufficiently large in order to obtain a critical point at a *finite* temperature. For an Ising model ($N = 1$) with nearest–neighbour interactions $d = 2$ is sufficient, and a Heisenberg model ($N = 3$) with nearest–neighbour interactions requires $d = 3$ [45]. The XY–model ($N = 2$) with nearest–neighbour interactions, however, is an interesting borderline case. On one hand it requires $d = 3$ like the Heisenberg model [45], on the other hand there is a phase transition already in $d = 2$. But contrary to the aforementioned examples this latter transition does *not* exhibit *long–ranged order* (Kosterlitz–Thouless transition [46]).

If a critical system is confined in a finite geometry, two distinct cases must therefore be considered. In the first case the "remaining" dimension is too small to sustain a sharp phase transition. For the above examples this is fulfilled for completely finite ("zero"–dimensional) and cylindrical (one–dimensional) geometries, where the *fractional shift* and the *fractional rounding* can be observed (see Refs.[48, 51, 52] and Sec.2.1). In the second case the remaining dimension is still sufficient to sustain a sharp phase transition. A *film geometry* in the embedding dimension $d = 3$ which confines an *Ising* model falls in this second category, otherwise $d = 3$ is *not* sufficient (apart from the Kosterlitz–Thouless transition in the XY–model). The crossover from d–dimensional critical behavior to the absence of a phase transition or critical behavior in reduced dimension has been termed the *dimensional crossover*. The efforts spent on a field–theoretical description of the dimensional crossover can be subdivided into the two categories described above.

5.5.1 *Completely Finite and One Dimensional Geometries*

In order to obtain a field–theoretical description of a critical system confined in a strictly finite model geometry the Ginzburg–Landau Hamiltonian (see Sec.1.4) has been studied in a cube L^d with periodic boundary conditions [106, 107]. More realistic cubic geometries involve mixed periodic and Dirichlet and pure Dirichlet boundary conditions [108, 109]. The latter ones have also been applied to spherical systems [63]. As a representative for one–dimensional samples the cylindrical geometry $L^{d-1} \times \infty$ has been investigated for periodic boundary conditions along the $d-1$ finite directions of the cylinder [106].

In a finite cube L^d with periodic boundary conditions the order parameter field $\Phi(\mathbf{r})$ in the Ginzburg–Landau Hamiltonian (see Sec.2.3) can be written in terms of a Fourier series [106]

$$\Phi(\mathbf{r}) = \sum_{\mathbf{q}} e^{i\mathbf{q}\mathbf{r}} \varphi_{\mathbf{q}}, \tag{5.58}$$

where the momentum spectrum is discrete according to $q_i = \frac{2\pi}{L} n_i$, $n_i = 0, \pm 1, \pm 2, \ldots$, and $1 \leq i \leq d$. Each number n_i stays below a cutoff n given by, e.g., the lattice spacing a. The $\mathbf{q} = 0$ Fourier component φ_0 in Eq.(5.58) which is occasionally referred to as the *zero mode* becomes massless at $T = T_{c,b}$, and therefore perturbation theory breaks down for φ_0 at $T = T_{c,b}$. The method presented in Refs.[106, 107] relies on the construction of an *effective action* (i.e., an *effective Hamiltonian*) for the zero mode φ_0 in terms of a loop expansion in the $\varphi_{\mathbf{q}}$–modes for $\mathbf{q} \neq \mathbf{0}$. To one–loop order the effective Hamiltonian of an $O(N = 1)$–symmetric system reads [106, 107]

$$\mathcal{H}_{eff}\{\varphi_0\} = L^d \left\{ \frac{\tau}{2}\varphi_0^2 + \frac{g}{4!}\varphi_0^4 \right\} + \frac{1}{2} \sum_{\mathbf{q} \neq 0} \ln\left(\mathbf{q}^2 + \tau + \frac{g}{2}\varphi_0^2 \right), \tag{5.59}$$

where $\tau = 0$ at $T = T_{c,b}$. Note, that $\mathcal{H}_{eff}\{\varphi_0\}$ given by Eq.(5.59) is only well defined for $\tau > -\left(\frac{2\pi}{L}\right)^2$, because the Fourier modes $\varphi_{\mathbf{q}}$ with $|\mathbf{q}| = \frac{2\pi}{L}$ become massless at

$\tau = -\left(\frac{2\pi}{L}\right)^2$. At the tree level ($d > 4$) the rescaling $\varphi = \left(gL^d\right)^{-1/4}\varphi_0$ suggests $x = g^{-1/2}\tau L^{4/2}$ as a scaling variable which already indicates the failure of the usual finite-size scaling *above* $d = 4$ (see Eq.(5.6) and Ref.[43] for $d < 4$) due to the breakdown of hyperscaling [106] (see also Refs.[54] and [231] and Eqs.(2.9) and (2.35))). The one–loop contribution to the effective Hamiltonian given by Eq.(5.59) leads to a renormalization of τ and g and a *shift* of τ in $d = 4 - \varepsilon$, where the renormalization factors Z_t and Z_u are the same as in the corresponding bulk system (see Eq.(5.2)), and the usual finite–size scaling (see Eq.(5.6)) is recovered in terms of an $\varepsilon^{1/2}$–expansion in $d < 4$ [106]. The effective Hamiltonian gives access to the specific heat of a cube in $d = 4$ and $d = 4 - \varepsilon$ [107] which displays a *rounded maximum* at a temperature $T_m < T_{c,b}$ and a negative slope at $T_{c,b}$ (see Fig.2.1 and Sec.6.1).

The cube L^d is a simple model for ^{4}He in pores near the λ–transition if the periodic boundary conditions are replaced by *Dirichlet* boundary conditions. In fact, the effective Hamiltonian given by Eq.(5.59) can be generalized to $N > 1$ ($N = 2$ for ^{4}He), where the zero mode φ_0 is replaced by the *lowest standing–wave mode* $\Phi_1(\mathbf{r}) = \phi_1 \prod_{j=1}^d \sin\frac{\pi x_j}{L}$ [108]. The specific heat which follows from the corresponding effective Hamiltonian at the one–loop level is in *quantitative* accordance with experimental data even at finite temperatures *below* bulk T_λ [108]. A further generalization to mixed periodic and Dirichlet boundary conditions for the cube L^d is given in Ref.[109], where the surface contributions to the internal energy and the specific heat are analyzed to two–loop order. For the special case of one Dirichlet and $d - 1$ periodic boundary conditions the geometry in Ref.[109] resembles the film geometry discussed in Secs.3.2, 5.3, and 5.4. Note, that as in Eq.(5.59) higher modes limit the range of validity of the effective Hamiltonian to some finite temperature below $T_{c,b}$. The analysis of the spherical geometry with Dirichlet boundary conditions in Ref.[63] does not make use of an effective zero–mode Hamiltonian and is therefore limited to temperatures well *above* the specific heat maximum, where a usual ε–expansion applies (see Sec.5.4). However, so far the field–theoretical treatment of the crossover from the bulk critical singularities to the rounded "critical" behavior in strictly finite geometries has reached the level of a quantitative theory.

In a cylindrical geometry $L^{d-1} \times \infty$ with $d - 1$ periodic boundary conditions the zero mode $\varphi_0 = \varphi_0(z)$ is still a function of the coordinate z along the axis of the cylinder. Therefore the effective zero–mode Hamiltonian is a *functional* of $\varphi_0(z)$

which at tree level for an $O(N)$–symmetric system is given by [106]

$$\mathcal{H}_{eff}\{\varphi_0\} = L^{d-1} \int dz \left\{ \frac{1}{2} \left(\varphi_0'(z)\right)^2 + \frac{\tau}{2}\varphi_0^2(z) + \frac{g}{4!} \left(\varphi_0^2(z)\right)^2 \right\}. \tag{5.60}$$

The effective Hamiltonian given by Eq.(5.60) can be treated along the lines of the Feynman path integral formulation of a N–dimensional quantum mechanical anharmonic oscillator in imaginary time with the Hamiltonian [106]

$$H(\mathbf{p},\mathbf{q}) = \frac{\mathbf{p}^2}{2L^{d-1}} + L^{d-1} \left(\frac{\tau}{2}\mathbf{q}^2 + \frac{g}{4!}(\mathbf{q}^2)^2 \right). \tag{5.61}$$

The correlation length ξ_L of the cylinder can be defined by the inverse separation of the lowest two energy levels of the anhamonic oscillator described by Eq.(5.61), i.e., $\xi_L = (E_1 - E_0)^{-1}$. As in the case of a cubic geometry finite–size scaling fails for $d > 4$. In $d = 4 - \varepsilon$ one–loop corrections give rise to an $\varepsilon^{1/3}$–expansion of the correlation length ξ_L and the usual finite–size scaling is recovered in $d < 4$ [106]. However, in comparison with the cubic geometry L^d the field–theoretical description of the crossover to the rounding of critical singularities in one–dimensional systems has not yet reached the same quantitative level. The above discussion can also be applied to the $O(N)$–nonlinear σ–model in $d = 2 + \varepsilon$ for both the cubic and the cylindrical geometry [106, 216], where finite–size scaling functions for the free energy and the correlation length ξ_L of the finite system can be studied at low temperatures.

Finally, we would like to point out that *first–order* phase transitions become rounded by finite size effects as well [231]. If, for example, a magnetic field H is applied to an Ising ferromagnet below $T_{c,b}$, the magnetization M jumps sharply from $M = -M_0(T)$ for $H < 0$ to $M = M_0(T)$ for $H > 0$. In a cubic geometry L^d with periodic boundary conditions the transition from $-M_0(T)$ to $M_0(T)$ becomes rounded according to [106, 231]

$$M(T, H, L) = M_0(T) \tanh \left(\frac{HL^d}{k_BT} M_0(T) \right), \tag{5.62}$$

where the magnetic field H is assumed to be small, but HL^d is taken to be finite. Note, that $M = 0$ for $H = 0$ in Eq.(5.62) which reflects the fact that there is no spontaneous magnetization in strictly finite systems. The finite–size scaling of the first–order transitions at $T < T_{c,b}$ merges with the usual finite–size scaling near critical points when $T \nearrow T_{c,b}$ [231]. A corresponding analysis has been performed for general $O(N > 1)$–symmetric systems in cubic L^d and cylindrical $L^{d-1} \times \infty$ geometries, where spin wave fluctuations must be considered [217]. The finite–size effect of a cubic L^d geometry with periodic boundary conditions on the temperature–driven first–order phase transition in a q–state Potts model has been studied in Ref.[232] especially for $q = 3$ and $d = 3$. Other investigations of the three–dimensional Heisenberg model indicate that the scaling theory of Ref.[217] must be modified near criticality or for small system sizes [233].

5.5.2 Geometries with Two or More Infinite Extensions

If the system is infinite in two or more directions, a critical point in its literal sense may still occur. The most prominent example for a corresponding geometry is given by a *film* or *slab* of thickness L which the preceding sections of this chapter have been devoted to. As already mentioned in Sec.5.4 the dimensional crossover from d–dimensional to $(d - 1)$–dimensional critical behavior cannot be captured in the usual subtraction scheme of field theory in $d = 4 - \varepsilon$. In order to find a field-theoretical description of this dimensional crossover for the Ising universality class an *L–dependent minimal subtraction* scheme has been suggested for the film geometry, where only periodic boundary conditions have been considered so far [234]. The perturbation expansion of the renormalization factors Z_Φ and Z_{Φ^2} is governed by coefficients which can be expressed in terms of a L–dependent Laurent series in $\varepsilon' = 3 - d$. The renormalization of the coupling constant is governed by a *crossover function* $h(L/L_0)$ with $h(x \to \infty) = 1$ and $h(x \to 0) \simeq x$ [234]. The ad hoc choice $h(x) = x/(x + 1)$ for the crossover function leads to an *effective* exponent $\gamma(t, L)$ of the static susceptibility $\chi(t, L)$ which interpolates between the three– and four-dimensional values for γ as L is varied between zero and infinity [234].

The basic ideas of Ref.[234] have been reconsidered and extended in Refs.[235, 236, 237] within the Ising universality class and for periodic boundary conditions for a film geometry. The *generalized minimal subtraction* [235] yields L–dependent Wilson functions which recover their known form for $L \to \infty$ and cross over smoothly to a finite limit as $L \to 0$. The renormalization of the coupling constant is uniquely given by the subtraction scheme so that the introduction of a crossover function $h(x)$ as in Ref.[234] is avoided. The effective critical exponents $\gamma_{eff}(t, L)$ and $\nu_{eff}(t, L)$ display the correct behavior in the limits $L \to 0$ and $L \to \infty$ in terms of an expansion in powers of $\varepsilon' = 3 - d$ and $\varepsilon = 4 - d$, respectively. This approach has recently been extended to a wider class of crossovers [238]. Moreover, the effective critical exponents β_{eff} and δ_{eff} have been calculated explicitly for the dimensional crossover from $d = 3$ to $d = 2$ within the Ising universality class [239]. However, for a comparison with experiments the consideration of other than periodic boundary conditions is required (see, e.g., Ref.[108]).

The dimensional crossover from three to two dimensions has especially been studied using lattice models confined to layered geometries or with spatially inhomogeneous coupling constants [76, 77, 78, 80]. In one case the dimensional crossover has even been observed experimentally in magnetic superlattices [75]. For details we refer to Sec.2.2.

6.   Experiments on Finite Size Scaling

6.1   The Specific Heat

The size dependence of the singular part of the free energy of a confined critical system can be expessed in a scaling form (see Chap.2). The consequence of scaling and especially finite–size scaling is data collapse, i.e., the singular parts of the thermodynamical quantities in the vicinity of a critical point are governed by *scaling functions* which depend on a reduced number of *scaling variables*. (For simple examples see Eqs.(1.3) and (1.4) in Sec.1.1.) In the case of short–ranged interactions in $d < d_c$ the finite–size scaling behavior of the *leading* singular part of the free energy is governed by *bulk* critical exponents. The finite–size scaling *functions* depend on the boundary conditions (surface universality classes) imposed at the surfaces of the system (see Eqs.(2.4) and (2.10) in Sec.2.1).

For a critical film the finite–size scaling behavior of the leading singular part of the specific heat is given by Eq.(2.8), which has been derived on phenomenological grounds in terms of an unspecified scaling function $C_{a,b}\left(|t|L^{1/\nu}, HL^{\Delta/\nu}\right)$. In the absence of external fields and for symmetry conserving boundary conditions the finite–size scaling function of the specific heat can be expressed in terms of the scaling function $\Theta_{\pm a,b}(y_\pm)$ (see Secs.5.3 and 5.4 and Eqs.(5.8), (5.15), and (5.18)). According to the construction of the *renormalized free energy* described in Sec.5.1 we obtain for a critical film the *renormalized specific heat* $c^R(t, L, u, \mu) \equiv -f_{tt}^R(t, L, u, \mu)$ per volume and in units of k_B in the following scaling form [173, 240]:

$$c_{a,b}^R(t, L) = \left(\xi_0^\pm\right)^{-d} \left[|t|^{-\alpha} J_{\pm a,b}(y_\pm) - J_{\pm a,b}(y_{1\pm})\right], \qquad (6.1)$$

where the variables u and μ have been omitted for simplicity. The finite–size scaling functions $J_{\pm a,b}(y_\pm)$ follow directly from the definition of the specific heat [173] (see Eq.(5.18)):

$$J_{\pm a,b}(y_\pm) = \frac{a_{bulk}^\pm}{\alpha} + \frac{a_{s,a}^\pm + a_{s,b}^\pm}{\alpha_s} y_\pm^{-1} - \nu y_\pm^{1-d} \left[\nu y_\pm \Theta_{\pm a,b}''(y_\pm) + (\nu - 1)\Theta_{\pm a,b}'(y_\pm)\right]. \quad (6.2)$$

Like the renormalized free energy (see Eq.(5.8)) the renormalized specific heat given by Eq.(6.1) subsumes the leading singular behavior of the specific heat in the vicinity of a critical point and additionally fulfills the normalization condition $c_{a,b}^R(t_1, L) = 0$ at the reference reduced temperature $t_1 = \mathrm{sgn} t$. According to Eqs.(5.7) and (5.8) an *unnormalized* singular specific heat $c_{a,b}(t, L)$ can be defined so that $c_{a,b}^R(t, L) = c_{a,b}(t, L) - c_{a,b}(\text{sgn} t, L)$, where $c_{a,b}(t, L)$ has the scaling form

$$c_{a,b}(t, L) = \left(\xi_0^\pm\right)^{-d} |t|^{-\alpha} J_{\pm a,b}(y_\pm) \tag{6.3}$$

which can be cast into the form of Eq.(2.8) using the substitution $|t| = \left(y_\pm \xi_0^\pm / L\right)^{1/\nu}$ with $H = 0$. Note, that for $t < 0$ the scaling argument $y_- = L/\xi_-$ is only defined for the Ising universality class (see Eq.(5.10)). Furthermore, Eqs.(6.1), (6.2), and (6.3) are only valid for short–ranged interactions in $d < d_c$ dimensions. Above the upper critical dimension or for long–ranged interactions finite–size scaling takes a different form (see Eqs.(2.9), (2.35), and (2.42)).

At the bulk critical point ($t = 0$) the renormalized specific heat of the film simplifies according to [173]

$$c_{a,b}^R(0, L) = \left(\xi_0^+\right)^{-d} \left[-2\Delta_{2(a,b)}^+ y_{1+}^{\alpha/\nu} - J_{+a,b}(y_{1+})\right] \tag{6.4}$$

which is a direct consequence of the analytic properties of the finite–size scaling functions $\Theta_{\pm a,b}(y_\pm)$ given by Eq.(5.43). The amplitudes $\Delta_{2(a,b)}^+$ (see Eq.(5.48)) are *universal*, and therefore Eq.(6.4) provides a link between universal finite–size properties and an experimentally testable quantity. The film thickness L is usually much larger than ξ_0^+ so that $y_{1+} = L/\xi_0^+ \gg 1$. From the asymptotic behavior of the finite–size scaling function $\Theta_{+a,b}(y_+)$ for $y_+ \to \infty$ (see Eq.(5.36)) we find

$$J_{+a,b}(y_+ \to \infty) = \frac{a_{bulk}^+}{\alpha} + \frac{a_{s,a}^+ + a_{s,b}^+}{\alpha_s} y_+^{-1} + \ldots, \tag{6.5}$$

where the dots stand for exponentially small corrections. The bulk contribution to the renormalized specific heat given by Eq.(6.1) can be cast into the form $c_{bulk}^R(t) = C_{bulk}^{(0)\pm}(|t|^{-\alpha} - 1)$, where $C_{bulk}^{(0)\pm} = \left(\xi_0^\pm\right)^{-d} a_{bulk}^\pm/\alpha$ is a nonuniversal bulk amplitude. In the limit $L/\xi_0^+ \to \infty$ the renormalized specific heat of a film at bulk criticality further simplifies to [173, 188, 240]

$$c_{a,b}^R(0, L \to \infty)/C_{bulk}^{(0)+} = \omega_{a,b} \left(L/\xi_0^+\right)^{\alpha/\nu} - 1 + \ldots, \tag{6.6}$$

where only the leading terms are displayed. Surface contributions to $J_{\pm a,b}(y_{1+})$ decay as L^{-1} (see Eq.(6.5)) and are therefore omitted in Eq.(6.6). The amplitudes

$$\omega_{a,b} = -2\alpha\Delta_{2(a,b)}^+/a_{bulk}^+ \tag{6.7}$$

Table 6.1:   Universal amplitudes $\omega_{O,O}$, $\omega_{O,SB}$, and ω_{aper} of the size dependence of the renormalized specific heat at $T = T_{c,b}$ (see Eq.(6.8)) for $N = 1, 2, 3$ in $d = 3$ ($\varepsilon = 1$) and $d = 4$ ($\varepsilon = 0$) [188].

Amplitude	$d = 3$			$d = 4$
	$N = 1$	$N = 2$	$N = 3$	$N = 1, 2, 3$
$\omega_{O,O}$	0.58	0.75	0.89	1
$\omega_{O,SB}$	1.04	1.03	1.01	1
ω_{aper}	0.81	0.89	0.95	1

are positive and universal [188]. Using the ε-expansion of $\Delta_{2(a,b)}^{+}$ (see Eq.(5.48)) and the well known universal bulk amplitude a_{bulk}^{+} [38] we obtain [188]

$$\omega_{O,O} = 1 + \varepsilon(\gamma - \ln 2\pi)\left(1 - 2\frac{N+2}{N+8}\right),$$

$$\omega_{O,SB} = 1 + \varepsilon\left(\gamma - \ln\frac{\pi}{2}\right)\left(1 - 2\frac{N+2}{N+8}\right),$$

$$\omega_{aper} = 1 + \varepsilon(\gamma - \ln\pi)\left(1 - 2\frac{N+2}{N+8}\right) \tag{6.8}$$

to two–loop order. Which of the two terms shown in Eq.(6.6) is the dominating one depends on the sign of α. In the Ising universality class ($N = 1$) α is positive (see Table 1.5), and therefore $c_{a,b}^{R}(0, L)$ *diverges* as $L^{\alpha/\nu}$ for $L \to \infty$. In the XY ($N = 2$) and Heisenberg ($N = 3$) universality classes α is negative (see Table 1.5), and thus $c_{a,b}^{R}(0, L)$ approaches its bulk value $-C_{bulk}^{(0)+} > 0$ in a cusplike singularity from below, because in this case $L^{\alpha/\nu} \to 0$ for $L \to \infty$.

The nonuniversal amplitudes $C_{bulk}^{(0)+}$ and ξ_0^{+} and the exponents α and ν are bulk quantities which can be determined independently. If these bulk data are known, Eqs.(6.6) and (6.8) yield a *quantitative* prediction for the leading singular behavior of the film specific heat at bulk criticality as a function of L. The numerical values of $\omega_{O,O}$, $\omega_{O,SB}$, and ω_{aper} according to Eq.(6.8) are summarized in Table 6.1 for $N = 1, 2, 3$ in $d = 3$ and $d = 4$. For the specific heat of liquid ^{4}He films at $T = T_\lambda$ the amplitude $\omega_{O,O}$ for $N = 2$ is relevant. However, in view of a comparison with experimental data the extrapolation of the ε-expansion for the amplitudes $\omega_{a,b}$ and the scaling functions $J_{a,b\pm}(y_\pm)$ to $\varepsilon = 1$ suffers from considerable ambiguities. To illustrate this we return to Eq.(6.7) for $(a,b) = (O,O)$ and use the representation of the amplitude $\Delta^+_{2(O,O)}$ according to Eq.(5.48) instead of Eq.(6.8) for a numerical evaluation of $\omega_{O,O}$. The universal bulk amplitude a^+_{bulk} is known beyond the ε–expansion used here, and one finds $a^+_{bulk} = 0.020,\ 0.047,\ 0.074$ for $N = 1, 2, 3$, respectively [38]. The exponent α cancels in the expression for $\omega_{a,b}$, and with ν taken from Table 1.5 we obtain for $\varepsilon = 1$ and $N = 1, 2, 3$ [173]

$$\omega_{O,O} \simeq 0.38, \quad \omega_{O,O} \simeq 0.38, \quad \omega_{O,O} \simeq 0.42, \tag{6.9}$$

respectively. The discrepancy between Eq.(6.9) and Table 6.1 demonstrates the deficiencies of the first–order ε–expansion in view of a quantitative analysis of the specific heat in $d = 3$ (see below). For a numerical evaluation of the renormalized specific heat on the basis of the ε–expansion of the scaling function $J_{+O,O}(y_+)$ and Eq.(6.9) we refer to Ref.[173].

The renormalized specific heat for a *tricritical* film is also given by Eq.(6.1) except that the finite–size scaling functions $J_{\pm a,b}(y_\pm)$ have a different form. For $t \geq 0$ the scaling functions $J_{+a,b}(y_+)$ can be determined from the tricritical scaling functions $\Theta_{+a,b}(y_+)$ (see Eqs.(5.40) and (5.50)) and Eq.(6.2). The results read [240]:

$$J_{+O,O}(y_+) = \frac{N}{16\pi}\left[1 - \frac{1}{y_+} + \frac{2}{e^{2y_+} - 1}\right],$$

$$J_{+O,SB}(y_+) = \frac{N}{16\pi}\left[1 - \frac{2}{e^{2y_+} + 1}\right],$$

$$J_{+aper}(y_+) = \frac{N}{16\pi}\left[1 - \frac{2}{e^{y_+} + 1}\right]. \tag{6.10}$$

At the bulk tricritical temperature $T = T_t$ the leading singular behavior of the specific heat is always governed by $L^{\alpha/\nu}$, and in analogy with Eq.(6.11) we find [188]

$$c^R_{a,b}(0, L \rightarrow \infty) = C^{(0)+}_{bulk}\omega_{a,b}L/\xi^+_0 + \ldots, \tag{6.11}$$

where $\alpha = \nu = \frac{1}{2}$ for a tricritical point in $d = 3$ has been used. The universal amplitudes $\omega_{a,b}$ are given by [188]

$$\omega_{O,O} = \tfrac{1}{3}, \quad \omega_{O,SB} = 1, \quad \omega_{aper} = \tfrac{1}{2}. \tag{6.12}$$

The scaling function $J_{+O,O}(y_+)$ in Eq.(6.10) and the amplitude $\omega_{O,O}$ are particularly interesting for films of liquid ^{3}He –^{4}He mixtures in the vicinity of the tricritical end point T_t of the line L of the tricritical superfluid transitions (see Fig.1.9).

6.1.1 ^{4}He in Critical Films

At bulk criticality the leading singular part of the specific heat of a fluid confined in a film geometry displays a power–law behavior as a function of the film thickness L (see Eq.(6.6)). The *universal* amplitudes $\omega_{a,b}$ which govern this size dependence are experimentally testable quantities. However, Eq.(6.6) only yields the asymptotic behavior of the *singular* part of the specific heat so that additional information on the *background* contribution to the measured total specific heat $C_{a,b}(t,L)$ of the film is needed. In fact, for liquid ^{4}He *above* the λ–transition this information is available [241]. First, we note that the bulk specific heat of ^{4}He at the λ–transition is *finite* due to $\alpha < 0$ (see Table 1.5). Therefore $\lim_{L\to\infty} C_{a,b}(0,L)$ is finite and equals the bulk specific heat of ^{4}He at the λ–transition. With regard to the *leading* L–dependence of the difference $C_{a,b}(0,L) - C_{a,b}(0,\infty)$ between the total film specific heat and the total bulk specific heat of ^{4}He Eq.(6.6) can be employed in the form [242, 243]

$$C_{a,b}(0,L) - C_{a,b}(0,\infty) = C_{bulk}^{(0)+}\omega_{a,b}\left(L/\xi_0^+\right)^{\alpha/\nu}. \tag{6.13}$$

Second, we note that for ^{4}He films the combination $(a,b) = (O,O)$ of surface universality classes seems to be adequate (see also Refs.[108, 109]). The bulk specific heat $C_{bulk}(t) = C_{O,O}(t,\infty)$ at saturated vapor pressure is identified with the constant pressure specific heat $C_p(t)$ which reads to leading order in the reduced temperature t [241]

$$C_{O,O}(t,\infty) = C_p(t) = A\left[(t^{-\alpha} - 1)/\alpha + \tilde{B}\right] \tag{6.14}$$

which leads to $C_{bulk}^{(0)+} = A/\alpha$ and $C_{O,O}(0,\infty) = A(\tilde{B} - 1/\alpha)$. Therefore the leading L–dependence of the specific heat $C_{O,O}(0,L)$ of a ^{4}He film at the λ–transition is

$$C_{O,O}(0,L) = A\left[\left(\omega_{O,O}\left(L/\xi_0^+\right)^{\alpha/\nu} - 1\right)/\alpha + \tilde{B}\right]. \tag{6.15}$$

At saturated vapor pressure the amplitudes A and $\tilde{B}$ and the correlation length amplitude ξ_0^+ take the values [241]

$$A = 6.156\,\mathrm{J\,mole^{-1}K^{-1}}, \quad \tilde{B} = -1.826, \quad \xi_0^+ = 1.432\,\text{Å}. \tag{6.16}$$

In comparison with a recent specific heat measurement for ^{4}He at saturated vapor pressure (see Fig.6 in Ref.[244]) Eq.(6.15) considerably overestimates the experimental value of the specific heat, if the data from Eq.(6.16) and $\omega_{O,O}$ from Table 6.1

or Eq.(6.9) are used. If the amplitude $\omega_{O,O}$ is adjusted to the experimental values presented in Fig.6 of Ref.[244], one finds $\omega_{O,O} = 1.0$ in contrast to the values given in Table 6.1 and Eq.(6.9). Moreover, the specific heat exponent α for ^{4}He is very small (see Table 1.5) so that the values of $C_{O,O}(0, L)$ predicted by Eq.(6.15) are very sensitive to the choice of $\omega_{O,O}$. In view of the considerable spread in the values for $\omega_{O,O}$ we refrain from evaluating Eq.(6.15) any further.

For $t > 0$ the universal scaling function $J_{+O,O}(y_+)$ (see Eq.(6.2)) governs the renormalized specific heat $c_{O,O}^R(t, L)$ (see Eq.(6.1)). In the spirit of Eq.(6.15) one can write the *total* specific heat $C_{O,O}(t, L)$ of a film of liquid ^{4}He at saturated vapor pressure and for $T > T_\lambda$ in the form

$$C_{O,O}(t, L) = A \left[\left(t^{-\alpha} J_{+O,O}(y_+) - J_{+O,O}(y_{1+}) \right) / a_{bulk}^+ + \tilde{B} \right], \qquad (6.17)$$

where A and $\tilde{B}$ are given by Eq.(6.16). Note, that the reference value $J_{+O,O}(y_{1+})$ is governed by the bulk contribution a_{bulk}^+/α, because $y_{1+} = L/\xi_0^+ \gg 1$. However, if the ε-expansion of the scaling function $J_{+O,O}(y_+)$ is used to evaluate Eq.(6.17) to $\mathcal{O}(\varepsilon)$, one again finds a gross overestimation of the specific heat as compared to experimental data [244].

The basic reason why the ε-expansion of the specific heat up to —indexspecific heat!of ^{4}He!ε-expansion $\mathcal{O}(\varepsilon)$ fails to yield reasonable agreement with experimental specific heat data for liquid ^{4}He is given by the misrepresentation of the specific heat exponent α for the XY universality class ($N = 2$) to order ε. For $N = 2$ one has $\alpha = \varepsilon/10 + \mathcal{O}(\varepsilon^2)$ which yields $\alpha = 0.1$ in $d = 3$ compared to $\alpha \simeq -0.01$ according to Table 1.5. Thus the critical behavior of the specific heat of ^{4}He is much closer to the logarithmic (one–loop) behavior than predicted by the ε-expansion. A proper field–theoretical treatment of the specific heat of ^{4}He to one–loop order directly in $d = 3$ should therefore yield far better agreement with the experimental data.

This improved agreement has in fact been obtained very recently by a field–theoretical analysis of $C_{O,O}(t, L) - C_{O,O}(t, \infty)$ to one–loop order in a three–dimensional film geometry [242, 243]. The finite–size scaling function $J_{+O,O}(y_+)$ in this case is basically given by the corresponding expression for a tricritical film in $d = 3$ (see Eq.(6.10)). After a suitable renormalization procedure [243] the specific heat $C_{O,O}(t, L)$ is obtained for temperatures T above T_λ and slightly below T_λ, where Eq.(6.17) cannot be applied. The agreement with experimental data for $T = T_\lambda$ [244, 246] as a function of the film thickness L as well as for $L = 19\mu$m [244] as a function of temperature is very convincing.

However, up to now most experimental efforts have been spent on the analysis of the film specific heat near its maximum (see Fig.2.1(b)) which is located *below* T_λ [26]. The advantage of this temperature regime of course is that temperature gradients in the ^{4}He samples are effectively suppressed by superfluidity. Moreover, the film thicknesses that have been reached in, e.g., Nuclepore filters are in the range of 20–50Å [26, 246, 247] which is presumably too small to be in the asymptotic regime. For further experimental tests of field–theoretical predictions for the specific heat of ^{4}He films measurements especially for thick films ($L \geq 1\mu$m) in the vicinity of T_λ would be desirable.

6.1.2 ^{4}He in Capillaries and Pores

A large number of experiments on the specific heat of finite systems near bulk criticality have been performed with liquid ^{4}He confined to *Nuclepore filters*. These filters are polycarbonate membranes of thickness $l = 5 - 10\mu$m with capillaries (cylindrical channels) at a density of the order 10^8cm^{-2} [247]. The diameter of these channels typically takes the values 300Å, 800Å, or 1000Å for $l = 5\mu$m and 2000Å for $l = 10\mu$m. Due to the typical angle between the channel axis and the normal of the filter surface intersections between different channels are very likely (see Fig.6.1) [248]. Basically two different types of geometries can be realized with liquid ^{4}He confined to Nuclepore filters. In the first case a layer of liquid ^{4}He is formed on the inner walls of the cylindrical channels. The layer thickness which typically attains values in the range 20–50Å is much smaller than the diameter of a channel so that this type of confinement resembles a *film* geometry. If the layer thickness grows beyond the capillary condensation threshold (56Å for 2000Å filters [246, 247]) the channels are filled with liquid ^{4}He thus providing a *cylindrical* confinement. With regard to finite–size scaling the fractional shift, the fractional rounding, and the specific heat maximum have been studied as functions of the channel diameter [26, 245, 246, 247]. The specific heat maximum can be fitted to a logarithmic law as a function of the channel diameter, because the critical exponent α of the specific heat is very small (see Eqs.(6.13) and (6.14)). Rather surprisingly a first analysis of the experimental data indicated a violation of the fractional shift and the fractional rounding prediction (see Eqs.(2.5) and (2.7)) in the sense that the shift exponent Λ and the rounding exponent θ dif-

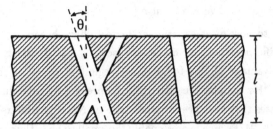

Fig. 6.1: Schematic cross section of a Nuclepore filter (taken from Ref.[248]). The filter thickness l is
5 or 10μm with channel diameters between 300 and
2000Å. The angle θ between the channel axis and the
filter surface normal is typically not larger than 34°
[248]. Many channels intersect at least one other channel along their length.

fer from the predicted value $1/\nu$ outside the error bars [246]. Moreover, the surface
contribution to the total specific heat near T_λ (see Eqs.(6.1) and (6.2)) seemed to be
governed by an exponent α_s which *violated* the scaling law $\alpha_s = \alpha + \nu$ (see Eq.(1.13))
[247]. It has been only very recently that the apparent violation of the scaling prediction $\theta = \Lambda = 1/\nu$ could be explained by errors in the first analysis of the data
[249]. In fact, a reanalysis of the original data of Refs.[245] and [246] confirmed the
scaling prediction $1/\Lambda = \nu \simeq 0.67$ for ^{4}He confined to cylindrical channels [249]. The
observed exponent α_s, however, should be interpreted as an *effective* exponent due
to the strong scatter of the data. In order to obtain a reliable estimate of α_s from
experimental data of the specific heat the size L of the samples should be sufficently
large ($L \geq 1\mu$m) in order to reach the scaling region [249]. The present status of the
theory concerning finite–size scaling of the specific heat of ^{4}He in a uniform type of
confinement as provided by Nuclepore filters has recently been reviewed in Ref.[242].

Apart from Nuclepore filters there are other types of porous media like Vycor glass
or pressed powders which provide a *nonuniform* confinement for liquid ^{4}He. The pore
size inside such media is typically of the order of 100Å [26]. The first measurements
of the specific heat of liquid ^{4}He confined in such a medium were performed for ^{4}He in
Vycor glass with a pore size of about 60Å [250]. The temperature dependence of the
specific heat in this case closely follows that of a 32.8Å film for temperatures T in the

range $2.12K \leq T \leq 2.17K$ (see p.51 in Ref.[26]). However, the height of the specific heat maximum and its position do not match those of the 32.8Å film. Moreover, it seems to be impossible to account for the observed scaling of the specific heat maximum and the fractional shift (see Eq.(2.5)) with one and the same confining pore size. These findings may be explained by the complicated topology of the confinement which is generated by the nonuniformly shaped pores in Vycor. To present knowledge it is impossible to characterize such topologies by a single length scale so that a theoretical explanation for the experimental findings must certainly go beyond the standard finite–size scaling [26]. An analogous behavior of the specific heat can be observed for ^{4}He confined to pressed carbon black powder of 120Å grain size. The pore–size distribution for this kind of confinement is presumably not reproducible from one experiment to another, and the topology seems to be even more complicated than the one in Vycor. The scaling of the specific heat maximum and the fractional shift again yield different confining sizes [26]. A third interesting kind of nonuniform complete confinement for liquid ^{4}He is provided by the implantation of ^{4}He atoms in copper followed by a high–temperature annealing process. During annealing cavities filled with ^{4}He gas form, where the cavity–size distribution depends on the annealing temperature. A typical cavity size is 100Å. Upon cooling the confined ^{4}He liquifies, and specific heat measurements can be performed. Compared to the specific heat of ^{4}He in Vycor or pressed carbon black powder the specific heat curves also display a clear maximum below T_λ, but their overall shape is different. In view of this difference it is interesting to note that the pressure in a cavity *varies* with the size of the cavity. The current experimental data for this type of complete confinement do therefore not yet permit a quantitative finite–size scaling analysis [26].

6.1.3 ^3He – ^4He Mixtures in Tricritical Films

The renormalized specific heat of a *tricritical* ^{3}He –^{4}He film (see Eq.(6.1)) is governed by the universal finite–size scaling function $J_{+0,0}(y_+)$ (see Eq.(6.10)) which is exact in $d = 3$. In the spirit of Eq.(6.17) the knowledge of the nonuniversal bulk amplitude A and the background constant $B = A\bar{B}$ gives the opportunity to calculate the *total* specific heat of a ^{3}He –^{4}He film in the vicinity of the tricritical point at, e.g., saturated vapor pressure. However, the observed critical behavior of a system near a tricritical (multicritical) point depends on the *path* in the phase diagram along which it is approached. Besides the temperature T (or the reduced temperature t) and the pressure p the composition x or the chemical potential difference $\phi = \mu_{^3He} - \mu_{^4He}$ of the ^{3}He $-^4$He mixture can be chosen as thermodynamical variables (see Fig.1.9). As a consequence there are basically two choices for a definition of a molar constant pressure specific heat, namely $C_{p,x}$ (constant x) and $C_{p,\phi}$ (constant ϕ). These two specific heats are related via the thermodynamical relation [251]

$$C_{p,\phi} = C_{p,x} - T \left.\frac{\partial \phi}{\partial T}\right|_{p,x} \left.\frac{\partial x}{\partial T}\right|_{p,\phi}, \tag{6.18}$$

where $t = (T - T_t)/T_t$ and $T_t = 0.867$K is the tricritical temperature for a ^{3}He $-^4$He mixture at saturated vapor pressure. From a *bulk* measurement of $C_{p,x} = T\partial S/\partial T|_{p,x}$ the quantity $\partial \phi/\partial T|_{p,x} = -\partial S/\partial x|_{T,p}$ can be derived so that the construction of $C_{p,\phi}$ according to Eq.(6.18) is possible, where $\partial x/\partial T|_{p,\phi}$ is given by the slope of the constant ϕ curves in the constant pressure surfaces of the phase diagram [251]. At saturated vapor pressure and for ϕ fixed to its tricritical value ϕ_t the specific heat $C_{p,\phi}$ of bulk ^{3}He $-^4$He behaves as [251, Eq.18]

$$C_{p,\phi}^{bulk} \propto |t|^{-1/2}, \tag{6.19}$$

for $t \to 0$ in accordance with the value $\alpha = \frac{1}{2}$ for the specific heat of tricritical $O(N)$-symmetric systems. Here it is interesting to note that the specific heat $C_{p,x}$ at $x = x_t$ is *linear* in t for $t \searrow 0$ which corresponds to an exponent $\alpha_x = -1 = -\alpha/(1 - \alpha)$ in accordance with the "renormalization" of the critical exponent α due to the constraint $x = x_t$ imposed on the *extensive* variable x [15, 251] (see Eq.(1.2)). Using the data displayed in Fig.7 of Ref.[251] for $t > 0$ we approximate the bulk specific heat $C_{p,\phi}^{bulk}$ at saturated vapor pressure by

$$C_{p,\phi}^{bulk}(t) = At^{-1/2} + B \quad \text{with}$$
$$A = 1.1 \,\text{J mole}^{-1}\text{K}^{-1}, \quad B = -0.8 \,\text{J mole}^{-1}\text{K}^{-1} \tag{6.20}$$

for $0 < t < 2 \cdot 10^{-3}$ and $\phi = \phi_t$. In the spritit of Eq.(6.17) the total specific heat $C_{p,\phi}(t, L)$ of a ^{3}He $-^4$He film of thickness L at saturated vapor pressure and $\phi = \phi_t$ can be written as (see Eq.(6.10))

$$C_{p,\phi}(t, L) = At^{-1/2}\left(1 - \frac{1}{y_+} + \frac{2}{e^{2y_+} - 1}\right) + B, \tag{6.21}$$

where A and B are taken from Eq.(6.20), $y_+ = L/\xi_+$, and the bulk correlation length

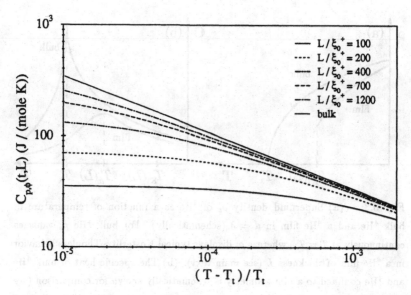

Fig. 6.2: Constant pressure specific heat $C_{p,\phi}(t, L)$ of a ^{3}He –^{4}He film of thickness L at saturated vapor pressure and $\phi = \phi_t$ above the tricritical temperature T_t. The dotted line shows the bulk behavior according to Eq.(6.20). For any finite L the specific heat levels off at a finite value for $t \to 0$ (see main text).

ξ_+ is given by $\xi_+(t) = \xi_0^+ t^{-1/2}$. The shape of $C_{p,\phi}(t, L)$ in the range $10^{-5} < t < 2 \cdot 10^{-3}$ is displayed in Fig.6.2 for several values of the dimensionless ratio L/ξ_0^+. The specific heat reaches its bulk limit for $L/\xi_0^+ \to \infty$ much more rapidly than for pure ^{4}He due to the much larger value of the specific heat exponent α. As in the case of pure ^{4}He the limiting bulk specific heat $C_{p,\phi}^{bulk}(t) = C_{p,\phi}(t, \infty)$ is approached from *below* for $L \to \infty$. At the tricritical point ($t = 0, \phi = \phi_t$) we conclude from Eq.(6.21) (see also Eqs.(6.11) and (6.12))

$$C_{p,\phi}(0, L) = \frac{A}{3} \frac{L}{\xi_0^+} + B \tag{6.22}$$

for $L \gg \xi_0^+$ which completes our analysis of $C_{p,\phi}(t, L)$ near the tricritical point of a ^{3}He –^{4}He mixture. Note, that Eq.(6.22) also gives access to the correlation length amplitude ξ_0^+ by a measurement of the specific heat $C_{p,\phi}(0, L)$.

Experimentally the specific heat of ^{3}He –^{4}He mixtures confined to Vycor has been probed at various compositions away from the tricritical point [252]. However,

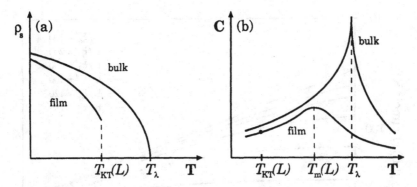

Fig. 6.3:   (a) Superfluid density ρ_s of ^{4}He as a function of temperature in
bulk ^{4}He and a ^{4}He film in $d = 3$ (schematically). For bulk ^{4}He ρ_s vanishes
continuously at $T = T_\lambda$, whereas ρ_s displays typical Kosterlitz–Thouless behavior
in a ^{4}He film of thickness L (see main text). (b) The specific heat of bulk ^{4}He
and ^{4}He confined to a film geometry is schematically shown for comparison (see
Fig.2.1).

experimental data for the specific heat of liquid ^{3}He –^{4}He *films* near the tricritical
point at saturated vapor pressure (see Fig.1.9(b)) seemingly do not exist.

6.2   The Superfluid Density in Confined ^{4}He

The most striking property of liquid ^{4}He is its superfluid phase at low temperatures
which leads to a wide spectrum of rather unusual physical properties. Superfluidity
itself is characterized by the superfluid density ρ_s which measures the fraction of
^{4}He in the superfluid state. The superfluid density is closely related to the *helicity
modulus* $\Upsilon(t)$ of an $O(2)$-symmetric vector model [44] which has already been investigated from the point of view of finite–size scaling (see Eqs.(2.18)–(2.22)). Therefore
Eq.(2.20) can be understood as a prediction of the finite–size scaling behavior of the
superfluid density in ^{4}He which therefore allows experimental tests.

The theoretical expectation concerning the global behavior of the superfluid density of ^{4}He confined to a film geometry is summarized in Fig.6.3(a) [26]. In bulk
^{4}He the superfluid density $\rho_s^{bulk}(t)$ vanishes identically for $T > T_\lambda$ and displays an
algebraic growth of the form $\rho_s^{bulk}(t) = \rho_0(-t)^\upsilon$ (see Eq.(2.19)) for $T < T_\lambda$ which is

reminiscent of the power–law behavior of the order parameter. The exponent v is given by $v = 2 - \alpha - 2\nu = \nu$ in $d = 3$ [44]. In a film of thickness L, however, the behavior of the superfluid density $\rho_s(t, L)$ of ^{4}He is effectively two–dimensional as expected from the XY model in $d = 2$, where long–ranged order is absent [45]. At a (Kosterlitz–Thouless) transition temperature $T_{KT}(L)$ the superfluid density $\rho_s(t, L)$ jumps to a finite value which approaches a certain limit as $L \rightarrow 0$. This limit is given by the value [46]

$$\Upsilon(T_{KT}) = \tfrac{2}{\pi} k_B T_{KT} \tag{6.23}$$

of the helicity modulus $\Upsilon(T)$ of the two–dimensional XY model at $T = T_{KT}(L = 0) \equiv T_{KT}$. Conversely, the jump in $\rho_s(t, L)$ vanishes in the limit $L \rightarrow \infty$. In fact, this kind of behavior has been found in a recent Monte–Carlo analysis of the helicity modulus in a XY model confined to a slab geometry with free boundary conditions in $d = 3$ [79] (see Sec.2.2). The specific heat of confined ^{4}He displays an essential singularity at $T = T_{KT}(L)$ corresonding to the Kosterlitz–Thouless transition which is invisible in the specific heat curve (see Fig.6.3(b) and Sec.2.1). Conversely, the position $T_m(L)$ of the specific heat maximum has no effect on the superfluid density as a function of temperature (see Fig.6.3(a)). Finally, we note that ρ_s in Fig.6.3(a) is not normalized. If ρ_s is taken as the superfluid *fraction*, ρ_s^{bulk} and $\rho_s(t, L)$ are both equal to unity at zero temperature.

6.2.1   *The Superfluid Density in Films*

The close relationship between the helicity modulus of $O(2)$–symmetric systems and the superfluid density in ^{4}He immediately yields a finite–size scaling form of the superfluid density $\rho_s(t, L)$ in a film geometry (see Eq.(2.20)) which is usually written in the normalized form [242]

$$\rho_s(t, L)/\rho = |t|^\nu g\left(|t|L^{1/\nu}\right) \tag{6.24}$$

or, alternatively, with the bulk contribution separated [253]

$$\rho_s(t, L)/\rho = \rho_s^{bulk}(t)/\rho \left[1 - f\left(|t|L^{1/\nu}\right)\right], \tag{6.25}$$

where $\rho_s^{bulk}(t)/\rho = k|t|^\nu$, $k =$const in the vicinity of the λ–transition. From the theoretical side there are ambiguities in the definition of the superfluid density via the helicity modulus in a finite system due to some freedom in the choice of the boundary conditions (see the discussion of the helicity modulus in Secs.2.1 and 2.2 for the spherical model). However, these ambiguities should not affect the finite–size scaling property of the helicity modulus, and therefore a scaling plot of $|t|^{-\nu}\rho_s(t, L)$ versus the scaling argument $|t|L^{1/\nu}$ with the correlation length exponent $\nu \simeq 0.67$ for ^{4}He is expected to yield data collapse onto a single curve. A corresponding experiment has been performed on superfluid ^{4}He confined between two silicon wafers providing a film geometry of thickness $L = 0.106, 0.519, 2.8$, and 3.9μm [253]. The silicon wafers form a flat cell for the confinement of ^{4}He which is attached to a Be–Cu torsional element to form a high–Q oscillator. The superfluid density is measured by recording the change of the period of the oscillations as a function of temperature with high resolution [253]. A scaling plot of the superfluid density shows that the data do *not* collapse onto a single curve which is in substantial disagreement with Eqs.(6.24) and (6.25). A reasonable data collapse can only be achieved if the exponent $\nu \simeq 0.67$ in the scaling argument of Eq.(6.25) is replaced by an *effective* exponent $\nu' = 1.14$ [253]. Several attempts have been made to reconcile the experimental findings with scaling theory, e.g., by including a logarithmic correction term according to Eqs.(2.20) and (2.21) in the finite–size scaling law for the superfluid density. However, the results are not convincing (see Ref.[242]). The apparent failure of finite–size scaling for the superfluid density still poses a major challenge for both experiment and theory.

Finally, we note that other theoretical definitions of the superfluid density $\rho_s(t, L)$ than the one in terms of the helicity modulus may be used [242]. It is not obvious which definition is most adequate for the description of current experiments. However, in view of the aforementioned difficulties with finite–size scaling the precise definition of the superfluid density of ^{4}He seems to be of minor importance.

6.2.2 The Superfluid Density in Capillaries and Pores

Channels in Nuclepore filters provide a uniform confinement of ^{4}He in a cylindrical geometry (capillaries) as shown in Fig.6.1. The superfluid density of ^{4}He confined to Nuclepore filters of thickness $l = 5\mu$m and with channel diameters of 300, 500, 800, and 1000Å has been measured in Ref.[254] by means of the Helmholtz–resonator technique. As expected from Fig.6.3 the superfluid density ρ_s for confined ^{4}He is smaller than ρ_s^{bulk} in bulk ^{4}He, where the transition temperature $T_{KT}(L)$ (see Fig.6.3) decreases with decreasing channel diameter. The temperature range of the Helmholtz–

resonator technique is limited to about $(T_\lambda - T)/T_\lambda \geq 10^{-3}$ due to dissipative signal losses in the resonator very close to T_λ. The data of Ref.[254] have been reanalyzed in the light of the finite–size scaling prediction given by Eq.(6.25) [26, 254], but as in the case of planar confinement in a film [253] a scaling plot of the superfluid density does *not* yield data collapse [26]. The deviations from scaling are found to be particularly strong for small confining sizes. Reasonable data collapse can only be obtained if the scaling prediction $1/\theta = \nu \simeq 0.67$ for the exponent θ in the scaling argument $|t|L^\theta$ is replaced by the smaller value $1/\theta \simeq 0.53$ [254]. However, in this case one may argue that due to the restriction $(T_\lambda - T)/T_\lambda \geq 10^{-3}$ the data points obtained in Ref.[254] are too remote from the λ–transition to give clear evidence of the leading finite–size effects.

For nonuniform confinements of ^{4}He as provided by packed powders or Vycor the experimental data cannot be scaled by a unique confining size (see Sec.6.1). However, a superfluid density $\langle \rho_s \rangle$ which has been averaged over the size distribution of the pores can still be analyzed with regard to its temperature dependence [26]. This has been done with various data sets from experiments with ^{4}He confined to Metricel (cellulose) or Millipore (similar to Metricel) filters for example. As a result the data for the temperature dependence of the fractional difference $(\rho_s^{bulk} - \langle \rho_s \rangle)/\rho_s^{bulk}$ which should be governed by the power law $|t|^{-\nu}/L$ for large pore sizes L and $t < 0$ in fact favor an exponent close to 1.0 instead of $\nu \simeq 0.67$ [26, p.61].

In summary we remark that although the initial problems with the finite–size scaling behavior of the specific heat of confined ^{4}He have been resolved [242] the finite–size scaling behavior of the superfluid density still poses major challenges for theory and experiment. For a recent review of experimental results on confined ^{4}He we refer the reader to Ref.[26]. The most recent developments in this field especially from the theoretical side have been summarized in Ref.[242].

6.3 *Wetting Phenomena near Criticality*

Liquids confined to finite geometries are formed in a natural way in the course of a *wetting transition* [25]. There are basically two physical situations in which wetting transitions can occur. In the first case a *substrate*, e.g., a container wall, is exposed to the vapor of a fluid which forms a liquidlike layer on the wall. In the vicinity of liquid–vapor *coexistence* the liquidlike layer can become macroscopically thick (see below). In the second case a fluid layer forms at the *interface* between two other fluids, where the layer thickness can become large in the vicinity of *coexistence* of all three fluids. This second kind of wetting is called *interfacial wetting* [6], and it occurs in binary liquid mixtures.

The shape of the wetting layer on a substrate depends on the spatial and chemical structure of the substrate [255]. If the substrate is flat and chemically homogeneous the wetting layer resembles a *film* as in the case of interfacial wetting. However, one boundary of this wetting induced film geometry is no longer provided by an external wall but by a fluid–fluid *interface*. In order to define a film thickness L we assume in the following that these interfaces are *sharp* and that *capillary waves* can be ignored (see below). For the wetting of a substrate by the liquid phase of a simple fluid this is justified for temperatures sufficiently below the critical temperature T_c (see Fig.1.1). In the case of interfacial wetting this is justified if the wetting transition occurs sufficiently below the critical end point of the line L_2 of critical demixing transitions as indicated by the point W in Fig.6.4. In this case the interfacial width is set by the bulk correlation length ξ_- which is small compared to the wetting layer thickness.

Upon approaching coexistence at constant temperature the wetting layer thickness L either levels off at some finite value (incomplete wetting) or diverges in the thermodynamic limit (complete wetting) [25]. In the latter case with which we will be concerned in the following the wetting film has become *macroscopically* thick in contrast to the former case in which the film remains *microscopically* thin. The onset of complete wetting, i.e., the *wetting transition* to a layer of macroscopic thickness, is marked by the *wetting temperature* T_w. The wetting transition itself may be first or second order, where the first order transition is accompanied by a *prewetting transition* (for details see Ref.[25]). For the following considerations we assume $T_3 < T_w < T_c$ for simple fluids (see Fig.1.1) and $T_w < T_{cep}$ for binary mixtures as indicated by the point W in Fig.6.4.

The thickness of the wetting layer is governed by two opposing contributions to the so–called *effective interface potential* [25]. In the case of complete wetting a substrate acts as an external potential on the fluid particles which is more attractive than the pair interaction potential in the vapor so that "potential energy" can be gained by the formation of a thick wetting layer. On the other hand the cost in free energy for building up a layer of a thermodynamically unfavorable phase grows with

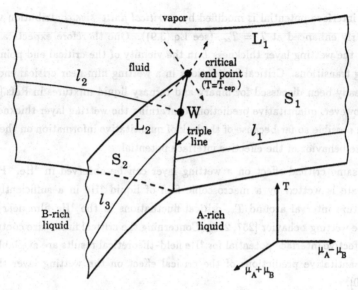

Fig. 6.4: Schematic phase diagram of a binary liquid mixture (see Fig.1.5). The lines l_1, l_2, and l_3 and the point W indicate possible positions of wetting transitions, where prewetting sheets and lines have been omitted for simplicity. The A–rich and the B–rich liquid wet the wall–vapor interface along l_1 and l_2, repectively, and the A–rich liquid wets the interface between a wall and the B–rich liquid along l_3. The B–rich liquid wets the interface between the vapor and the A–rich liquid at the point W. The complete wetting regime is located above the lines l_1, l_2, l_3, and the point W, respectively (taken from Ref.[25, Fig.4.7(a)]). The dashed arrows indicate paths on the vapor side of S_1 which pass close to the critical end point ($T = T_{cep}$).

the layer thickness L. In thermodynamical equilibrium the thickness L of the wetting layer takes a value which *minimizes* the effective interface potential.

If a substrate is exposed to the vapor of a binary liquid mixture a wetting layer of an A–rich liquid, a B–rich liquid, or the AB mixture may be formed depending on the values of the chemical potential difference $\mu_A - \mu_B$ and the temperature T relative to the lines l_1 and l_2 of wetting transitions shown in Fig.6.4. If the critical end point of the second–order demixing transitions ($T = T_{cep}$) lies in the complete wetting regime (see Fig.6.4), a macroscopic *critical wetting film* forms upon approaching the vapor pressure surface S_1 from the vapor side at $T = T_{cep}$. In this case the effective interface potential is modified by a *critical finite–size* contribution which is particularly enhanced at $T = T_{cep}$ (see Eq.(3.9)). One therefore expects a critical effect on the wetting layer thickness L in the vicinity of the critical end point of the demixing transitions. Critical fluctuations in a wetting film near critical end points have already been discussed for binary and ternary liquid mixtures in Refs.[99] and [256]. However, quantitative predictions concerning the wetting layer thickness have not been possible so far because of the lack of quantitative information on the critical finite–size behavior of the effective interface potential.

The same critical effect on a wetting layer can be observed in ^{4}He. Provided a substrate is wetted by a macroscopic film of liquid ^{4}He in a sufficiently broad temperature interval around T_λ, critical fluctuations in the ^{4}He film near T_λ will affect the wetting behavior [257, 258]. Concerning the critical finite–size contribution to the effective interface potential for ^{4}He field–theoretical results are available which allow quantitative predictions of the critical effect on the wetting layer thickness [173, 240].

With ^{4}He one can also study an "inverted" situation in which the bulk fluid becomes critical instead of the wetting layer [257, 258]. Consider liquid ^{4}He slightly below the melting pressure confined between parallel plates at a distance D. For a suitably chosen plate material a macroscopic wetting layer of *solid* ^{4}He of thickness L is formed on the inside walls of this slab geometry. In the vicinity of the *upper* λ–point $(T_\lambda^+, p_\lambda^+)$ (see Fig.1.6) the remaining liquid layer of thickness $D - 2L$ becomes critical which imposes a change on the thickness L of the solid wetting layers due to the finite separation D between the plates [257, 258]. However, in practice *capillary condensation* severely limits the growth of the solid ^{4}He layers on the inside walls of a slab or a capillary.

Analogous to liquid ^{4}He below saturated vapor pressure near the lower λ–point a ^{3}He $-^4$He mixture below saturated vapor pressure near the tricritical end point of the tricritical superfluid transitions (line L in Fig.1.9(a)) gives the opportunity to study a tricritical wetting film, provided complete wetting occurs in the neighborhood of this point [257, 258].

At least on one side wetting films are bounded by a fluid–fluid *interface* rather than a solid wall. Especially for macroscopic films *capillary waves* (see Eqs.(2.52) and (2.53) in Sec.2.3) occur in the interface, because the interface becomes *rough* in the limit of large film thicknesses and in the absence of gravity. In general a rough interface is characterized by a height–height correlation function of the form [259]

$$\left\langle (h(\mathbf{r}) - h(\mathbf{r'}))^2 \right\rangle = A_s \left| \mathbf{r} - \mathbf{r'} \right|^{2\zeta_s}, \tag{6.26}$$

where A_s is a constant, $h(\mathbf{r})$ is the height of the interface over the substrate at the position $\mathbf{r}$ in the substrate, and ζ_s is the roughness exponent. Stochastic roughness in comparison is defined by a Gaussian height–height correlation function [260]. According to Eq.(6.26) a rough boundary of a wetting film leads to large spatial fluctuations of the film thickness L over macroscopic distances which raises the question, how these fluctuations modify the critical finite–size behavior of the effective interface potential. A rigorous answer to this question is not yet available. However, in the case of *quenched* roughness the height–height correlations given by Eq.(6.26) lead to two L–dependent contributions to the free energy which are proportional to $A_s L^{-4}$ and $A_s L^{-4+2\zeta_s}$, respectively [259]. In $d = 3$ one has $\zeta_s = 0$, i.e., a logarithmic law in Eq.(6.26) indicating that roughness of a film boundary only leads to subdominant terms in the free energy compared to the leading L^{-2}–behavior of the finite–size part at $T = T_{c,b}$ in $d = 3$ (see Eq.(3.9)). The same holds for stochastic roughness which only yields a contribution proportional to L^{-4} to the free energy in $d = 3$ [260]. At present it is not known what happens slightly away from the critical point and if roughness is part of the relevant degrees of freedom which usually is the case for fluid–fluid interfaces. However, due to $\zeta_s = 0$ in $d = 3$ the growth of the height–height correlation function is rather slow, and we therefore expect that roughness of interfaces can be ignored in the discussion of the *leading* critical effect on the wetting layer thickness L.

We now turn to a more detailed analysis of critical ^{4}He wetting films under the assumption that these films exist in the complete wetting regime.

6.3.1 *Critical ^{4}He on Substrates*

In experiments the distinction between complete and incomplete wetting behavior of a fluid on a given substrate is a rather delicate problem. Especially this aspect of the wetting behavior of ^{4}He in the vicinity of the lower λ–point has received a considerable amount of interest. In order to give an impression of the efforts spent on this subject we quote a few examples.

The vibrating–wire microbalance technique provides a useful experimental tool for the characterization of wetting behavior (see Sec.1.3 and Ref.[27]). The frequency shift of the wire oscillation depends on the mass load of the wire due to adsorption of the wetting agent and therefore measures the thickness of the liquid film on the wire. The typical diameter of a single wire is in the range of 10μm. Using graphite and platinum fibers complete wetting of ^{4}He has first been found only below T_λ, where the wetting transition occurs close to T_λ [27]. However, in a later study with graphite fibers of 9.5μm average thickness complete wetting of ^{4}He is found both *above* ($T = 2.251$K) and *below* T_λ ($T = 1.987$K) [261]. This discrepancy can be explained by small temperature gradients in normal ^{4}He which easily lead to an incomplete wetting behavior of ^{4}He [261] so that graphite is in fact wetted completely by ^{4}He in a whole temperature interval around T_λ.

Silver and gold exhibit a strong substrate potential on ^{4}He so that one expects complete wetting of ^{4}He on these materials. For silver evaporated on borosilicate glass this can be confirmed both above and below T_λ by means of a capacitance measurement [262]. A parallel plate capacitor made of the substrate material is placed in the experimental chamber filled with ^{4}He vapor close to saturated vapor pressure. The inner surfaces of the capacitor provide the substrate for the wetting agent so that the capacitance increases with the wetting layer thickness. For ^{4}He a smooth increases of the capacitance with the vapor pressure along the vapor pressure curve is observed so that wetting remains complete [262]. Rather interestingly a small reduction in the film thickness (i.e., the capacitance) has been observed in a small intervall $\sim \pm 2$mK around T_λ which may be due to critical fluctuations [173, 257] (see below). For gold plated quartz stable films of ^{4}He up to 600Å thickness have been observed both above and below T_λ [263] which indicate complete wetting of gold by ^{4}He as well. In this experiment the frequency shift of a gold–plated quartz–crystal resonator provides the measure of the film thickness as a function of the vapor pressure [263]. Here one should note that in an earlier study of the wetting behavior of ^{4}He on Ag and Au(111) surfaces using the same quartz–crystal resonator technique incomplete wetting has been reported [264] in contrast to Refs.[262] and [263]. This discrepancy may be due to an additional mode of crystal vibration which couples to sound waves in the vapor and leads to larger frequency shifts than expected for a film alone [263]. The observed finite–frequency intercept at saturated vapor pressure which serves as an indicator for incomplete wetting [264] may thus erroneously be attributed to an incomplete wetting film. Finally, we mention that incomplete wetting of normal ^{4}He on copper has been obtained from a capacitance measurement [265].

In the course of the experimental verification of the Lifshitz theory for the van-der-Waals potential [143, 144] in Ref.[266] complete wetting of ^{4}He has also been observed on weak substrates like SrF_2, NaF, and Ne at $T = 1.38$K. Moreover, complete wetting of ^{4}He occurs on H_2 crystals [267, 268] which also belongs to the weaker substrates. Concerning the strength of Ne as a substrate for ^{4}He additional data are given in Ref.[269].

While the wetting behavior of normal ^{4}He depends on the substrate material it has been believed that *superfluid* ^{4}He is a universal wetting agent. However, after theoretical indications that superfluid ^{4}He does *not* wet Cs, Rb, and K at $T=0$ [270] first experimental confirmations were obtained for Cs substrates from the disruption of superfluid flow in the ^{4}He film [271] and from the suppression of third sound propagation [272] which limits the film thickness to about nine layers. Later experimental investigations provided first evidence for a *prewetting* transition of ^{4}He on Cs above T_λ [273] by means of the quartz microbalance technique. The observation of prewetting is supported further by theoretical [274] and experimental studies [275] which yield a wetting temperature $T_w = 1.95$K [274, 275] and a prewetting line which can be observed up to the critical prewetting temperature $T_c^{pw} = 2.5$K [275]. The present status of theory and experiment concerning this unexpected wetting behavior of ^{4}He on Cs is summarized in Ref.[276] to which we refer the reader for further details. We close this digression on the wetting behavior of ^{4}He with the remark that very recently it has become possible to "tune" the substrate potential of gold electrodes by adsorbing monolayers of cesium on their surface [277]. By varying the thickness of the cesium layer from 1 to 25 monolayers the wetting temperature changes from $T_w = 0$ to $T_w \simeq 2.0$K. This is especially interesting for the enhancement of the critical effect on a complete wetting layer of ^{4}He near the lower λ–point to which we turn now.

The equilibrium thickness L of a complete wetting film is a function of the reduced temperature t and the undersaturation $\delta\mu = \mu_0 - \mu \geq 0$ which measures the deviation of the chemical potential μ of the ^{4}He vapor from its value $\mu_0(T)$ at liquid–vapor coexistence. Since the undersaturation $\delta\mu$ is assumed to be small we employ the

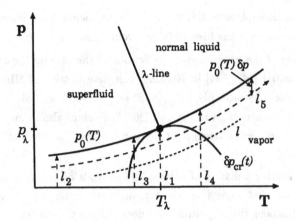

Fig. 6.5: Phase diagram of ^{4}He in the vicinity of the lower λ–point (see Fig.1.6). The paths l_1, l_2, l_3, and l_4 correspond to isotherms inside, outside, and passing through the critical region, respectively. The path l_5 runs parallel to the vapor pressure curve $p_0(T)$ at a distance $p_0(T)\delta p$ with $\delta p = \text{const}$. The domain in which wetting is influenced by the lower λ–point is bounded by the parabola shaped curve $\delta p_{cr}(t)$ and the short–dashed line l (see main text).

Gibbs–Duhem relation in the form

$$\delta\mu \simeq (p_0(T) - p)/\rho_v = p_0(T)/\rho_v\delta p, \tag{6.27}$$

where ρ_v denotes the vapor density of ^{4}He at liquid–vapor coexistence, and $\delta p = 1 - p/p_0(T) \geq 0$ provides a dimensionless measure of the undersaturation with respect to the saturated vapor pressure $p_0(T)$. The region of the phase diagram of ^{4}He which will be considered for the investigation of complete wetting films is shown in Fig.6.5.

The effective interface potential $\omega(l)$ for a liquid layer of a prescribed thickness l has the form [173]

$$\omega(l) = l\left(\rho_l/\rho_v - 1\right)p_0(T)\delta p + \sigma_{wl} + \sigma_{lv} + \delta\omega(l), \tag{6.28}$$

where ρ_l is the density of liquid ^{4}He at liquid–vapor coexistence. The term linear in l in Eq.(6.28) corresponds to the "penalty" for the formation of a liquid layer in the vapor region of the phase diagram, σ_{wl} and σ_{lv} denote the wall–liquid and the liquid–vapor surface tensions, respectively, and $\delta\omega(l)$ is the finite–size contribution due to the finite separation l between the substrate and the liquid–vapor interface. Besides the decomposition of $\omega(l)$ into bulk, surface, and finite–size contributions shown in Eq.(6.28) the effective interface potential furthermore decomposes into a singular and a regular part according to $\omega(l) = \omega^{sing}(l) + \omega^{reg}(l)$. The bulk and surface contributions to $\omega^{sing}(l)$ exhibit the well–known $t^{2-\alpha}$ and $t^{2-\alpha_s}$ singularities, respectively. However, since the surface tensions σ_{wl} and σ_{lv} do not depend on l we disregard these in the following. Moreover, the expansion $p_0(T) = p_\lambda + p_1 t + p_2 t^2 + P_\pm|t|^{2-\alpha} + \ldots$ [220] of the saturated vapor pressure near $t = 0$ $(T = T_\lambda)$ will be truncated after the constant term p_λ. According to Eqs.(2.4) and (5.15) the leading singular part of $\delta\omega(l)$ in $d = 3$ is given by

$$\delta\omega^{sing}(l) = k_B T_\lambda l^{-2}\Theta_{\pm 0,0}(l/\xi_\pm), \tag{6.29}$$

where $\xi_\pm$ can be identified with the bulk correlation length of ^{4}He *only* for $t > 0$ (see Eq.(5.10)). The universal finite–size scaling function $\Theta_{\pm 0,0}$ is discussed in Chap.5. The regular part of $\delta\omega(l)$ has the asymptotic form

$$\delta\omega^{reg}(l) = \begin{cases} \mathcal{W}\,l^{-2} + \mathcal{V}\,l^{-3} + \ldots & \text{(nonretarded)} \\ \mathcal{V}_r\,l^{-3} + \mathcal{U}_r\,l^{-4} + \ldots & \text{(retarded)} \end{cases} \tag{6.30}$$

for large l, where $\mathcal{W}$ is the so–called Hamaker constant, and the coefficient $\mathcal{V}$, inter alia, depends on the structure of the substrate along the direction of the surface normal [25]. In ^{4}He the crossover from the nonretarded form of the van–der–Waals interaction (top line in Eq.(6.30)) to the retarded form (bottom line in Eq.(6.30)) sets in for separations $l \simeq 100\text{Å}$. For $l \geq 0.1\mu m$ the interaction is fully retarded. The correction coefficient $\mathcal{V}$ and its retarded counterpart $\mathcal{U}_r$ can be used to describe the effect of preplating the substrate by some other material (see Ref.[277]). The temperature dependence of the coefficients $\mathcal{W}$, $\mathcal{V}$, $\mathcal{V}_r$, and $\mathcal{U}_r$ has been disregarded in Eq.(6.30).

According to the minimum principle for the effective interface potential the thickness $L(t, \delta p)$ of the complete wetting layer is obtained from the condition $\partial\omega/\partial l = 0$ evaluated at $l = L(t, \delta p)$, i.e., [173]

$$(\rho_l/\rho_v - 1)\, p_\lambda \delta p = k_B T_\lambda L^{-3}\vartheta_{\pm 0,0}(L/\xi_\pm) - \delta\omega'^{reg}(L), \tag{6.31}$$

Table 6.2:   Hamaker constants $\mathcal{W}$ for ^{4}He on SrF$_2$
and NaF [266], Ne [266, 269], and H$_2$ [267, 268] in
units of $k_B T_\lambda$.

	SrF$_2$	NaF	Ne	H$_2$
$\mathcal{W}/k_B T_\lambda$	5.3 [a]	3.6 [a]	1.5 [a]	4.6±1.6 [b]
$\mathcal{W}/k_B T_\lambda$			2.1±0.3 [c]	

[a]Ref.[266], [b]Refs.[267, 268], [c]Ref.[269]

where $\vartheta_{\pm 0,o}(y_\pm) = 2\Theta_{\pm 0,o}(y_\pm) - y_\pm \Theta'_{\pm 0,o}(y_\pm)$. In order to account for the crossover to retarded van–der–Waals forces in ^{4}He we adopt an empirical crossover law which has been determined in Ref.[266] for ^{4}He on SrF$_2$ in the form [173, 266]

$$\delta\omega'^{reg}(L) = - \left(2\mathcal{W}L^{-3} + 3\mathcal{V}L^{-4}\right)(1 + BL(L+c))^{-1/2}$$

$$\text{with} \quad B = 5.41 \cdot 10^{-5} \text{Å}^{-2} \quad \text{and} \quad c = 132 \text{Å}. \tag{6.32}$$

Note, that according to Eq.(6.32) $\mathcal{V}_r = 2\mathcal{W}/(3\sqrt{B})$. The values for the Hamaker constant $\mathcal{W}$ in units of $k_B T_\lambda$ are listed in Table 6.2 for ^{4}He on SrF$_2$, NaF, Ne, and H$_2$. For ^{4}He itself we use the following bulk data (see, e.g., Refs.[9] and [241])

$$T_\lambda = 2.17 \text{K}$$
$$\rho_l^{-1} = 27.4 \text{cm}^3/\text{mole}$$
$$\rho_v^{-1} = 3400 \text{cm}^3/\text{mole}$$
$$p_\lambda = 0.050 \text{bar} = 1.7 \cdot 10^{-4} k_B T_\lambda \text{Å}^{-3}$$
$$\xi_0^+ = 1.43 \text{Å}. \tag{6.33}$$

The critical effect on the wetting layer thickness becomes transparent if $L(t, \delta p)$ is compared to the corresponding thickness $L_0(\delta p)$ outside the critical region (e.g., along path l_2 in Fig.6.5) which is most conveniently achieved by considering the ratio [173]

$$R(t, \delta p) \equiv L(t, \delta p)/L_0(\delta p). \tag{6.34}$$

For $t = 0$, i.e., along path l_1 in Fig.6.5 one obtains in the *nonretarded* regime [173]

$$R(0, \delta p \to 0) = (1 + k_B T_\lambda \Delta_{0,o}/\mathcal{W})^{1/3}\left(1 + \mathcal{O}\left(\delta p^{1/3}\right)\right). \tag{6.35}$$

Table 6.3:   Relative thinning $R(0,0)$ of a complete wetting
layer of ^{4}He on SrF$_2$, NaF, Ne, and H$_2$ according to Eq.(6.35).
The Hamaker constants W are taken from Table 6.2.

	SrF$_2$	NaF	Ne	H$_2$
$R(0,0)$	0.999	0.998	0.995	0.998±0.001
$R(0,0)$			0.997±0.001	

The numerical values for $R(0,0)$ are summarized in Table 6.3 using the Hamaker
constants given by Table 6.2 and $\Delta_{o,o} \simeq -0.022$ (see Table 3.1). The experimental
detection of such small thinning effects on a complete wetting film seems to be a rather
demanding task. However, one should note that it suffices to detect the *change* of
the wetting layer thickness instead of its absolute value. A first experimental hint for
a thinning of a ^{4}He wetting film near T_λ has been reported for silver evaporated on
borosilicate glass [262].

In the retarded regime the critical thinning of a ^{4}He wetting film becomes significantly enhanced. In this case $\delta\omega''^{reg}(L) = -2W/\sqrt{B}L^{-4}$, and Eq.(6.31) implies for
$T = T_\lambda$

$$L_{max} \equiv L(0,0) = -W/\left(\sqrt{B}k_B T_\lambda \Delta_{o,o}\right) \tag{6.36}$$

which means that the critical fluctuations prevent complete wetting. For Ne ($W \simeq$
$1.5k_B T_\lambda$) for example one has $L_{max} \simeq 0.9\mu m$. Outside the critical region we find
$L_0(\delta p) \sim \delta p^{-1/4}$ in the retarded regime which leads to [173]

$$R(0,\delta p) = \frac{\rho_l - \rho_v}{2K_B T_\lambda(-\Delta_{o,o})}\left[\frac{2W}{\sqrt{B}(\rho_l - \rho_v)}\right]^{3/4}\left[\frac{p_\lambda}{\rho_v}\right]\delta p^{1/4} + \cdots \tag{6.37}$$

so that $R(0,\delta p)$ *vanishes* at coexistence.

For $t > 0$ the size of the critical effect on $L(t,\delta p)$ depends on the value of the scaling argument y_+ of the scaling function $\vartheta_{+0,o}(y_+)$ (see Eq.(6.31)). The exponential
decay of $\vartheta_{+0,o}(y_+)$ sets in at $y_+ \simeq 1$ (see Eq.(5.36)) so that the critical regime can
be characterized by the requirement $y_+ < 1$ (for an alternative characterization see
Ref.[173]). As long as $L(t,\delta p) \ll L_{max}$ one has $y_+ \simeq L_0(\delta p)/\xi_+$ so that the crossover
line $\delta p_{cr}(t)$ shown in Fig.6.5 can be obtained from the condition $L_0(\delta p_{cr}) = \xi_0^+ t^{-\nu}$
which leads to [173]

$$\delta p_{cr}(t) = \frac{2W\rho_v}{p_\lambda \xi_0^{+3}(\rho_l - \rho_v)}t^{3\nu} \tag{6.38}$$

in the nonretarded regime and

$$\delta p_{cr}(t) = \frac{2\mathcal{W}\rho_v}{p_\lambda \zeta_0^{+4}\sqrt{B}\,(\rho_l - \rho_v)} t^{4\nu} \tag{6.39}$$

in the retarded regime. For ^{4}He on Ne ($\mathcal{W} \simeq 1.5k_BT_\lambda$) for example we obtain from the bulk data given by Eq.(6.33) $\delta p_{cr}(t) \simeq 100\,t^{3\nu}$ in the nonretarded regime. Consequently the critical region ("cone of influence") in Fig.6.5 is given by the inequality $\delta p > \delta p_{cr}(t)$, where an *upper* bound for δp is set by the condition that the complete wetting layer must not become too thin to resemble a critical slab. This upper bound is indicated by the short–dashed line l in Fig.6.5. Note, that $\delta p_{cr}(t)$ as a function of t crosses over from the nonretarded form according to Eq.(6.38) to the retarded form given by Eq.(6.39) as t decreases. Furthermore, the Hamaker constant $\mathcal{W}$ determines the *width* of the critical region in the sense that for a weaker substrate (smaller $\mathcal{W}$) the temperature interval over which the critical reduction of the wetting layer thickness $L(t,\delta p)$ can be observed along the path l_5 (see Fig.6.5) becomes larger. For the following investigation we therefore choose Ne as a substrate, where we use the Hamaker constant $\mathcal{W}$ and Eq.(6.32) according to Ref.[266] (see Table 6.2).

For $t > 0$ the wetting layer thickness $L(t,\delta p)$ is determined by solving Eq.(6.31) numerically, where $\vartheta_{+0,0}(y_+)$ is calculated using Eq.(5.34) for $N = 2$ and $\varepsilon = 1$. The thickness $L_0(\delta p)$ of the *noncritical* wetting film fulfills Eq.(6.31) with $\vartheta_{+0,0}(y_+) = 0$. For $\mathcal{V} = 0$ the absolute layer thickness $L_0(\delta p)$ as a function of δp outside the critical region and the relative critical thinning $R(t,\delta p)$ as a function of t along path l_5 for various values of δp (see Fig.6.5) is displayed in Fig.6.6. The solid line in Fig.6.6(top) shows $L_0(\delta p)$ according to Eq.(6.32) ($\mathcal{V} = 0$), and the dashed line displays $L_0(\delta p)$ without retardation, i.e., according to Eq.(6.32) with $B = 0$ for comparison. Retardation leads to thinner films for finite δp, because the $\delta p^{-1/3}$–growth for non-retarded van–der–Waals interactions crosses over to the slower $\delta p^{-1/4}$–growth in the retarded regime for $\delta p \to 0$. The relative thinning $R(t,\delta p)$ shown in Fig.6.6(bottom) and $L_0(\delta p)$ shown by the solid line in Fig.6.6(top) yield the wetting layer thickness $L(t,\delta p) = R(t,\delta p)L_0(\delta p)$. If retardation were absent the curves $R(t,\delta p)$ in Fig.6.6(bottom) would meet at the same value $R(0,\delta p) = (1 + k_BT_\lambda\Delta_{0,0}/\mathcal{W})^{1/3}$ at $t = 0$ (see Eq.(6.35)). Since retardation leads to thinner films the critical effect on $R(t,\delta p)$ becomes enhanced for decreasing δp (see Eq.(6.37)). If δp grows, i.e., the path l_5 is taken further away from the vapor pressure curve, the temperature interval in which the critical thinning of the wetting film can be observed increases.

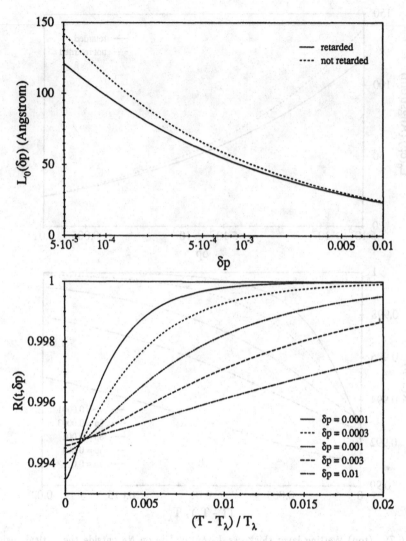

Fig. 6.6:   (top) Wetting layer thickness $L_0(\delta p)$ of ^{4}He on Ne outside the critical region (path l_2 in Fig.6.5) according to Eq.(6.31) for $\vartheta_+ = 0$ and $\mathcal{V} = 0$ with retardation (solid line) and without retardation (dashed line). (bottom) Relative thinning of a wetting layer of ^{4}He on Ne along path l_5 in Fig.6.5 for $\mathcal{V} = 0$, various values of δp, and with retardation according to Eq.(6.32). The region in which $R(t, \delta p)$ is suppressed widens as δp increases.

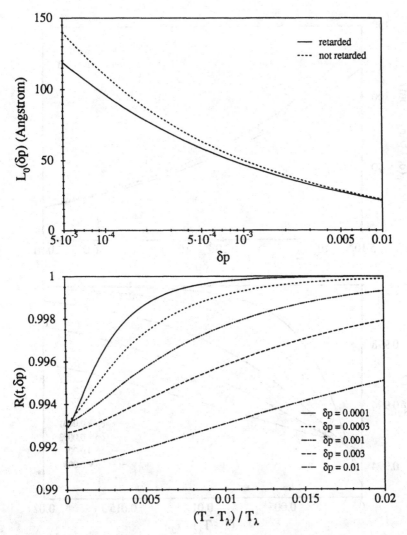

Fig. 6.7:   (top) Wetting layer thickness $L_0(\delta p)$ of ^{4}He on Ne outside the critical region (path l_2 in Fig.6.5) according to Eq.(6.31) for $\vartheta_+ = 0$ and $\mathcal{V} = -3\xi_0^+\mathcal{W}$ with retardation (solid line) and without retardation (dashed line). (bottom) Relative thinning of a wetting layer of ^{4}He on Ne along path l_5 in Fig.6.5 for $\mathcal{V} = -3\xi_0^+\mathcal{W}$, various values of δp, and with retardation according to Eq.(6.32). The relative thinning of the wetting layer is enhanced compared to Fig.6.6.

This is in qualitative accordance with the shape of the crossover line $\delta p_{cr}(t)$ given by Eqs.(6.38) and (6.39) (see also Fig.6.5). In contrast to previous conjectures concerning the shape of $L(t, \delta p)$ near $t = 0$ [257, 258] there is no cusp singularity in the wetting layer thickness L at $t = 0$ [173].

Preplating the substrate by a weaker material, e.g., cesium as suggested by the experiment presented in Ref.[277] weakens the substrate potential and thus enhances critical effects. However, the Hamaker constant W of the substrate cannot be changed this way. Inhomogeneities in the substrate along the direction of the surface normal (like adsorbed layers of another material) are subsumed in the next–to–leading term in the van–der–Waals potential which is governed by the coefficient V (see Eq.(6.32)) [25]. Here we use V as a model parameter to describe preplating of Ne by a weaker material. For $V = -3\xi_0^+ W < 0$ the results are summarized in Fig.6.7, where apart from V the same parameters as in Fig.6.6 have been used. The negative value of V leads to slightly thinner films which like retardation enhances the critical thinning of the wetting layer. Here one should note that the thickness of the liquid portion of a ^{4}He film on a substrate may be reduced due to layering effects which produce solid–like layers of ^{4}He on the surface of the substrate. For ^{4}He on H_2 for example a pronounced layering up to nine layers has been observed [268] which reduces the thickness of the superfluid film considerably.

The wetting layer thicknesses $L_0(\delta p)$ and $L(t, \delta p)$ are invertible functions of their argument δp, i.e., $L_0(\delta p)$ posesses the inverse $\delta p_0(L)$, and from $L(t, \delta p)$ one can construct the inverse $\delta p(t, L)$ [173]. The minimum principle for the effective interface potential (see Eq.(6.31)) can now be solved for the scaling function $\vartheta_{+0,0}$ in the form

$$\vartheta_{+0,0}(L/\xi_+) = \frac{L^3 p_\lambda}{k_B T_\lambda} \left(\frac{\rho_l}{\rho_v} - 1 \right) (\delta p(t, L) - \delta p_0(L)) . \qquad (6.40)$$

If the right hand side of Eq.(6.40) is plotted versus the scaling variable L/ξ_+ one obtains data collapse onto a universal curve which displays the scaling function $\vartheta_{+0,0}$ for $t > 0$. For $t < 0$ a corresponding scaling plot versus the scaling variable $L(-t)^\nu$ shows the scaling function $\vartheta_{-0,0}$ for which there is as yet no theoretical prediction. An experiment in which the critical thinning of a ^{4}He wetting film is measured as a function of t and δp therefore probes the universal finite–size scaling function $\Theta_{\pm 0,0}$ [173]. However, the first–order ε–expansion of $\Theta_{\pm 0,0}$ which has been used here in order to obtain quantitative predictions suffers from significant ambiguities (see

Sec.6.1). The critical effect on the thickness of complete wetting layers may therefore be stronger than predicted here.

6.3.2 Tricritical ^3He – ^4He Mixtures on Substrates

The phase diagram of a ^{3}He and ^{4}He mixture (see Fig.1.9) offers the opportunity to study both critical and tricritical behavior. The superfluid transition in ^{3}He –^{4}He mixtures becomes first–order along the line L of tricritical points which terminates in a tricritical end point on the vapor pressure surface. In analogy with the lower λ–point in pure ^{4}He the presence of the tricritical end point affects the thickness $L(t, \delta p)$ of the complete wetting layer.

Concerning the wetting behavior of weakly undersaturated ^{3}He –^{4}He vapor on a given substrate material much less seems to be known than for pure ^{4}He. In the course of an investigation of copper as a substrate for pure ^{4}He by capacitance measurements the superfluid transition in ^{3}He –^{4}He has been found to be the wetting transition for ^{3}He –^{4}He which is in accordance with the behavior of pure ^{4}He [265]. Interestingly, a thick superfluid film of ^{4}He or a ^{3}He –^{4}He mixture on copper which is heated up to the normal fluid state relaxes only very slowly to a thin film. The relaxation time for this process is between two and three orders of magnitude larger than the typical diffusion time [265].

As a model assumption for tricritical ^{3}He –^{4}He wetting films we adopt Eq.(6.32) for the interaction part of the effective interface potential, where the Hamaker constant $\mathcal{W}$ has been chosen as for ^{4}He on Ne. Preplating will not be considered so that we have $\mathcal{V} = 0$ in Eq.(6.32). Furthermore, we assume Dirichlet boundary conditions $((a, b) = (O, O))$ for the superfluid order parameter in liquid ^{3}He –^{4}He so that the finite–size scaling function in Eq.(6.29) is given by $\Theta_{+O,O}(y_+)$ according to Eq.(5.40) ($t > 0$ and $d = 3$). The corresponding scaling function $\Theta_{-O,O}(y_-)$ for $t < 0$ is not yet available.

The tricritical Casimir amplitude $\Delta_{O,O} \simeq -0.048$ (see Table 3.5) for tricritical ^{3}He –^{4}He is about twice as large as the corresponding amplitude for ^{4}He, and it has the same sign. Therefore the relative thinning of a wetting film of liquid ^{3}He –^{4}He shown in Fig.6.8 is about twice as large as for ^{4}He. The wetting layer thickness $L_0(\delta p)$ away from the tricritical end point for the model parameters used in Fig.6.8 is displayed in Fig.6.6(top). A mixture of ^{3}He and ^{4}He provides an alternative system in which

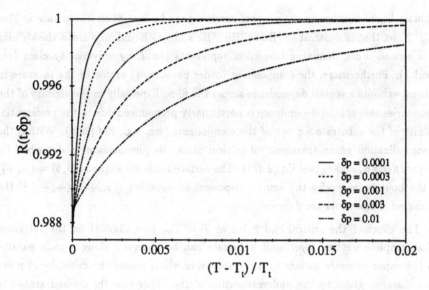

Fig. 6.8: Model calculation for the relative thinning $R(t, \delta p)$ of a complete wetting layer of a ^{3}He –^{4}He mixture along a path analogous to l_5 in Fig.6.5 parallel to the vapor pressure surface passing the tricritical end point $(T, z) = (T_t, z_t)$ of the line L in Fig.1.9 at a distance $p(T_t, z_t)\delta p$ on the vapor side. The parameters are the same as in Fig.6.6. The critical reduction of R is approximately twice as large as for pure ^{4}He. The curves do not meet at $t = 0$ due to retardation.

the effect of criticality on a complete wetting layer can be detected and measured in an experiment. For a direct comparison of the size of the critical effect in pure ^{4}He and ^{3}He –^{4}He the corresponding Hamaker constants must be known in addition to the Casimir amplitudes. Moreover, ^{3}He –^{4}He mixtures offer the opportunity to study both critical and tricritical effects on the wetting layer and the crossover between them.

6.3.3 *Binary Liquid Mixtures on Substrates*

In binary liquid mixtures a considerable variety of wetting transitions is possible (see Fig.6.4 and Refs.[6] and [25]). If the wetting transitions are located as shown schematically in Fig.6.4, the critical end point of the line L_2 of second–order demixing transitions lies in the complete wetting regime. In this case a critical wetting layer of a binary liquid mixture on the verge of demixing can be formed on a substrate at $T = T_{cep}$. Note, that in contrast to ^{4}He or ^{3}He –^{4}He a binary liquid mixture in the vicinity of a second–order demixing transition represents the *Ising* universality class (see Sec.1.1). Furthermore, the composition (order parameter) profile of the mixture in general exhibits a spatial dependence across the film. Especially in the vicinity of the boundaries this spatial dependence is particularly pronounced due to the preferential affinity of the substrate for one of the components (see, e.g., Ref.[205]). Within the renormalization group treatment of critical films this phenomenon is described by *surface fields* h_a and h_b (see Eq.(2.48)). The surface fields are aligned $((a, b) = (+, +))$ if the boundaries prefer the same component or opposing $((a, b) = (+, -))$ if the preferred components are different.

The effect of the critical end point at $T = T_{cep}$ (see Fig.6.4) on the thickness of a complete wetting layer on a substrate can be observed along a path parallel to the vapor pressure surface on the vapor side which passes the critical end point at a distance given by the undersaturation of the vapor (see the dashed arrows in Fig.6.4). There are many choices for such a path so that a second parameter, e.g., the chemical potential difference $\mu_A - \mu_B$ (see Fig.6.4) is needed in addition to the reduced temperature $t = (T - T_{cep})/T_{cep}$ in order to describe the full spectrum of possibilities. If $\mu_A - \mu_B$ is fixed to its value at the critical end point and only t is varied around $t = 0$, the universal finite–size scaling functions $\Theta_{\pm(+,+)}(y_\pm)$ or $\Theta_{\pm(+,-)}(y_\pm)$ for the Ising universality class are needed in order to solve Eq.(6.31) for the wetting layer thickness $L(t, \delta p)$. However, at present these scaling functions are unavailable so that a quantitative analysis of $L(t, \delta p)$ as a function of t is not possible.

At $T = T_{cep}$ ($t = 0$) the scaling functions $\Theta_{\pm(+,+)}$ and $\Theta_{\pm(+,-)}$ reduce to the corresponding Casimir amplitudes $\Delta_{+,+}$ and $\Delta_{+,-}$, respectively, which in three dimensions have been estimated by means of a Migdal–Kadanoff renormalization group ananlysis (see Ref.[190] and Table 3.4). The result $\Delta_{+,+} = 0$ in $d = 3$ may not be reliable because this analysis yields $\Delta_{+,+} = 0$ also in $d = 2$ which disagrees with the exact result $\Delta_{+,+} = -\pi/48$ (see Table 3.3). For opposing surface fields the result $\Delta_{+,-} \simeq 0.279$ indicates that for large layer thicknesses L the critical contribution to Eq.(6.31) *dominates* over the van–der–Waals contribution due to retardation (see Eq.(6.32)). The change of the relative layer thickness $R(0, \delta p)$ in this case is governed by [173]

$$R(0, \delta p \to 0) = \left[\frac{2k_B T_{cep}\Delta_{+,-}}{\rho_l - \rho_v}\right]^{1/3} \left[\frac{2W}{\sqrt{B}(\rho_l - \rho_v)}\right]^{-1/4} \left[\frac{p_0}{\rho_v}\right]^{-1/12} \delta p^{-1/12} + \ldots, \quad (6.41)$$

where p_0 is the saturated vapor pressure at the critical end point of the demixing transitions. As in the case of a negative Casimir amplitude (see Eq.(6.37)) retardation of the van–der–Waals potential leads to a substantial enhancement of the critical effect on the wetting layer thickness. Moreover, an experimental investigation of $L(t, \delta p)$ for binary liquid mixtures probes the universal finite–size scaling functions $\Theta_{\pm(+,+)}$ and $\Theta_{\pm(+,-)}$ for which at present no theoretical analysis exists in $d = 3$.

6.4 The Casimir Force

The well known Casimir force between perfectly conducting plates given by Eqs.(3.1), (3.3), or (3.4) for $T = 0$, low temperatures, and high temperatures, respectively, has a direct analogue in a critical film (see Chap.3 and Chap.5). The spectrum of the critical fluctuations in a film depends on the film thickness L which leads to a L–dependent *singular* part of the free energy (see Eqs.(2.1) and (2.4)). The derivative of the singular part of the free energy with respect to L defines the *critical* or *singular* Casimir force $K_C^{sing}(t, L)$ [173]. However, critical fluctuations are not the only source of a L–dependence of the free energy. According to the Lifshitz theory [143, 144] (see also Sec.3.1) thermal fluctuations of the electromagnetic field in a dielectric film impose an L–dependence on the *regular* part of the free energy. If the total free energy of the film is denoted by $\mathcal{F}(T, L)$, the Casimir force K_C per unit area is given by [173]

$$K_C(T, L) = -\lim_{A \to \infty} \frac{1}{A} \frac{\partial}{\partial L} \mathcal{F}(T, L) = K_C^{reg}(T, L) + K_C^{sing}(t, L), \quad (6.42)$$

where $K_C^{reg}(T, L)$ denotes the regular van–der–Waals part of the Casimir force (see Refs.[143] and [144]), and $K_C^{sing}(t, L)$ is the singular (critical) part of the Casimir force. The Casimir force furthermore decomposes into a bulk pressure contribution and a finite–size contribution $\delta K_C(T, L)$ [173], where the latter will be analyzed in the following.

The functional form of $\delta K_C(T, L)$ which sometimes called *solvation force* [225] can be read off from Eq.(6.31) which in the light of the above definition of the Casimir force can be interpreted as a *force balance* which determines the equilibrium thickness $L(t, \delta p)$ of a complete wetting film. If $\delta K_C(T, L)$ is represented as $\delta K_C(T, L) = \delta K_C^{reg}(T, L) + \delta K_C^{sing}(t, L)$ (see Eq.(6.42)), one finds

$$\delta K_C^{reg}(T, L) = \left(2\mathcal{W}L^{-3} + 3\mathcal{V}L^{-4}\right)(1 + BL(L + c))^{-1/2}$$
$$\delta K_C^{sing}(t, L) = k_B T_{c,b} L^{-d} \vartheta_{\pm a,b}(L/\xi_\pm),$$ (6.43)

where $\vartheta_{\pm a,b}(y_\pm) = (d-1)\Theta_{\pm a,b}(y_\pm) - y_\pm \Theta'_{\pm a,b}(y_\pm)$ (see also Eq.(6.31)) is the universal finite–size scaling function of the critical Casimir force. The critical thinning of a wetting layer in the vicinity of a critical end point (see, e.g., Figs.6.5, 6.6, and 6.7) therefore also provides direct evidence for the critical Casimir force $\delta K_C^{sing}(t, L)$ in a film.

Apart from wetting experiments the singular contribution δK_C^{sing} to the Casimir force in a critical fluid confined between parallel plates should be measurable by a suitably adapted version of the atomic force microscope (AFM) [278]. The AFM gives access to the mechanical properties of the surface of a sample like elasticity and hardness on a nanometer scale and to the van–der–Waals force between a tip on a cantilever beam and the surface with high force and spatial resolutions (1nN and 0.2Å, respectively) [279]. The surface *imaging* of the AFM without contact between the tip and the surface dependes on both the topography of the surface and the surface energy [280]. The AFM therefore allows a clear distinction between chemically different adsorbates on the surface of a sample [280]. Using an optical detection system for the position of the lever arm the force between the tip on the cantilever and the sample surface can be measured in a medium like air or water [281]. The minimum force between the tip and the surface neccessary for AFM imaging can be reduced considerably which is especially important for the non–destrutive investigation of surfaces. In another experiment the tip on the cantilever has been replaced by a silica sphere [282]. The force exerted on the sphere upon approaching a flat silica surface in an aqeous sodium–chloride solution could therefore be measured directly by the AFM [282]. Concerning the direct measurement of the (Casimir) force between two planar surfaces the recently developed surface force apparatus in which a piezoelectric bimorph is used as a cantilever [283] looks very promising. The voltage produced by the piezo is proportional to the deflection of the cantilever which provides a measure of the force between the surfaces and their separation [283].

For the measurement of the force between two parallel planar surfaces in a liquid medium two configurations must be distinguished. In the first configuration the amount of fluid confined between the surfaces is fixed, i.e., the measurement is performed in the *canonical* ensemble. In this case the definition of the Casimir force according to Eq.(6.42) holds, and $\delta K_C^{sing}(t, L)$ is given by Eq.(6.43). In the second configuration the fluid between the surfaces is connected to a reservoir so that the force measurement is performed at fixed chemical potential, i.e., in the *grand canonical* ensemble. The pressure p between the surfaces in this case is given by $p(T, L, \mu) = -\omega(T, L, \mu)/L$ rather than Eq.(6.42). The grand canonical potential per area $\omega(T, L, \mu)$ decomposes into bulk, surface, and finite–size contributions so that the pressure $p(T, L, \mu)$ can be written as [173]

$$p(T, L, \mu) = p_{bulk}(T, \mu) - L^{-1} \left(\omega_{s,a}(T, \mu) + \omega_{s,b}(T, \mu) + \delta\omega_{a,b}(T, L, \mu) \right). \qquad (6.44)$$

A convenient experimental realization of the grand canonical ensemble is obtained by immersing two parallel plates into a large trough filled with the fluid. The net force per area $K_{C,net}$ between these plates is given by [173]

$$K_{C,net}(T, L, \mu) = -L^{-1} \left(\omega_{s,a}(T, \mu) + \omega_{s,b}(T, \mu) + \delta\omega_{a,b}(T, L, \mu) \right) \qquad (6.45)$$

so that the measurement of $K_{C,net}$ near a critical end point probes the scaling function $\Theta_{\pm a,b}(L/\xi_\pm)$ (see Eq.(5.34)) instead of $\vartheta_{\pm a,b}(L/\xi_\pm)$ (see Eq.(6.43)). Since the scaling functions $\Theta_{\pm a,b}(L/\xi_\pm)$ have already been discussed in Chap.5 in great detail we restrict the following analysis to force measurements in the *canonical* ensemble. From the field–theoretical point of view it should be noted that subtractions according to additive renormalization (see Eq.(5.20)) have been neglected in $\delta K_C^{sing}(t, L)$, because these subtractions are exponentially small [173].

6.4.1 *Critical ^{4}He between Plates*

The second–order superfluid transition in liquid ^{4}He offers the opportunity to investigate the Casimir force between surfaces in the presence of a critical liquid. The leading singular finite–size contribution to the Casimir force $\delta K_C^{sing}(t, L)$ for a film of liquid ^{4}He of thickness L confined between parallel plates reads for $t \geq 0$ in $d = 3$ (canonical ensemble, see Eq.(6.43))

$$\delta K_C^{sing}(t, L) = k_B T_\lambda L^{-3} \vartheta_{+0,0}(L/\xi_+), \qquad (6.46)$$

where Dirichlet boundary conditions have been assumed. Note, that the thickness of the liquid ^{4}He film may be considerably reduced due to layering effects on the surfaces of the plates (see Ref.[268]) so that the liquid ^{4}He is in contact with a few layers of *solid* ^{4}He rather than with the surface. In this case we still assume Eq.(6.46) to be correct for the *liquid* portion of the ^{4}He layer. In order to evaluate Eq.(6.46) for a film of liquid ^{4}He in the vicinity of the lower λ–point we use the bulk data given in Eq.(6.33).

In the nonretarded regime the regular van–der–Waals contribution $\delta K_C^{reg}(T, L)$ to the Casimir force has the form (see Eq.(6.43) for $\mathcal{V} = 0$)

$$\delta K_C^{reg}(T, L) = 2\mathcal{W}L^{-3} \qquad (6.47)$$

so that for $T = T_\lambda$ δK_C^{sing} and δK_C^{reg} are governed by the same power law as functions of the film thickness L. The critical and the noncritical contribution to δK_C can therefore be subsumed in an effective Hamaker constant $\mathcal{W}_{eff}$ which is defined by [173]

$$\mathcal{W}_{eff} = \mathcal{W} + k_B T_\lambda \Delta_{0,0}. \qquad (6.48)$$

Typical values for the Hamaker constants of relatively weak substrate materials are summarized in Table 6.2 in units of $k_B T_\lambda$. A direct comparison with $\Delta_{0,0} \simeq -0.022$ (see Table 3.1) shows that the critical contribution to the force between parallel planar surfaces is about 100 times weaker than the van–der–Waals contribution. However, in view of the force resolution achieved by an atomic force microscope [279, 280, 283] an effect of this size is sufficiently large to be measurable. In absolute units $\delta K_C^{sing}(t, L)$ is shown in Fig.6.9 as a function of L for various values of $t \geq 0$. The scaling function $\vartheta_{+0,0}(y_+)$ (see Eq.(6.46)) has been determined from the ε–expansion of $\Theta_{+0,0}(y_+)$ for $N = 2$ and $\varepsilon = 1$ (see Eqs.(5.34) and (6.43)). The curves shown in Fig.6.9 do not intersect, because the scaling function $\vartheta_{+0,0}(y_+)$ is *monotonic* (see also Fig.5.1). If Neumann boundary conditions $((a, b) = (SB, SB))$ can be realized for an experimental setup, the effect of the *nonmonotonic* scaling function $\vartheta_{+SB,SB}(y_+)$ can be studied [173] (see also Fig.5.1). In this case curves for different t do intersect [173].

In the retarded regime the regular van–der–Waals contribution $\delta K_C^{reg}(T, L)$ to $\delta K_C(T, L)$ can be written as

$$\delta K_C^{reg}(T, L) = 3\mathcal{V}_r L^{-4} \qquad (6.49)$$

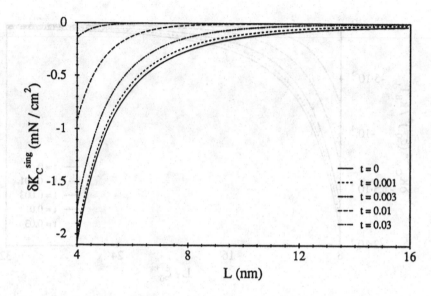

Fig. 6.9:   Singular Casimir force per unit area $\delta K_C^{sing}(t, L)$ for a ^{4}He film according to Eq.(6.46) in $d = 3$ for various values of $t = (T - T_\lambda)/T_\lambda$. Curves for different t do not intersect.

so that for $T = T_\lambda$ the leading L–dependence of $\delta K_C(T, L)$ is governed by $\delta K_C^{sing}(t, L)$, and one has

$$\delta K_C(T_\lambda, L) = 2k_B T_\lambda \Delta_{0,0} L^{-3} + 3\mathcal{V}_r L^{-4}. \tag{6.50}$$

For $\mathcal{V}_r > 0$ $\delta K_C(T_\lambda, L)$ changes its sign at a finite separation L_{max} which is the limiting thickness of a ^{4}He wetting layer on a substrate for $\delta p = 0$ and $T = T_\lambda$ (see Eq.(6.36) for $\mathcal{V}_r = 2W/(3\sqrt{B})$).

6.4.2   Tricritical ^{3}He – ^{4}He Mixtures between Plates

Apart from critical contributions to the Casimir force the influence of a tricritical point can be studied in a liquid ^{3}He –^{4}He mixture along the line L of tricritical superfluid transitions (see Fig.1.9). The functional form of the tricritical contribution $\delta K_C^{sing}(t, L)$ to the Casimir force has the same structure as Eq.(6.46), where T_λ is replaced by the tricritical temperature T_t, and for $t = (T - T_t)/T_t > 0$ the scaling function $\vartheta_{+0,0}$ is replaced by its tricritical counterpart. As in the case of pure ^{4}He we

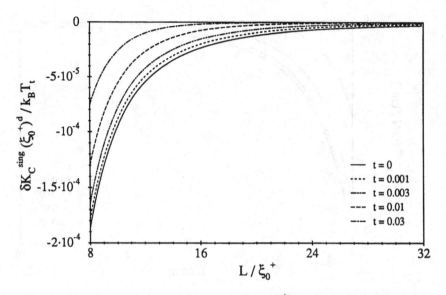

Fig. 6.10: Singular Casimir force per unit area $\delta K_C^{sing}(t, L)$ for a ^{3}He –^{4}He film
according to Eqs.(6.46) and (6.51) in units of $k_B T_t (\xi_0^+)^{-d}$ in $d = 3$ for various
values of $t = (T - T_t)/T_t$. Curves for different t do not intersect.

assume Dirichlet boundary conditions for the superfluid order parameter at the two
surfaces, i.e., $(a, b) = (O, O)$. From Eq.(5.40) one obtains for the tricritical finite–size
scaling function $\vartheta_{+O,O}(y_+)$ for $N = 2$ and $d = 3$ [173]

$$\vartheta_{+O,O}(y_+) = -\frac{1}{4\pi} \left[\mathcal{L}_3 \left(e^{-2y+} \right) + 2y_+ \mathcal{L}_2 \left(e^{-2y+} \right) - 2y_+^2 \ln \left(1 - e^{-2y+} \right) \right]. \qquad (6.51)$$

The corresponding scaling function $\vartheta_{-O,O}$ for $t < 0$ is not yet available. At the
tricritical point one has $\vartheta_{+O,O}(0) = 2\Delta_{O,O} \simeq -0.096$ which is about twice as large as
the corresponding critical value for pure ^{4}He. In comparison with typical values for
the Hamaker constants (see Table 6.2) the strength of the tricritical contribution to
the Casimir force can be expected to be of the order of 1...2% of the van–der–Waals
force. This effect is large enough to be measurable by an atomic force microscope
(see Refs.[281], [282], and [283]).

A graphical representation of the tricritical Casimir force given by δK_C^{sing} in units
of $k_B T_t (\xi_0^+)^{-d}$ in $d = 3$ is shown in Fig.6.10, where ξ_0^+ is the amplitude of the correlation length near the tricritical superfluid transition in ^{3}He –^{4}He (line L in Fig.1.9).

The scaling function $\vartheta_{+0,0}(y_+)$ (see Eq.(6.51)) is monotonic so that curves for different values of t do not intersect [173]. By varying the concentration of ^{3}He in the ^{3}He $-^4$He mixture it is possible to study the crossover of the singular Casimir force δK_C^{sing} from usual critical to tricritical behavior. Furthermore, the tricritical scaling function $\vartheta_{+0,0}(y_+)$ (see Eq.(6.51)) is exact in $d=3$ so that the numerical accuracy of the field–theoretical result for the tricritical contribution to the Casimir force is not limited by the well known shortcomings of the ε–expansion. Finally, we note that layering effects similar to those in pure ^{4}He [268] may reduce the effective film thickness.

6.4.3 Binary Liquid Mixtures between Plates

Along the line L_2 of second–order demixing transitions (see Figs.1.5 and 6.4) critical fluctuations in a binary liquid mixture confined between parallel plates generate a critical contribution δK_C^{sing} to the Casimir force between the plates. In contrast to pure ^{4}He and ^{3}He $-^4$He mixtures a critical binary liquid mixture on the verge of demixing represents the Ising universality class, and, furthermore, the surfaces are characterized by surface fields (extraordinary surface universality class) with the combinations $(a,b) = (+,+)$ and $(a,b) = (+,-)$. At present the available information about the corresponding finite–size scaling functions in $d=3$ is limited to Migdal-Kadanoff estimates for the Casimir amplitudes $\Delta_{+,+}$ and $\Delta_{+,-}$ (see Table 3.4) [190], where only $\Delta_{+,-} \simeq 0.279$ seems to be reliable (see above). This value, however, indicates that the critical contribution to the Casimir force for $(a,b) = (+,-)$ is one order of magnitude larger than for ^{4}He at the superfluid transition. The effective Hamaker constant (see Eq.(6.48)) is *enhanced* at critical demixing points, because the Casimir amplitude $\Delta_{+,-}$ is positive. If the Hamaker constants for binary liquid mixtures measured in units of $k_B T_{c,b}$ are comparable to those given in Table 6.2, the critical contribution to the force between the plates becomes as large as 10% of the van–der–Waals contribution and should therefore be easily detectable by, e.g., the surface force apparatus described in Ref.[283].

For $T < T_{c,b}$ opposing surface fields generate an *interface* between the A–rich liquid and the B–rich liquid (see Sec.2.1). The associated interfacial tension $\sigma_{AB}(t) = \sigma_0(-t)^{2-\alpha_s}$ contributes to the universal finite–size scaling function $\Theta_{-(+,-)}(y_-)$ (see Eq.(5.18)) as an extra surface term which is proportional to y_-^{d-1}. Whether $\sigma_{AB}(t)$

contributes to the critical Casimir force depends on the experimental setup. In the canonical ensemble (fixed amount of liquid between the plates) δK_C^{sing} is independent of $\sigma_{AB}(t)$ (see Eq.(6.42)). In the grand canonical ensemble, however, Eq.(6.45) holds, and therefore $\sigma_{AB}(t)$ can be detected experimentally by a force measurement at temperatures slightly *below* the bulk critical demixing temperature $T_{c,b}$.

7. Summary and Outlook

7.1 The Casimir Effect

The Casimir effect is a phenomenon common to all physical systems in which relevant degrees of freedom are characterized by fluctuating fields. In Chap.3 we have demonstrated that critical systems fit into this scheme due to the presence of a fluctuating order parameter. Critical fluids confined to a film geometry exhibit a particularly pronounced finite–size contribution to the free energy which is governed by the *universal Casimir amplitudes* $\Delta_{a,b}$. The *force* associated with this size dependence is the analogue of the well known electromagnetic *Casimir force* which two parallel conducting plates experience in vacuum (see Secs.3.1 and 6.4). The critical finite–size contribution to the effective interface potential changes the ratio $R = L/L_0$ of the equilibrium thickness L of a critical complete wetting layer and its thickness L_0 outside the critical region by an amount which is governed by the ratio of the Casimir amplitude and the Hamaker constant (see Sec.6.3).

In two dimensions a considerable variety of Casimir amplitudes for the Ising and 3–state Potts universality class is known exactly from conformal invariance considerations (see Sec.3.4). However, in the experimentally relevant case $d = 3$ the knowledge of the Casimir amplitudes is limited in two respects. First, field–theoretical estimates of these amplitudes are limited to symmetry conserving boundary conditions, and, second, their numerical reliability is still unknown which is in part due to discrepancies between the field–theoretical and Migdal–Kadanoff renormalization group estimates (see Sec.3.2). Therefore independent tests of the Casimir amplitudes in $d = 3$ by additional real–space renormalization and Monte–Carlo techniques are desirable.

In order to obtain experimentally relevant results which are applicable beyond ^{4}He and ^{3}He –^{4}He mixtures the field–theoretical analysis of the Casimir amplitudes should be extended to symmetry breaking boundary conditions. Apart from films of binary liquid mixtures near their critical demixing transition Casimir amplitudes for symmetry breaking boundary conditions are interesting for the lamellar phase of microemulsions, where the oil phase has become critical [284]. Furthermore, the numerical reliability of field–theoretical estimates can be improved by employing the minimal subtraction scheme directly in $d = 3$ (see, e.g., Refs.[110], [111], and [112]). One should therefore consider this approach for future investigations of the Casimir effect in critical films.

7.2   Wall Effects and Profiles

In the vicinity of boundaries the spatial dependence of scaling density profiles in critical systems is particularly pronounced. Especially near a planar wall the dominating
contribution to the profile is given by the scaling density profile in the corresponding
semi–infinite system. The presence of other boundaries leads to *distant wall corrections* to the scaling density profile which are given by the *short–distance expansion*
of the corresponding *scaling operator*. The short–distance expansion of the energy
density scaling operator near planar O and SB walls has been discussed in Chap.4
for $T = T_{c,b}$ (see Sec.4.3). It turns out that the *surface stress tensor* at the near
wall always governs one of the contributions to the short–distance expansion. As a
consequence a term proportional to $(z/L)^d$ occurs as a distant wall correction to the
energy density profile in a *film* of thickness L at a distance z from the near wall. The
coefficient of $(z/L)^d$ is closely related to the *Casimir amplitude* [128, 129, 192]. In
$d = 2$ this relation involves the *central charge* which can be defined by the *hyperuniversal* amplitude of the stress–tensor stress–tensor correlation function. However, the
hyperuniversality of this amplitude is lost for an $O(N)$–symmetric $(\Phi^2)^2$–model with
surfaces in $d = 4 - \varepsilon$ (see Sec.4.4). The structure of the short–distance expansion
of the energy density in the vicinity of an E wall resembles that for an O wall (see
Sec.4.3) [285]. In the context of critical adsorption (see Ref.[22]) the short–distance
expansion of the *order parameter* near an E wall would be particularly interesting
(see especially Ref.[203]).

The energy density profiles for critical films at $T = T_{c,b}$ with symmetry conserving
boundary conditions have been discussed in Sec.4.2. In view of experimental tests
one should also consider symmetry breaking boundary conditions. The order parameter profile in a film geometry is particularly interesting, because it gives access to,
e.g., the composition profile in a binary liquid mixture between plates which can be
tested experimentally (see Ref.[286]). Furthermore, critical adsorption on spatially or
chemically inhomogeneous substrates (see Ref.[255]), for which exact results in $d = 2$
can be obtained using conformal invariance [61, 131, 133], should be considered in
future investigations.

7.3 The Finite Size Scaling Functions

The universal scaling functions $\Theta_{\pm a,b}(y_\pm)$ of the singular finite-size part of the free energy of a critical film provide the extension of the critical Casimir effect to temperatures slightly away from $T_{c,b}$. These scaling functions give access to the temperature dependence of the thickness of a critical wetting layer near $T_{c,b}$ (see Sec.6.3) and to the Casimir force between plates in a critical medium (see Sec.6.4). Furthermore, the specific heat of a critical film can be calculated in the vicinity of $T_{c,b}$ if the finite-size scaling functions $\Theta_{\pm a,b}(y_\pm)$ are known (see Sec.6.1). However, in three dimensions the knowledge about these scaling functions is again limited in two respects. First, field-theoretical estimates for $\Theta_{\pm a,b}(y_\pm)$ in $d = 3$ exist only for symmetry conserving boundary conditions and $T \geq T_{c,b}$ (see Sec.5.3). For $T_c(L) < T < T_{c,b}$ the first terms of an expansion of $\Theta_{-O,O}$, $\Theta_{-O,SB}$, and Θ_{aper} in powers of the reduced temperature are known (see Sec.5.4). Second, scaling functions for symmetry breaking boundary conditions are still unavailable in $d = 3$. Therefore only ^{4}He and ^{3}He $-^4$He films near the superfluid transition offer the opportunity for quantitative experimental tests of the known finite-size scaling functions (see Chap.6). However, the numerical uncertainty of the ε-expansion of the scaling functions for $\varepsilon = 1$ poses a major problem. For the specific heat of a critical ^{4}He film the strict ε-expansion fails to give reasonable agreement with experimental data (see Sec.6.1). This gap has recently been narrowed considerably by a one-loop calculation of the specific heat of a critical ^{4}He film using minimal renormalization directly in $d = 3$ [242].

For future investigations independent tests of the finite-size scaling functions by field-theoretical calculations in $d = 3$, real-space renormalization group analyses, or Monte-Carlo studies would be desirable. Furthermore, symmetry breaking boundary conditions should be considered in order to extend the spectrum of applications to simple liquids and binary liquid mixtures near criticality. The *dimensional crossover* in a critical film from $d = 3$ to $d = 2$ is still a demanding task from the point of view of *quantitative* predictions (see Sec.5.5). At present only a real-space renormalization analysis or a Monte-Carlo simulation seem to be capable of handling the dimensional crossover quantitatively (see also Ref.[75]).

7.4 Other Geometries

Silica spheres of 1600Å diameter immersed in the binary liquid mixture water+2-6-lutidine aggregate in the vicinity of the lower consolute point of the mixture due to the formation of a thick lutidine layer on the spheres [287, 288]. It is very tempting to assume that the attractive force which causes the aggregation of these colloid particles originates from the *Casimir force* between spherical particles in a critical medium. In order to obtain a theoretical understanding of this phenomenon in a first step critical adsorption on the surface of a single sphere should be studied by means of a field–theoretical renormalization group analysis of the order parameter profile in presence of a surface field. In a second step the Casimir effect between two spheres in a critical medium should be analyzed by the same method in order to obtain an estimate for the strength of the critical contribution to the force between the spheres in comparison with the van–der–Waals contribution.

Another application of a spherical geometry is provided by AFM measurements of the force between a silica sphere and a planar silica surface in an aqueous sodium-chloride solution [282]. In a critical medium one again expects a critical contribution to this force which can be studied by field–theoretical methods.

The vibrating wire microbalance technique, which is often used to investigate the wetting behavior of ^{4}He on substrates in the vicinity of the lower λ–point (see Sec.6.3), provides a cylindrical surface on which ^{4}He is adsorbed. A field–theoretical study of the critical contribution to the effective interface potential of a cylindrical wetting layer on a substrate would therefore be interesting.

7.5 Correlation Functions and Critical Dynamics

Scattering experiments give access to correlation functions in critical systems. For example the surface critical exponents $\eta_{\parallel}$, γ_{11}, and β_1 (see Table 1.4) can be probed by X–ray and neutron scattering under grazing incidence (see Refs.[289] and [290]). Corresponding scattering experiments for critical films give the opportunity to study the effect of the second boundary on, e.g., correlations parallel to the surfaces. A field–theoretical analysis of two–point correlation functions in critical films would therefore be very interesting, in particular in the presence of symmetry breaking boundary conditions. At $T = T_{c,b}$ and in $d = 2$ exact results for correlation functions in strips can be obtained from conformal invariance considerations [39, 122]. However, in three dimensions the theoretical investigation of the two–point correlation function for critical fluids confined in films and other geometries is still a widely open field.

Concerning dynamical critical behavior in finite systems relaxational dynamics without conserved order parameter (model A) and with conserved order parameter (model B) (see also Ref.[291]) is currently under investigation. Critical relaxation and the universal short–time behavior of the order parameter has recently been analyzed within model A in a finite cube L^d for periodic boundary conditions at $T = T_{c,b}$ [292]. In general the presence of surfaces leads to a subdivision of the dynamical bulk universality classes (A,B,...) into dynamical surface universality classes (e.g., B_A and B_B), where the subscripts A and B indicate the absence or presence of a conservation law for the *surface* magnetization (see Ref.[293]). However, these dynamical models must first be studied for semi–infinite geometries so that their extension to critical films leads to future research.

8.   Acknowledgements

I would like to express my warmest thanks to Prof. S. Dietrich for a very enjoyable and fruitful collaboration and for the stimulating atmosphere in his group. Furthermore, I have been very fortunate to have collaborated with Prof. E. Eisenriegler to whom I am very grateful for many helpful and stimulating discussions.

This work has profited very much from correspondence and discussions with Prof. H. W. Diehl and his group, Prof. V. Dohm and his group, Prof. M. E. Fisher, Prof. J. O. Indekeu, Prof. D. P. Landau and his group, Prof. P. Leiderer, Prof. P. S. Pershan, Prof. V. Privman, and Prof. J. V. Sengers and his group. The major part of this work has been supported by the *Deutsche Forschungsgemeinschaft* through Sonderforschungsbereich 237 *Unordnung und große Fluktuationen* which I gratefully acknowledge.

References

1. H. E. Stanley, *Introduction to Phase Transitions and Critical Phenomena* (Clarendon, Oxford, 1971).

2. W. Gebhardt and U. Krey, *Phasenübergänge und Kritische Phänomene* (Vieweg, Braunschweig, 1980).

3. J.-C. Tolédano and P. Tolédano, *The Landau Theory of Phase Transitions* (World Scientific, Singapore, 1987).

4. M. Schick, in *Progress in Surface Science*, edited by S. G. Davison and W. K. Liu (Pergamon, New York, 1982), Vol. 11, p. 245.

5. E. H. Hauge, *Phase Transitions and Critical Phenomena*, preprint 1991.

6. S. Dietrich and A. Latz, Phys. Rev. B **40**, 9204 (1989).

7. K. R. Atkins, *Liquid Helium* (Cambridge University Press, Cambridge, 1959), p. 19.

8. M. E. Fisher, in *Proceedings of the Gibbs Symposium, Yale University, May 15–17, 1989* (American Mathematical Society, 1990), p. 39.

9. G. Ahlers, in *The Physics of Liquid and Solid Helium, Part I*, edited by K. H. Bennemann and J. B. Ketterson (Wiley, New York, 1978).

10. M. E. Fisher, J. Math. Phys. **5**, 944 (1964).

11. F. J. Wegner, Phys. Rev. B **5**, 4529 (1972).

12. R. B. Griffith, Phys, Rev. Lett. **24**, 1479 (1970).

13. M. E. Fisher, S. K. Ma, and G. Nickel, Phys. Rev. Lett. **29**, 917 (1972).

14. J. C. Le Guillou and J. Zinn–Justin, J. Physique Lett. **46**, L137 (1985).

15. M. E. Fisher, Phys. Rev. **176**, 257 (1968).

16. R. B. Griffith, Phys. Rev. Lett. **14**, 623 (1965).

17. B. Widom, J. Chem. Phys. **43**, 3898 (1965).

18. L. P. Kadanoff in *Phase Transitions and Critical Phenomena*, edited by C. Domb and M. S. Green (Academic, London, 1976), Vol. 5A, p.1.

19. K. Binder in *Phase Transitions and Critical Phenomena*, edited by C. Domb and J. L. Lebowitz (Academic, London, 1983), Vol. 8, p. 2.

20. H. W. Diehl in *Phase Transitions and Critical Phenomena*, edited by C. Domb and J. L. Lebowitz (Academic, London, 1986), Vol. 10, p. 76.

21. K. Binder and P. C. Hohenberg, Phys. Rev. B 9, 2194 (1974).

22. H. W. Diehl and M. Smock, Phys. Rev. B 47, 5841 (1993).

23. K. Binder and P. C. Hohenberg, Phys. Rev. B 6, 3461 (1972).

24. S. Dietrich and H. W. Diehl, Z. Phys. B 43, 315 (1981).

25. S. Dietrich, in *Phase Transitions and Critical Phenomena*, edited by C. Domb and J. L. Lebowitz (Academic, New York, 1988), Vol. 12, p. 1.

26. F. M. Gasparini and I. Rhee, in *Progress in Low Temperature Physics*, edited by D. F. Brewer (North–Holland, Amsterdam, 1992), Vol.XIII, p. 1.

27. P. Taborek and L. Senator, Phys. Rev. Lett. 57, 218 (1986).

28. R. Evans, U. M. B. Marconi, and P. Tarazona, J. Chem Phys. 84, 2376 (1986).

29. R. J. Baxter, *Exactly Solved Models in Statistical Mechanics*, (Academic, New York, 1982).

30. D. P. Landau, in *Monte Carlo Methods in Statistical Physics*, edited by K. Binder (Springer, Berlin, 1979), p. 121 and p. 337.

31. F. Y. Wu, Rev. Mod. Phys. 54, 235 (1982).

32. A. Ciach and H. W. Diehl, Europhys. Lett. 12, 635 (1990).

33. H. W. Diehl and E. Eisenriegler, Europhys. Lett. 4, 709 (1987).

34. D. J. Amit, *Field Theory, the Renormalization Group, and Critical Phenomena*, (McGraw–Hill, New York, 1978).

35. G. Parisi, *Statistical Field Theory* (Addison–Wesley, Wokingham, 1988).

36. J. Zinn–Justin, *Quantum Field Theory and Critical Phenomena* (Clarendon Press, Oxford, 1989).

37. J. J. Binney, N. J. Dowrick, A. J. Fisher, and M. E. J. Newman, *The Theory of Critical Phenomena, An Introduction to the Renormalization Group*, (Clarendon Press, Oxford, 1992).

38. V. Privman, P. C. Hohenberg, and A. Aharony, in *Phase Transitions and Critical Phenomena*, edited by C. Domb and J. L. Lebowitz (Academic, New York, 1991), Vol. 14, p. 1.

39. J. L. Cardy in *Phase Transitions and Critical Phenomena*, edited by C. Domb and J. L. Lebowitz (Academic, London, 1987), Vol. 11, p. 55.

40. M. E. Fisher and H. Nakanishi, J. Chem. Phys. **75**, 5857 (1981).

41. H. Nakanishi and M. E. Fisher, J. Chem. Phys. **79**, 3279 (1983).

42. M. N. Barber, in *Phase Transitions and Critical Phenomena*, edited by C. Domb and J. L. Lebowitz (Academic, New York, 1983), Vol. 8, p. 145.

43. V. Privman and M. E. Fisher, Phys. Rev. B **30**, 322 (1984).

44. M. E. Fisher, M. N. Barber, and D. Jasnow, Phys. Rev. A **8**, 1111 (1973).

45. N. D. Mermin and H. Wagner, Phys. Rev. Lett. **17**, 1133 (1966).

46. J. M. Kosterlitz and D. J. Thouless, J. Phys. C **6**, 1181 (1973).

47. K. Binder and D. P. Landau, Physica A **177**, 483 (1991).

48. M. E. Fisher and A. E. Ferdinand, Phys. Rev. Lett. **19**, 169 (1967).

49. A. E. Ferdinand and M. E. Fisher, Phys. Rev. **185**, 832 (1969).

50. H. Au-Yang and M. E. Fisher, Phys. Rev. B **11**, 3469 (1975).

51. M. E. Fisher, in *Critical Phenomena*, Proceedings of the 1970 International School of Physics, "Enrico Fermi", Course LI, edited by M. S. Green (Academic, New York, 1971), p. 1.

52. M. E. Fisher and M. N. Barber, Phys. Rev. Lett. **28**, 1516 (1972).

53. M. N. Barber and M. E. Fisher, Phys. Rev. A **8**, 1124 (1973).

54. E. Brezin, J. de Phys. **43**, 15 (1982).

55. K. Binder, M. Nauenberg, V. Privman, and A. P. Young, Phys. Rev. B **31**, 1498 (1985).

56. V. Privman, in *Finite Size Scaling and Numerical Simulation of Statistical Systems*, edited by V. Privman (World Scientific, Singapore, 1990), p. 1.

57. J. G. Brankov and N. S. Tonchev, Physica A **189**, 583 (1992).

58. V. Privman, Phys. Rev. B **38**, 9261 (1988).

59. J. L. Cardy and I. Peschel, Nucl. Phys. **B300**, 377 (1988).

60. M. P. Gelfand and M. E. Fisher, Physica A **166**, 1 (1990).

61. P. Kleban and I. Vassileva, J. Phys. A: Math. Gen. **25**, 5779 (1992).

62. V. Privman, J. Phys. A: Math. Gen. **23**, L711 (1990).

63. E. Eisenriegler, Z. Phys. B **61**, 299 (1985).

64. H. Au–Yang and M. E. Fisher, Phys. Rev. B **21**, 3956 (1980).

65. H. Li, M. Paczuski, M. Kadar, and K. Huang, Phys. Rev. B **44**, 8274 (1991).

66. L. V. Mikheev and M. E. Fisher, Phys. Rev. Lett. **70**, 186 (1993).

67. L. V. Mikheev and M. E. Fisher, Phys. Rev. B **49**, 378 (1994).

68. L. V. Mikheev and M. E. Fisher, J. Stat. Phys. **66**, 1225 (1992).

69. D. P. Landau, Phys. Rev. B **13**, 2997 (1976).

70. D. P. Landau, Phys. Rev. B **14**, 255 (1976).

71. K. Binder and D. P. Landau, J. Chem. Phys. **96**, 1444 (1992).

72. C. Ruge, S. Dunkelmann, and F. Wagner, Phys. Rev. Lett. **69**, 2465 (1992).

73. D. P. Landau and K. Binder, Phys. Rev. B **41**, 4633 (1990).

74. W. Selke, N. M. Švrakić, and P. J. Upton, Z. Phys. B **89**, 231 (1992).

75. D. Lederman, C. A. Ramos, V. Jaccarino, and J. L. Cardy, Phys. Rev. B **48**, 8365 (1993).

76. W. Janke and T. Matsui, Phys. Rev. B **42**, 10673 (1990).

77. W. Janke and K. Nather, Nucl. Phys. B (Proc. Suppl.) **30**, 834 (1993).

78. A. Schmidt and T. Schneider, Z. Phys. B **87**, 265 (1992).

79. T. Schneider and A. Schmidt, J. Phys. Soc. Jpn. **61**, 2169 (1992).

80. D. J. Bukman and J. M. J. van Leeuwen, J. Phys. Cond. Matter **3**, 9995 (1991).

81. M. P. Nightingale and H. W. J. Blöthe, Phys. Rev. Lett. **60**, 1562 (1988).

82. C. Holm and W. Janke, Phys. Lett. A **173**, 8 (1993).

83. K. Chen, A. M. Ferrenberg, and D. P. Landau, J. App. Phys. **73**, 5488 (1993).

84. K. Binder and H. P. Deutsch, Europhys. Lett. **18**, 667 (1992).

85. G. An and M. Schick, J. Phys. A: Math. Gen. **21**, L213 (1988).

86. H. W. Blöthe and M. P. Nightingale, Physica A, **112**, 405 (1982).

87. H. Park and M. den Nijs, Phys. Rev. B **38**, 565 (1988).

88. M. Fukugita, H. Mino, M. Okawa, and A. Ukawa, J. Stat. Phys. **59**, 1397 (1990).

89. M. Bernaschi, M. Guagnelli, E. Marinari, and S. Patarnello, Nucl. Phys. **B360**, 283 (1991).

90. C. Borgs, Nucl. Phys. **B384**, 605 (1992).

91. S. Allen and R. K. Pathria, Can. J. Phys. **67**, 952 (1989).

92. S. Allen and R. K. Pathria, Can. J. Phys. **69**, 753 (1991).

93. M. Henkel and R. A. Weston, J. Phys. A **25**, L207 (1992).

94. J. G. Brankov and N. S. Tonchev, J. Stat. Phys. **60**, 519 (1990).

95. J. G. Brankov and D. M. Danchev, J. Math. Phys. **32**, 2543 (1991).

96. J. G. Brankov and D. M. Danchev, J. Stat. Phys. **71**, 775 (1993).

97. D. M. Danchev, J. Stat. Phys. **73**, 267 (1993).

98. M. Napiórkowski and S. Dietrich, Phys. Rev. E **47**, 1836 (1993).

99. M. P. Nightingale and J. O. Indekeu, Phys Rev. Lett. **54**, 1824 (1985); **55**, 1700 (1985).

100. M. R. Swift, A. L. Owczarek, and J. O. Indekeu, Europhys. Lett. **14**, 475 (1991).

101. A. O. Parry and R. Evans, Phys. Rev. Lett. **64**, 439 (1990).

102. A. O. Parry, J. Phys. A: Math. Gen. **25**, 257 (1992).

103. A. O. Parry and R. Evans, Physica A **181**, 250 (1992).

104. E. Cheng, M. R. Swift, and M. W. Cole, J. Chem. Phys. **99**, 4064 (1993).

105. A. O. Parry and R. Evans, J. Phys A: Math. Gen. **25**, 275 (1992).

106. E. Brézin and J. Zinn-Justin, Nucl. Phys. **B257**, 867 (1985).

107. J. Rudnick, H. Guo, and D. Jasnow, J. Stat. Phys. **41**, 353 (1985).

108. W. Huhn and V. Dohm, Phys. Rev. Lett. **61**, 1368 (1988).

109. V. Dohm, Z. Phys. B **75**, 109 (1989).

110. H. J. Krause, R. Schloms, and V. Dohm, Z. Phys. B **79**, 287 (1990).

111. R. Schloms and V. Dohm, Phys. Rev. B **42**, 6142 (1990).

112. F. J. Halfkann and V. Dohm, Z. Phys. B **89**, 79 (1992).

113. J. J. Morris, J. Stat. Phys. **69**, 539, (1992).

114. A. M. Polyakov, Sov. Phys. JETP Lett. **12**, 381 (1970).

115. A. A. Belavin, A. M. Polyakov, and A. B. Zamolodchikov, Nucl. Phys. **B241**, 333 (1984).

116. A. A. Belavin, A. M. Polyakov, and A. B. Zamolodchikov, J. Stat. Phys. **34**, 763 (1984).

117. J. L. Cardy, Nucl. Phys. **B240**, 514 (1984).

118. J. Cardy, Nucl. Phys. **B275**, 200 (1986).

119. D. Friedan, Z. Qui, and S. Shenker, Phys. Rev. Lett. **52**, 1575 (1984).

120. H. W. Blöthe, J. L. Cardy, and M. P. Nightingale, Phys. Rev. Lett. **56**, 742 (1986).

121. I. Affleck, Phys Rev. Lett. **56**, 746 (1986).

122. J. L. Cardy, J. Phys. A: Math. Gen. **17**, L385 (1984).

123. T. W. Burkhardt and I. Guim, Phys. Rev. B. **36**, 2080 (1987).

124. T. W. Burkhardt, E. Eisenriegler, and I. Guim, Nucl. Phys. **B316**, 559 (1989).

125. T. W. Burkhardt and J.-Y. Choi, Nucl. Phys. **B376**, 447 (1992).

126. P. Frojdh and H. Johannesson, Nucl. Phys. **B366**, 429 (1991).

127. T. W. Burkhardt and E. Eisenriegler, J. Phys. A: Math. Gen. **18**, L83 (1985).

128. T. W. Burkhardt and T. Xue, Phys. Rev. Lett. **66**, 895 (1991).

129. T. W. Burkhardt and T. Xue, Nucl. Phys. **B354**, 653 (1991).

130. I. Peschel, L. Turban, and F. Iglói, J. Phys. A: Math. Gen. **24**, L1229 (1991).

131. P. Kleban and I. Vassileva, J. Phys. A: Math. Gen. **24**, 3407 (1991).

132. P. Kleban, Phys. Rev. Lett. **67**, 2799 (1991).

133. T. W. Burkhardt and I. Guim, Phys. Rev. B **47**, 14306 (1993).

134. H. B. G. Casimir, Proc. K. Ned. Akad. Wet. **51**, 793 (1948).

135. H. B. G. Casimir, Physica **19**, 846 (1953).

136. H. B. G. Casimir and D. Polder, Phys. Rev. **73**, 360 (1948).

137. D. Tabor and R. H. S. Winterton, Nature **219**, 1120 (1968).

138. D. Tabor and R. H. S. Winterton, Proc. R. Soc. London, Ser. A: **312**, 435 (1969).

139. I. I. Abrikosava and B. V. Deriagin, Dokl. Akad. Nauk. SSSR, **90**, 1055 (1953).

140. M. J. Sparnaay, Physica **24**, 751 (1958).

141. V. M. Mostepanenko and N. N. Trunov, Sov. Phys. Usp. **31**, 965 (1988).

142. E. Elizalde and A. Romeo, Am. J. Phys. **59**, 711 (1991).

143. E. M. Lifshitz, Sov. Phys. JETP **2**, 73 (1956).

144. I. E. Dzyaloshinskii, E. M. Lifshitz, and L. P. Pitaevskii, Adv. Phys. **10**, 165 (1961).

145. J. Schwinger, L. L. DeRaad, and K. A. Milton, Ann. Phys. (N.Y.) **115**, 1 (1978).

146. L. S. Brown and G. L. Maclay, Phys. Rev. **184**, 1272 (1969).

147. I. H. Duru and M. Tomak, Phys. Lett. A **176**, 265 (1993).

148. G. Plunien, B. Müller, and W. Greiner, Phys. Rep. **134**, 87 (1986).

149. P. W. Milonni and M.-L. Shih, Contemp. Phys. **33**, 313 (1992).

150. I. Brevik and R. Sollie, J. Math. Phys. **31**, 1445 (1990).

151. I. Brevik and G. Einevoll, Phys. Rev. D **37**, 2977 (1988).

152. D. Kupiszewska, Phys. Rev. A **46**, 2286 (1992).

153. I. Brevik and J. S. Høye, Physica A **173**, 583 (1991).

154. P. W. Milonni and M.-L. Shih, Phys. Rev. A **45**, 4241 (1992).

155. C. Wotzasek, J. Phys. A: Math. Gen. **23**, 1627 (1990).

156. S. Hacyan, R. Jáuregui, F. Soto, and C. Villarreal, J. Phys. A: Math. Gen. **23**, 2401 (1990).

157. V. Sandoghdar, C. I. Sukenik, E. A. Hinds, and S. Haroche, Phys. Rev. Lett. **68**, 3432 (1992).

158. C. I. Sukenik, M. G. Boshier, D. Cho, V. Sandoghdar, and E. A. Hinds, Phys. Rev. Lett. **70**, 560 (1993).

159. S. Hacyan, R. Jáuregui, and C. Villarreal, Phys. Rev. A **47**, 4204 (1993).

160. O. Panella, A. Widom, and Y. N. Srivastava, Phys. Rev. B **42**, 9790 (1990).

161. Y.- C. Cheng and J. S. Yang, Phys. Rev. B **41**, 1196 (1990).

162. G. Barton, J. Phys. A: Math. Gen. **24**, 991 (1990).

163. G. Barton, J. Phys. A: Math. Gen. **24**, 5533 (1991).

164. C. Eberlein, J. Phys. A: Math. Gen. **25**, 3015 (1992).

165. C. Eberlein, J. Phys. A: Math. Gen. **25**, 3039 (1992).

166. V. V. Dodonov, A. B. Klimov, and V. I. Man'ko, Phys. Lett. A **142**, 511 (1989).

167. G. Calucci, J. Phys. A: Math. Gen. **25**, 3873 (1992).

168. P. Candelas, Ann. Phys. (N.Y.) **143**, 241 (1982).

169. J. Baacke and G. Krüsemann, Z. Phys. C **30**, 413 (1986).

170. E. D'Hoker and P. Sikivie, Phys. Rev. Lett. **71**, 1136 (1993).

171. C. L. Adler and N. M. Lawandy, Phys. Rev. Lett. **66**, 2617 (1991).

172. K. Symanzik, Nucl. Phys. **B190**, 1 (1981).

173. M. Krech and S. Dietrich, Phys. Rev. A **46**, 1922 (1992).

174. H. Li and M. Kadar, Phys. Rev. A **46**, 6491 (1992).

175. A. Ajdari, L. Peliti, and J. Prost, Phys. Rev. Lett. **66**, 1481 (1991).

176. A. Ajdari, B. Duplantier, D. Hone, L. Peliti, and J. Prost, J. Phys. II France **2**, 487 (1992).

177. E. Elizalde, Il Nuovo Cim. B **104**, 685 (1989).

178. E. Elizalde and A. Romeo, J. Math. Phys. **30**, 1133 (1989).

179. E. Elizalde and A. Romeo, Phys. Rev. D **40**, 436 (1989).

180. E. Elizalde and A. Romeo, Int. J. Mod. Phys. A **5**, 1653 (1990).

181. E. Elizalde, J. Math. Phys. **31**, 170, (1990).

182. K. Kirsten, J. Phys. A: Math. Gen. **24**, 3281 (1991).

183. B. P. Dolan and C. Nash, Commun. Math. Phys. **148**, 139 (1992).

184. N. F. Svaiter and B. F. Svaiter, J. Math. Phys. **32**, 175 (1990).

185. N. F. Svaiter and B. F. Svaiter, J. Phys. A: Math. Gen. **25**, 979 (1992).

186. A. M. Cetto and L. de la Peña, Il Nuovo Cim. B **108**, 447 (1993).

187. P. Attard, C. P. Ursenbeck, and G. N. Patey, Phys. Rev. A **45**, 7621 (1992).

188. M. Krech and S. Dietrich, Phys. Rev. A **46**, 1886 (1992).

189. M. Krech and S. Dietrich, Phys. Rev. Lett. **66**, 345 (1991); Phys. Rev. Lett. **67**, 1055 (1991).

190. J. O. Indekeu, M. P. Nightingale, and W. V. Wang, Phys. Rev. B **34**, 330 (1986).

191. H. W. Blöthe and M. P. Nightingale, Physica A, **129**, 1 (1984).

192. J. L. Cardy, Phys. Rev. Lett. **65**, 1443 (1990).

193. J. C. Le Guillou and J. Zinn-Justin, Phys. Rev. B **21**, 3976 (1980).

194. G. Gumbs, J. Math. Phys. **24**, 202 (1983).

195. A. Benyoussef, N. Boccara, and M. Saber, J. Phys. C: Solid State Phys. **19**, 1983 (1986).

196. D. Friedan, Z. Qui, and S. Shenker, Phys. Lett. B **151**, 37 (1985).

197. C. Itzykson and J.-M. Drouffe, *Statistical Field Theory*, Vol. 2, (Cambridge University Press, Cambridge, 1992).

198. M. R. Spiegel in *Schaum's Outline Series, Theory and Problems of Complex Variables* (Schaum Publishing Company, New York, 1964).

199. E. Eisenriegler, *Polymers Near Surfaces* (World Scientific, Singapore, in press).

200. L. Schäfer, J. Phys. A **9**, 377 (1976).

201. J. C. Collins, Phys. Rev. D **14**, 1965 (1976).

202. L. S. Brown, Ann. of Phys. **126**, 135 (1980).

203. M. E. Fisher and P.-G. de Gennes, C. R. Acad. Sc. Paris B **287**, 207 (1978).

204. H. W. Diehl and A. Ciach, Phys. Rev. B **44**, 6642 (1991).

205. M. E. Fisher and H. Au-Yang, Physica **101A**, 255 (1980).

206. J. Rudnick and D. Jasnow, Phys. Rev. Lett. **49**, 1595 (1982).

207. E. Eisenriegler, M. Krech, and S. Dietrich, Phys. Rev. Lett. **70**, 619 (1993); Phys. Rev. Lett. **70**, 2051 (1993).

208. A. O. Parry, R. Evans, and D. B. Nicolaides, Phys. Rev. Lett. **67**, 2978 (1991).

209. M. Krech, E. Eisenriegler, and S. Dietrich, to be published.

210. J. Spanier and K. B. Oldham, *An Atlas of Functions* (Springer, Berlin, 1987).

211. H. W. Diehl and S. Dietrich, Z. Phys. B **42**, 65 (1981).

212. E. Eisenriegler, M. Krech, and S. Dietrich, to be published.

213. H. W. Diehl, S. Dietrich, and E. Eisenriegler, Phys. Rev. B **27**, 2937 (1983).

214. A. Capelli, J. I. Latorre, and X. Vilasis-Cardona, Nucl. Phys. **B376**, 510 (1992).

215. D. M. McAvity and H. Osborn, Nucl. Phys. **B406**, 655 (1993).

216. E. Brézin, E. Korutcheva, Th. Jolicoeur, and J. Zinn-Justin, J. Stat. Phys. **70**, 583 (1993).

217. M. E. Fisher and V. Privman, Phys. Rev. B **32**, 447 (1985).

218. C. Bervillier, Phys. Rev. B **14**, 4964 (1976).

219. T. H. Magerlein and T. M. Sanders, Jr., Phys. Rev. Lett. **36**, 258 (1976).

220. M. E. Fisher and P. J. Upton, Phys. Rev. Lett. **65**, 2402 (1990).

221. M. E. Fisher and P. J. Upton, Phys. Rev. Lett. **65**, 3405 (1990).

222. E. Eisenriegler, J. Chem. Phys. **81**, 4666 (1984); E. Eisenriegler and P. J. Upton, J. Chem. Phys. **98**, 3582 (1993).

223. L. V. Mikheev and M. E. Fisher, J. Low Temp. Phys. **90**, 119 (1993).

224. P. J. Upton, Phys. Rev. Lett. **45**, 8100 (1992).

225. R. Evans and J. Stecki, preprint (1993).

226. E. Elizalde, J. Phys. A: Math. Gen. **22**, 931 (1989).

227. K. Kirsten, J. Math. Phys. **32**, 3008 (1991).

228. K. Kirsten, J. Phys. A: Math. Gen. **25**, 6297 (1992).

229. F. J. Wegner and E. K. Riedel, Phys. Rev. B **7**, 248 (1973).

230. E. Eisenriegler and H. W. Diehl, Phys. Rev. B **37**, 5257 (1988).

231. V. Privman and M. E. Fisher, J. Stat. Phys. **33**, 385 (1983).

232. K. Vollmayr, J. D. Reger, M. Scheucher, and K. Binder, Z. Phys. B **91**, 113 (1993).

233. K. Chen and D. P. Landau, Phys. Rev. B **46**, 937 (1992).

234. D. Schmeltzer, Phys. Rev. B **32**, 7512 (1985).

235. D. O'Connor and C. R. Stephens, Nucl. Phys. **B360**, 297 (1991).

236. Y. Kubyshin, D. O'Connor, and C. R. Stephens in *Renormalization Group '91, Second International Conference*, edited by D. V. Shirkov and V. B. Priezzhev (World Scientific, Singapore, 1992), p. 80.

237. D. O'Connor and C. R. Stephens, J. Phys. A: Math. Gen. **25**, 101 (1992).

238. D. O'Connor and C. R. Stephens, Phys. Rev. Lett. **72**, 506 (1994).

239. F. Freire, D. O'Connor, and C. R. Stephens, J. Stat. Phys. **74**, 219 (1994).

240. M. Krech and S. Dietrich, J. Low Temp. Phys. **89**, 145 (1992).

241. W. P. Tam and G. Ahlers, Phys. Rev. B **32**, 5932 (1985).

242. V. Dohm, Physica Scripta **T49**, 46 (1993).

243. P. Sutter and V. Dohm, Physica B, to appear (1993).

244. J. A. Nissen, T. C. P. Chui, and J. A. Lipa, J. Low Temp. Phys. **92**, 353 (1993).

245. T.-P. Chen, Ph.D. Thesis, State University of New York, Buffalo, NY (1977).

246. T.-P. Chen and F. M. Gasparini, Phys. Rev. Lett. **40**, 331 (1978).

247. F. M. Gasparini, T.-P. Chen, and B. Bhattacharyya, Phys. Rev. B **23**, 5797 (1981).

248. D. T. Smith, K. M. Godshalk, and R. B. Hallock, Phys. Rev. B **36**, 202 (1987).

249. A. Wacker and V. Dohm, Physica B, to appear (1993).

250. D. F. Brewer, D. C. Champeney, and K. Mendelssohn, Cryogenetics **1**, 1 (1960).

251. S. T. Islander and W. Zimmermann, Jr., Phys. Rev. A **7**, 188 (1973).

252. D. F. Brewer, J. Low Temp. Phys. **3**, 205 (1970).

253. I. Rhee and F. M. Gasparini, Phys. Rev. Lett. **63**, 410 (1989).

254. J. S. Brooks, B. B. Sabo, P. C. Schubert, and W. Zimmermann, Jr., Phys. Rev. B **19**, 4524 (1979).

255. M. Napiórkowski, W. Koch, and S. Dietrich, Phys. Rev. A **45**, 5760 (1992).

256. R. Lipowsky and U. Seifert, Phys. Rev. B 31, 4701 (1985); R. Lipowski, Phys. Rev. Lett. 55, 1699 (1985).

257. J. O. Indekeu, J. Chem. Soc., Faraday Trans. II 12, 1835 (1986).

258. J. O. Indekeu, Habilitation thesis, Katholieke Universiteit Leuven, 1990.

259. H. Li and M. Kadar, Phys. Rev. Lett. 67, 3275 (1991).

260. M. Yu. Novikov, A. S. Sorin, and V. Ya. Chernyak, Theor. Math. Phys. 92, 773 (1993).

261. G. Zimmerli and M. H. W. Chan, Phys. Rev. B 38, 8760 (1988).

262. R. J. Dionne and R. B. Hallock, in *Quantum Fluids and Solids - 1989*, Gainesville, FL, 1989, edited by G. Ikas and Y. Takano, AIP Conf. Proc. No. 194 (AIP, New York, 1989), p. 199.

263. M. J. Lea, D. S. Spencer, and P. Fozooni, Phys. Rev. B 35, 6665 (1987).

264. A. D. Migone, J. Krim, and J. G. Dash, Phys. Rev. B 31, 7643 (1985).

265. G. M. Graham and P. Taborek, Phys. Rev. B 40, 802 (1989).

266. E. S. Sabisky and C. H. Anderson, Phys. Rev. A 7, 790 (1973).

267. M. A. Paalanen and Y. Iye, Phys. Rev. Lett. 55, 1761 (1985).

268. D. Cieslikowski, A. J. Dahm, and P. Leiderer, Phys. Rev. Lett. 58, 1751 (1987).

269. P. J. Shirron, K. A. Gillis, and J. M. Mochel, J. Low Temp. Phys. 75, 349 (1989); 78, 157 (1989).

270. E. Cheng, M. W. Cole, W. F. Saam, and J. Treiner, Phys. Rev. Lett. 67, 1007 (1991).

271. P. J. Nacher and J. Dupont-Roc, Phys. Rev. Lett. 67, 2966 (1991).

272. K. S. Ketola, S. Wang, and R. B. Hallock, Phys. Rev. Lett. 68, 201 (1992).

273. P. Taborek and J. E. Rutledge, Phys. Rev. Lett. 68, 2184 (1992).

274. W. F. Saam, J. Treiner, E. Cheng, and M. W. Cole, J. Low Temp. Phys. **89**, 637 (1992).

275. J. E. Rutledge and P. Taborek, Phys. Rev. Lett. **69**, 937 (1992).

276. E. Cheng, M. W. Cole, J. Dupont-Roc, W. F. Saam, and J. Treiner, Rev. Mod. Phys. **65**, 557 (1993).

277. P. Taborek and J. E. Rutledge, Phys. Rev. Lett. **71**, 263 (1993).

278. J. N. Israelachvili and P. M. McGuggian, Science **241**, 795 (1988).

279. N. A. Burnham and R. J. Colton, J. Vac. Sci. Technol. **A7**, 2906 (1989).

280. N. A. Burnham, D. D. Dominguez, R. L. Mowery, and R. J. Colton, Phys. Rev. Lett. **64**, 1931 (1990).

281. A. L. Weisenhorn, P. K. Hansma, T. R. Albrecht, and T. F. Quate, Appl. Phys. Lett. **54**, 2651 (1989).

282. W. A. Ducker, T. J. Senden, and R. M. Pashley, Nature **353**, 239 (1991).

283. J. L. Parker, Langmuir **8**, 551 (1992).

284. W. Fenzl, private communication.

285. E. Eisenriegler, private communication.

286. A. J. Liu and M. E. Fisher, Phys. Rev. A **40**, 7202 (1989).

287. D. Beysens and D. Estève, Phys. Rev. Lett. **54**, 2123 (1985).

288. V. Gurfein, D. Beysens, and F. Perrot, Phys. Rev. A **40**, 2543 (1989).

289. G. P. Felcher, Phys. Rev. B **24**, 1595 (1981).

290. S. Dietrich and H. Wagner, Phys. Rev. Lett. **51**, 1469 (1983).

291. P. C. Hohenberg and B. I. Halperin, Rev. Mod. Phys. **49**, 435 (1977).

292. H. W. Diehl and U. Ritschel, J. Stat. Phys., to appear.

293. H. W. Diehl and H. K. Janssen, Phys. Rev. A **45**, 7145 (1992).

Index